Hydrogel for Sustained Delivery of Therapeutic Agents

Hydrogel for Sustained Delivery of Therapeutic Agents

Guest Editors

Adina Magdalena Musuc
Magdalena Mititelu
Mariana Chelu

MDPI

Basel • Beijing • Wuhan • Barcelona • Belgrade • Novi Sad • Cluj • Manchester

Guest Editors
Adina Magdalena Musuc
Department of Chemical Kinetics
Institute of Physical
Chemistry - Ilie Murgulescu
Bucharest
Romania

Magdalena Mititelu
Department of Clinical
Laboratory and Food Safety
University of Medicine and
Pharmacy Carol Davila
Bucharest
Romania

Mariana Chelu
Department of Chemical Kinetics
Institute of Physical
Chemistry - Ilie Murgulescu
Bucharest
Romania

Editorial Office
MDPI AG
Grosspeteranlage 5
4052 Basel, Switzerland

This is a reprint of the Special Issue, published open access by the journal *Gels* (ISSN 2310-2861), freely accessible at: www.mdpi.com/journal/gels/special_issues/5VJK569X52.

For citation purposes, cite each article independently as indicated on the article page online and using the guide below:

Lastname, A.A.; Lastname, B.B. Article Title. *Journal Name* **Year**, *Volume Number*, Page Range.

ISBN 978-3-7258-2738-1 (Hbk)
ISBN 978-3-7258-2737-4 (PDF)
https://doi.org/10.3390/books978-3-7258-2737-4

Contents

About the Editors

Adina Magdalena Musuc

Habil. Dr. Adina Magdalena Musuc is a senior researcher and Head of the Chemical Kinetics Department at the Institute of Physical Chemistry, Ilie Murgulescu of the Romanian Academy, Bucharest, Romania, with over 25 years of experience in the field of chemistry. She earned her B.Sc. in Physical Chemistry (1999), an M.Sc. in Environmental Quality Control (2001), a Ph.D. in Chemistry (2007) from the University of Bucharest, and habilitation in Chemistry (2023). She was awarded the "Gheorghe Spacu" medal in 2018 and 2024 from the Romanian Chemical Society and the "Nicolae Teclu" award in 2014 from the Romanian Academy. Furthermore, she received an honorary diploma from the Romanian Chemical Society in 2015 for contributing to the activities of the Romanian Chemical Society to promote chemistry in Romania. Dr. Musuc has published over 100 papers in various ISI journals and 115 papers presented at international conferences. Dr. Musuc holds three invention patents and two patent applications pending. She is a vice president of the Romanian Society of Chemistry, Bucharest. Dr. Musuc is the co-author of four book chapters published by international publishers. Her research interests include various domains of chemistry, physical chemistry, and green and sustainable chemistry involving biopolymers, polysaccharides, biofuels, thermal analysis, kinetics, explosion, combustion, risk analysis, nanomaterials, cyclodextrin inclusion complexes, drug delivery systems, biocomposites, biomass, environmental chemistry, and hydrogels for biomedical and water remediation applications.

Magdalena Mititelu

Associate Professor Dr. habil. Magdalena Mititelu at the Department of Clinical Laboratory and Food Safety, Faculty of Pharmacy, University of Medicine and Pharmacy Carol Davila, Bucharest, Romania, holds degrees in pharmacy, economics, and physical chemistry. She is a specialist in Intra-European Transactions, Condensed Matter Physics, and Atomic Systems. Her postgraduate studies include advanced courses in Intra-European Transactions, Condensed Matter Physics, and Atomic Systems, as well as postgraduate training as a Nutrition Consultant (1500 hours) and as a Nutritionist and Dietitian (2400 hours). She is an Internal Auditor for the Food Safety Management System according to ISO 22000:2018 and ISO 19011:2018, a senior pharmacist, and Vice President of the Praova College of Pharmacists.

With over 250 published articles, 11 patents, 10 patent applications, 3 gold medals, and one silver medal at international invention exhibitions, she is a recognized researcher. Dr. Mititelu is also the author and co-author of 13 specialized books in the fields of clinical biochemistry, immunology, nutrition, pharmaceutical technology, and pharmacy communication. Her research expertise includes the development, analysis, and control of innovative therapeutic formulas; marine biotechnology aimed at extracting nutraceuticals from industrial waste; toxicological analysis of active ingredients and food products with risk assessment for consumption; as well as clinical and nutritional evaluation studies.

Mariana Chelu

Dr. Mariana Chelu received her Ph.D. degree in Chemistry in 2021 at the Romanian Academy. Currently, she is a scientific researcher at the Institute of Physical Chemistry "Ilie Murgulescu" of the Romanian Academy, working in the Department of Chemical Kinetics. Her research interests cover several key areas in physical chemistry and sustainable chemistry, including green chemistry, sol-gel processes, nanostructured and piezoelectric materials, thin films, hydrogels, drug delivery systems, and advanced polymer composites. She is involved in the design, synthesis, and application of innovative nano- and bio-materials adapted for biomedical and environmental applications. Contributes to the development of functional materials aimed at regenerative medicine, thin films, sensors, and biosensors. Her dedication to green and sustainable design principles is evidenced by her pursuit of green materials and processes that align with modern sustainability requirements in chemistry.

gels

Editorial

Hydrogel for Sustained Delivery of Therapeutic Agents

Adina Magdalena Musuc [1,*], Magdalena Mititelu [2] and Mariana Chelu [1]

[1] Institute of Physical Chemistry—Ilie Murgulescu, Romanian Academy, 060021 Bucharest, Romania; mchelu@icf.ro
[2] Department of Clinical Laboratory and Food Safety, Faculty of Pharmacy, "Carol Davila" University of Medicine and Pharmacy, 020956 Bucharest, Romania; magdalena.mititelu@umfcd.ro
* Correspondence: amusuc@icf.ro

Citation: Musuc, A.M.; Mititelu, M.; Chelu, M. Hydrogel for Sustained Delivery of Therapeutic Agents. *Gels* **2024**, *10*, 717. https://doi.org/10.3390/gels10110717

Received: 4 November 2024
Accepted: 6 November 2024
Published: 7 November 2024

1. Introduction

In recent years, hydrogels have emerged as a highly promising platform for the sustained delivery of therapeutic agents, addressing critical challenges in drug delivery systems, from controlled release to biocompatibility. With their high-water content, biocompatibility, and tunable physical and chemical properties, hydrogels have enabled significant advancements in delivering a wide range of therapeutic agents, including small molecules, proteins, nucleic acids, and cells. This Special Issue, "Hydrogels for Sustained Delivery of Therapeutic Agents" of the journal *Gels* seeks to explore the latest innovations, challenges, and potential future directions in this field, highlighting the role of hydrogels in biomedicine (Figure 1).

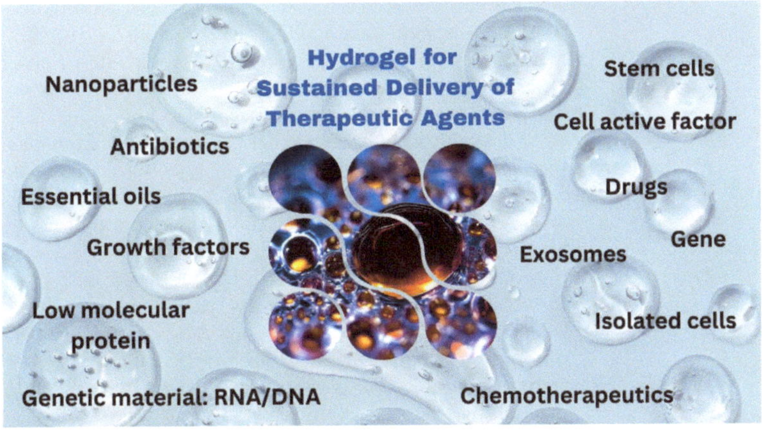

Figure 1. Various applications of hydrogels for sustained delivery of functional compounds.

Hydrogels are hydrophilic polymer networks that can absorb large amounts of water, resulting in a soft, tissue-like structure that can seamlessly interface with biological tissues. This unique feature allows hydrogels to deliver drugs in a more controlled manner than traditional drug delivery systems. By manipulating their molecular composition and structure, researchers have developed hydrogels with fine-tuned release kinetics, enabling the sustained delivery of therapeutic agents over extended periods. Furthermore, hydrogels can be engineered to respond to specific environmental cues—such as pH, temperature, or enzymes—making them ideal for targeted therapies in diseases such as cancer, cardiovascular diseases, and chronic inflammatory conditions.

This Special Issue aims to showcase cutting-edge research on the synthesis, characterization, and functionalization of hydrogels tailored for drug delivery applications. Novel hydrogel systems are increasingly designed for the co-delivery of multiple therapeutic agents, such as antibiotics with anti-inflammatory agents or chemotherapeutic drugs with immunomodulators. This strategy enables enhanced therapeutic efficacy while reducing side effects by maintaining localized, sustained release. Featuring a collection of ten papers, including six original research articles and four reviews, this Special Issue highlights the versatility of hydrogels in sustained drug delivery and explores how these systems are tailored for specific therapeutic challenges.

2. Contributions

The first research article published in this Special Issue is a study conducted by Pérez-González et al. [1]. This article focuses on caspofungin, an echinocandin-class antifungal used to treat severe and invasive fungal infections, and explores its application in treating cutaneous candidiasis—a challenging fungal skin infection. The study investigates the effect of two permeation enhancers, Azone and Transcutol-P, on caspofungin-loaded gels, evaluating the resulting formulations for their drug release profile, skin permeation, tolerability, and antimicrobial efficacy. In the treatment of fungal skin infections such as candidiasis, a significant challenge is achieving adequate drug retention and absorption in the skin layers while minimizing systemic absorption, which could lead to unwanted side effects. Pérez-González et al. [1] address this by incorporating Azone in a caspofungin gel (CPF-AZ-gel) to assess its potential for improved cutaneous application, comparing it to a standard caspofungin gel without permeation enhancers (CPF-gel). The study reports promising outcomes with the CPF-AZ-gel, which demonstrated superior skin retention and controlled drug release compared to the promoter-free formulation. Through in vitro release studies and ex vivo permeation testing on human skin, the authors showed that the CPF-AZ-gel provided an enhanced release profile while confining caspofungin's diffusion primarily within the targeted skin layers. Additionally, both formulations displayed pseudoplastic behavior, making them suitable for easy and uniform application, as well as excellent spreadability and compatibility with skin biomechanics. Notably, antimicrobial efficacy testing confirmed that both formulations were effective against *Candida glabrata*, *Candida parapsilosis*, and *Candida tropicalis*, while *Candida albicans* exhibited resistance. Histological analysis of skin samples confirmed that both gels were well tolerated, indicating their suitability for clinical use in patients with cutaneous candidiasis, particularly those who may not respond to or tolerate conventional antifungal therapies.

Petrini et al. explore [2] a novel application of photodynamic therapy (PDT) using an aminolevulinic acid-based gel in periodontal tissue repair. The study assesses how photodynamic therapy (ALAD-PDT) with red LED irradiation affects human gingival fibroblasts (hGFs) and osteoblasts (hOBs) cultured on porcine acellular dermal matrix membranes (PADMMs), a common material used in periodontal surgery. The findings offer promising insights into PDT's potential for accelerating healing and reinforcing the stability of membrane grafts in oral surgery. In this study, human gingival fibroblasts and osteoblasts obtained from dental patients were cultured on PADMMs and subjected to three treatment groups: a control group (CTRL) without exposure, a group receiving red LED irradiation only, and a group treated with ALAD-PDT (45 min of aminolevulinic acid incubation followed by 7 min of red LED exposure). The results show that ALAD-PDT significantly enhanced cellular proliferation and organization, forming a dense network of cells on PADMMs. Further assays—including MTT, histology, SEM, and mineralization assays—confirmed that ALAD-PDT significantly increased collagen and fibronectin production in fibroblasts and promoted bone marker expression in osteoblasts, suggesting that ALAD-PDT facilitates faster and more robust healing outcomes.

Hussain et al. [3] explore a cutting-edge, non-invasive approach to delivering valproic acid (VA) directly to the brain via nasal administration. This innovative study leverages computational predictive modeling and nanoemulsion gel formulation to overcome the challenges of conventional valproic acid delivery routes, which often result in systemic side effects, rapid hepatic metabolism, and low bioavailability in the brain. The study incorporated GastroPlus Version 9.8.3 and HSPiP (Hansen Solubility Parameters in Practice) programs to identify optimal excipients, evaluate formulation parameters, and predict the in vivo performance of valproic acid. GastroPlus simulations provided insights into drug absorption, distribution, and predicted the advantages of nasal administration over traditional oral and parenteral routes. Hansen solubility parameters helped select excipients that offered optimal miscibility, forming stable and effective nanoemulsion gels.

The development of advanced hydrogel systems for wound care has significant potential in managing chronic conditions, especially for diabetic patients facing slow or non-healing wounds. The research article by Aldakheel and colleagues [4] showcases a promising approach that merges green synthesis techniques with hydrogel technology to address this urgent healthcare need. Diabetic chronic wounds are prone to infection, poor healing outcomes, and an increased risk of complications like limb amputation, making effective, rapid-healing wound care solutions critical. In this study, the authors developed a polysaccharide-based hydrogel infused with silver nanoparticles (AgNPs) synthesized through an eco-friendly process using garlic extract, which acts as a reducing agent. The hydrogel matrix, composed of chitosan, starch, and alginate (PsB), was further modified with acrylamide to improve its mechanical and adhesive properties. The green-synthesized AgNPs demonstrated effective antibacterial activity against *Klebsiella pneumoniae* and *Staphylococcus aureus*, two bacterial strains commonly associated with wound infections. The choice of silver, known for its broad-spectrum antibacterial properties, enhances the hydrogel's functionality by providing infection control, while the polysaccharide matrix aids in creating a moist wound environment conducive to healing. By using green synthesis for AgNP production, this research aligns with the growing demand for sustainable medical materials and demonstrates a practical alternative to traditional, chemical-based wound care products.

In their paper, Ji and colleagues [5] present an innovative approach to anticancer drug delivery. This study addresses the significant challenge of safely delivering toxic anticancer agents, such as tamoxifen citrate, in a controlled and sustained manner to improve therapeutic efficacy and reduce side effects. Utilizing a modified coaxial electrospraying technique, the authors created microparticles composed of tamoxifen citrate (TC) within a matrix of ethylcellulose (EC), coated with stearic acid (SA) to regulate the drug's release. The use of ethylcellulose as the core matrix provided a robust foundation for the encapsulation, while the stearic acid layer functioned as an additional control mechanism to moderate the diffusion of the drug. The authors analyzed the morphology, structural compatibility, and physical state of the microparticles using advanced characterization techniques, including scanning and transmission electron microscopy (SEM and TEM), X-ray diffraction (XRD), and Fourier-transform infrared spectroscopy (FTIR). These methods validated the integrity and uniformity of the SA coating, which was integral to the observed sustained release profile.

In another research paper, Slavkova and colleagues [6] present a promising therapeutic approach for pediatric atopic dermatitis, utilizing budesonide-loaded nanoparticles in a hydrogel matrix to enhance treatment precision and efficacy. Budesonide, a corticosteroid often used in treating skin inflammation, has shown limited success in topical applications due to side effects and challenges related to its stability and permeability in the skin. This study leverages the pH differences observed in atopic dermatitis lesions to create a responsive nanocarrier, Eudragit L100, which enhances budesonide's release to the inflamed site. Nanoparticles were created via a nanoprecipitation method, producing particles with a mean size of 57 nm, a negative surface charge (−31.2 mV), and high drug encapsulation efficiency (~90%). Cytotoxicity assays on HaCaT keratinocyte cells indicated their safety for skin applications, making this a viable delivery method for young patients. The nanoparticles were subsequently incorporated into two types of hydrogels: methylcellulose or Pluronic F127, which were rigorously analyzed for characteristics like pH, occlusion, rheology, spreadability, and drug release profiles. These hydrogels demonstrated controlled, targeted release of budesonide, positioning them as an effective solution for treating atopic dermatitis in pediatric patients. This study underscores the potential of nanoparticle-infused hydrogels as advanced, localized treatment options for skin conditions, especially where traditional therapies may fall short.

In their comprehensive review, Chen et al. [7] examine recent innovations in hydrogel coatings applied to titanium and titanium alloy implants, materials widely used due to their mechanical properties and biocompatibility. However, titanium's performance within the physiological environment can be limited, particularly in promoting cellular interactions and biological integration. To address these challenges, hydrogel coatings offer a biochemical approach to enhance surface bioactivity by attaching functional biomolecules, including proteins, growth factors, and peptides. This biochemical strategy enables the implant surface to better support cell adhesion, proliferation, and differentiation, which are critical for improved biocompatibility and long-term implant success. The review highlights both natural polymers (e.g., collagen, gelatin, chitosan, alginate) and synthetic polymers (e.g., polyvinyl alcohol, polyacrylamide, polyethylene glycol, polyacrylic acid) used in hydrogel coatings. Various application methods for creating these coatings, such as electrochemical deposition, sol–gel processes, and layer-by-layer assembly, are thoroughly reviewed. Furthermore, the authors explore the five key benefits hydrogel coatings bring to titanium implants: enhanced osseointegration, improved angiogenesis, the modulation of macrophage responses (promoting an anti-inflammatory effect), antimicrobial properties, and the capability of localized drug delivery.

The review of Chelu et al. [8] delves into the growing field of *Aloe vera*-based hydrogels, emphasizing their advantages as biocompatible, therapeutic wound dressings. *Aloe vera* is renowned for its healing properties, and when integrated into hydrogels, it provides an ideal environment for the promotion of tissue repair, mitigating inflammation, and delivering bioactive agents directly to wound sites. This review discusses the synthesis techniques and structural characteristics of these hydrogels, examining how their properties support wound healing. Chelu et al. [8] explore the various mechanisms through which therapeutic agents are released from *Aloe vera* hydrogels, including diffusion, swelling, and degradation, which allow for controlled and sustained drug delivery. In addition to enhancing wound closure, these hydrogels offer significant antimicrobial and anti-inflammatory benefits due to *Aloe vera*'s natural compounds and the potential for incorporating additional therapeutic agents. The review covers different approaches for embedding antimicrobial and anti-inflammatory agents into these hydrogels, thus expanding their efficacy against infections and inflammation.

Villa et al. [9] examine the use of natural deep eutectic solvents (NaDESs) as eco-friendly, effective solvents in the cosmetic and pharmaceutical industries. Recognized as a new generation of green solvents, NaDESs provide a safer, non-flammable alternative to conventional ionic liquids and can be tailored for both lipophilic and hydrophilic molecules. This versatility makes them ideal for various applications, from sustainable extraction processes to biocompatible drug delivery systems. In pharmaceuticals, NaDESs are highlighted for their role as biopolymer modifiers, where they act as "therapeutic deep eutectic systems". These systems enhance the solubility and stability of active ingredients, offering potential improvements in drug delivery. In cosmetics, NaDESs show promise in forming more sustainable, efficient formulations, providing a means for the stable incorporation of bioactive ingredients in topical and dermal applications. This review synthesizes the current understanding of NaDES applications in these fields, discussing both their practical uses and the challenges ahead. By examining the multifunctionality of NaDESs, the review underscores their potential to transform formulation practices within cosmetics and pharmaceuticals, advocating for their broader adoption as a green solution in bioactive ingredient delivery.

Chelu [10] presents a comprehensive analysis of recent innovations combining essential oils with hydrogel technology for diverse applications such as biomedical, dental, cosmetic, food, packaging, and heritage restoration. This review explores the synthesis, polymeric sources, and cross-linking techniques used in these hydrogels, emphasizing their biocompatibility, non-toxicity, antibacterial properties, controlled release capabilities, and cytocompatibility. The unique properties of essential oils, including their bioactivity and aromatic potential, are examined alongside their extraction and encapsulation processes. The review delves into the benefits and challenges of these methods, addressing issues such as the volatility, solubility, and stability of essential oils within hydrogel matrices. The encapsulation of essential oils in hydrogels enhances both stability and biological efficacy, making these formulations viable for sustained release applications in health and cosmetic products as well as food preservation and cultural conservation. Chelu's review outlines the challenges and limitations faced in essential oil hydrogel technologies and discusses their promising future, noting significant potential across a broad spectrum of fields due to the multifunctionality of hydrogels and the enhanced delivery of natural bioactive compounds.

3. Future Directions in Hydrogel-Based Therapeutics

While the studies presented in this Special Issue illustrate the tremendous potential of hydrogel systems, challenges remain that require collaborative research efforts. The field is moving towards more personalized, patient-specific approaches, leveraging advanced techniques such as 3D printing and machine learning to create hydrogels tailored to individual needs. The integration of "smart" responsive materials that adjust drug release based on real-time feedback holds promise for next-generation, self-regulating delivery systems. Furthermore, the inclusion of both therapeutic and diagnostic functionalities within a single hydrogel platform—often referred to as "theranostic" applications—is an exciting development that could transform treatment paradigms for chronic and complex diseases.

4. Conclusions

This Special Issue captures both the depth and diversity of hydrogel-based drug delivery research. It reflects the field's journey from theoretical advancements to tangible applications and encourages continued exploration to overcome existing limitations. By integrating high-quality reviews and pioneering research, we aim to inspire innovation and collaboration among researchers and industry professionals working towards safer, more effective, and patient-centered drug delivery solutions. We hope that this collection will serve as a valuable resource for anyone interested in the field of hydrogels for sustained therapeutic delivery and look forward to witnessing the future impact of these advancements on modern medicine.

We would like to extend our deepest gratitude to the journal *Gels* for the invaluable opportunity to produce this Special Issue, and we express our heartfelt thanks to the editorial team for their tireless support, especially our managing editor Ms. Miranda Song, who provided continuous guidance throughout the submission and publication process. Finally, this Special Issue could not have been realized without the dedication of the contributing authors and the conscientiousness of our reviewers, whose critical insights and commitment have ensured the high quality and scientific rigor of this collection.

Author Contributions: All authors have read and agreed to the published version of the manuscript.

Acknowledgments: The authors of this Special Issue who contributed high-quality research are gratefully acknowledged.

Conflicts of Interest: The authors declare no conflicts of interest.

References

1. Pérez-González, N.; Espinoza, L.C.; Rincón, M.; Sosa, L.; Mallandrich, M.; Suñer-Carbó, J.; Bozal-de Febrer, N.; Calpena, A.C.; Clares-Naveros, B. Gel Formulations with an Echinocandin for Cutaneous Candidiasis: The Influence of Azone and Transcutol on Biopharmaceutical Features. *Gels* **2023**, *9*, 308. [CrossRef] [PubMed]
2. Petrini, M.; D'Amico, E.; Pierfelice, T.V.; Aceto, G.M.; Karaban, M.; Felice, P.; Piattelli, A.; Barone, A.; Iezzi, G. Photodynamic Therapy with Aminolevulinic Acid Enhances the Cellular Activity of Cells Cultured on Porcine Acellular Dermal Matrix Membranes Used in Periodontology. *Gels* **2023**, *9*, 584. [CrossRef] [PubMed]

3. Hussain, A.; Altamimi, M.A.; Ramzan, M.; Mirza, M.A.; Khuroo, T. GastroPlus- and HSPiP-Oriented Predictive Parameters as the Basis of Valproic Acid-Loaded Mucoadhesive Cationic Nanoemulsion Gel for Improved Nose-to-Brain Delivery to Control Convulsion in Humans. *Gels* **2023**, *9*, 603. [CrossRef] [PubMed]
4. Aldakheel, F.M.; Mohsen, D.; El Sayed, M.M.; Fagir, M.H.; El Dein, D.K. Green Synthesized Silver Nanoparticles Loaded in Polysaccharide Hydrogel Applied to Chronic Wound Healing in Mice Models. *Gels* **2023**, *9*, 646. [CrossRef] [PubMed]
5. Ji, Y.; Zhao, H.; Liu, H.; Zhao, P.; Yu, D.-G. Electrosprayed Stearic-Acid-Coated Ethylcellulose Microparticles for an Improved Sustained Release of Anticancer Drug. *Gels* **2023**, *9*, 700. [CrossRef] [PubMed]
6. Slavkova, M.; Lazov, C.; Spassova, I.; Kovacheva, D.; Tibi, I.P.-E.; Stefanova, D.; Tzankova, V.; Petrov, P.D.; Yoncheva, K. Formulation of Budesonide-Loaded Polymeric Nanoparticles into Hydrogels for Local Therapy of Atopic Dermatitis. *Gels* **2024**, *10*, 79. [CrossRef] [PubMed]
7. Chen, H.; Feng, R.; Xia, T.; Wen, Z.; Li, Q.; Qiu, X.; Huang, B.; Li, Y. Progress in Surface Modification of Titanium Implants by Hydrogel Coatings. *Gels* **2023**, *9*, 423. [CrossRef] [PubMed]
8. Chelu, M.; Musuc, A.M.; Popa, M.; Calderon Moreno, J. *Aloe vera*-Based Hydrogels for Wound Healing: Properties and Therapeutic Effects. *Gels* **2023**, *9*, 539. [CrossRef] [PubMed]
9. Villa, C.; Caviglia, D.; Robustelli della Cuna, F.S.; Zuccari, G.; Russo, E. NaDES Application in Cosmetic and Pharmaceutical Fields: An Overview. *Gels* **2024**, *10*, 107. [CrossRef] [PubMed]
10. Chelu, M. Hydrogels with Essential Oils: Recent Advances in Designs and Applications. *Gels* **2024**, *10*, 636. [CrossRef] [PubMed]

gels

MDPI

Article

Formulation of Budesonide-Loaded Polymeric Nanoparticles into Hydrogels for Local Therapy of Atopic Dermatitis

Marta Slavkova [1,*], Christophor Lazov [1], Ivanka Spassova [2], Daniela Kovacheva [2], Ivanka Pencheva-El Tibi [1], Denitsa Stefanova [1], Virginia Tzankova [1], Petar D. Petrov [3] and Krassimira Yoncheva [1,*]

[1] Faculty of Pharmacy, Medical University of Sofia, 1000 Sofia, Bulgaria; itibi@pharmfac.mu-sofia.bg (I.P.-E.T.); denitsa.stefanova@pharmfac.mu-sofia.bg (D.S.); vtzankova@pharmfac.mu-sofia.bg (V.T.)

[2] Institute of General and Inorganic Chemistry, Bulgarian Academy of Sciences, 1113 Sofia, Bulgaria; ispasova@svr.igic.bas.bg (I.S.); didka@svr.igic.bas.bg (D.K.)

[3] Institute of Polymers, Bulgarian Academy of Sciences, Akad. G. Bonchev Str. 103A, 1113 Sofia, Bulgaria; ppetrov@polymer.bas.bg

* Correspondence: mslavkova@pharmfac.mu-sofia.bg (M.S.); kyoncheva@pharmfac.mu-sofia.bg (K.Y.)

Abstract: Budesonide is a mineral corticoid applied in the local therapy of pediatric atopic dermatitis. Unfortunately, its dermal administration is hindered by the concomitant adverse effects and its physicochemical properties. The characteristic pH change in the atopic lesions can be utilized for the preparation of a pH-sensitive nanocarrier. In this view, the formulation of Eudragit L 100 nanoparticles as a budesonide delivery platform could provide more efficient release to the desired site, improve its penetration, and subsequently lower the undesired effects. In this study, budesonide-loaded Eudragit L100 nanoparticles were prepared via the nanoprecipitation method (mean diameter 57 nm, −31.2 mV, and approx. 90% encapsulation efficiency). Their safety was proven by cytotoxicity assays on the HaCaT keratinocyte cell line. Further, the drug-loaded nanoparticles were incorporated into two types of hydrogels based on methylcellulose or Pluronic F127. The formulated hydrogels were characterized with respect to their pH, occlusion, rheology, penetration, spreadability, and drug release. In conclusion, the developed hydrogels containing budesonide-loaded nanoparticles showed promising potential for the pediatric treatment of atopic dermatitis.

Keywords: budesonide; nanoparticles; Eudragit L100; hydrogels; atopic dermatitis

Citation: Slavkova, M.; Lazov, C.; Spassova, I.; Kovacheva, D.; Tibi, I.P.-E.; Stefanova, D.; Tzankova, V.; Petrov, P.D.; Yoncheva, K. Formulation of Budesonide-Loaded Polymeric Nanoparticles into Hydrogels for Local Therapy of Atopic Dermatitis. *Gels* **2024**, *10*, 79. https://doi.org/10.3390/gels10010079

Academic Editors: Adina Magdalena Musuc, Magdalena Mititelu and Mariana Chelu

Received: 12 November 2023
Revised: 17 January 2024
Accepted: 18 January 2024
Published: 20 January 2024

1. Introduction

Atopic dermatitis is a chronic, relapsing inflammatory skin disease. It has three stages: infantile, childhood, and adult. The onset could be as early as birth and can manifest with erythematous papules and vesicles on the cheeks, forehead, and/or scalp. It has a high prevalence and affects 15% to 30% of children [1]. The disease's pathogenesis is related to genetic predisposition, environmental factors, and immune dysregulation [2]. It is characterized by increased transepidermal water loss due to barrier dysfunction and a pH increase of up to 6 or even higher [3]. Therapy for atopic dermatitis depends on the manifestation of the disease, and in severe cases, it may require a systemic remedy. In mild to moderate conditions, usually topical therapies are sufficient for disease management [4]. Furthermore, topical application is a convenient and affordable method of administration with minimal systemic toxicity [5]. The pharmacological topical treatment consists mainly of glucocorticoids, calcineurin inhibitors, or topical crisaborole [6]. Current therapeutic strategies are focused on reducing inflammation, restoring the skin barrier, and antibacterial therapy [7].

Corticosteroids are used as a first-line treatment for many dermal conditions, from pruritic lesions to atopic dermatitis or psoriasis. They have anti-inflammatory, immunosuppressive, anti-proliferative, and vasoconstrictor effects [8]. Even though they have been intensively applied topically, they have been associated with local or systemic adverse effects such as cutaneous atrophy, telangiectasia, striae, skin infections, and hypothalamic-pituitary-adrenal axis suppression [9]. This is a drawback in their application, especially for long-term therapies for chronic conditions such as atopic dermatitis.

Budesonide is a potent synthetic nonhalogenated representative of the corticosteroid family with predominantly anti-inflammatory activity and a weak mineralocorticoid [10]. It is applicable in the inhaled therapy of asthma [11] and the targeted oral treatment of ulcerative colitis [12]. There are recent reports about the use of budesonide in the management of pediatric atopic dermatitis [4]. However, budesonide has poor aqueous solubility [13] and a partition coefficient (log P) of 2.32 [14]. These characteristics make it an unattractive drug for the dermal route of application, and different strategies have been proposed to improve its skin delivery, e.g., cyclodextrin inclusion complex-based hydrogels [4], PLGA-PVA nanoparticles [10], and PEO-PCL-PEO triblock nanoparticles [15].

Nowadays, scientific efforts are directed toward the development of innovative formulations like nanoparticles, liposomes, microemulsions, etc. for enhanced delivery of drug molecules into the skin [7]. The nanotechnological approach can boost the therapy of skin disorders. The penetration and transport of drugs from nanoparticles can be modified by the different chemical properties of the polymer used, the encapsulation mechanism, the size of the nanoparticles, and the viscosity of the formulations. The loading of the drug in nanocarriers improves the solubility of highly hydrophobic drugs, provides sustained and controlled release, increases drug stability, and provides site-specific delivery. Therefore, the adverse effects can be diminished [10]. Polymeric nanoformulations are sub-micrometric colloidal drug carriers prepared by biocompatible and biodegradable polymers. They vary in composition and structure and include such nanosystems as nanocapsules, nanospheres, nanofibers, etc. [7]. Natural and synthetic polymers can be used in their preparation. The second group is characterized by high purity and batch-to-batch reproducibility and is therefore suitable for more consistent drug release profiles. Some typical examples of synthetic polymers used for nanoparticle preparation are poly(lactic-co-glycolic acid), tyrosine-derived triblock polymer, poly(ε-caprolactone), and others [16].

Polymeric nanoparticles can be prepared by various methods, which can be generally classified as the application of preformed polymers or the direct polymerization of monomers. There are various techniques applied for the formation of nanoparticles with preformed polymers, such as solvent evaporation, salting-out, dialysis, supercritical fluid technology, and others [17]. Eudragit® is a manufactural name for a diverse range of synthetic polymethacrylate-based copolymers. They can be commercialized with different acidic or alkaline end groups, allowing pH-dependent drug release. Eudragits are functional polymers widely used in the development of polymeric nanoparticles. They have the potential to encapsulate and increase the solubility and bioavailability of poorly soluble drugs [18], as well as control drug delivery [3]. Eudragit® L100 is an anionic representative with a mean molecular mass of approximately 135,000 Da and an apparent viscosity of 50–200 mPas [19]. It is soluble at $pH \geq 6$ and is generally used in the preparation of enteric coatings. A current review of its applications shows that Eudragit® L100 can be utilized in the preparation of microspheres, microsponges, nanoparticles, liposomes, tablets, etc. in order to achieve sustained release or bioavailability improvement [20]. Nanoparaticles based on Eudragit L100 were also proposed for dermal drug delivery [21]. It has also been suggested that those nanoparticles possess negative zeta potential and remain on the epidermis surface, limiting systemic absorption as well as side effects, which could be especially useful in the pediatric population [22].

Topical corticosteroids are available in different conventional dosage forms, including creams, ointments, gels, sprays, foams, and others [8]. The main barrier for the topical and transdermal routes of administration happens to be the stratum corneum (SC) layer of the skin. It has a two-compartment structure, often referred to as a "brick and mortar system.0". The corneocytes are stacked in up to 20 layers and play the physical barrier role of SC. The spaces between them are occupied by mortar lipids. This is a complex mixture of around 13 lipid types, including ceramides, cholesterol, and free fatty acids, which play the permeability barrier role of SC [23]. The substances that are capable of diffusion through the intracellular route of the SC are small (molecular weight $\leq$ 400 Da) and lipophilic in nature (log P > 3) [24]. The main issue regarding the application of the classic semi-solid formulations is the enhancement of drug penetration and the simultaneous minimization of the risks of percutaneous absorption. The choice of vehicle can significantly affect the potency of the corticosteroid applied. Ointments, for example, exert more pronounced occlusion, which promotes their absorption and reach into the bloodstream [25]. In addition, the pediatric population, which is the most common end user of topical steroid therapy, is characterized by considerable differences in skin structure and thickness. The drug permeability is significant in children due to the thinner skin and the high skin-to-body-weight ratio. Thus, topical steroid application in children is prone to more systemic side effects such as growth retardation, Cushing disease, hyperglycemia, Addisonian crises upon cessation, and others [26]. Therefore, significant attention should be paid to the choice of vehicle for corticosteroid topical delivery. Simultaneously, more effective therapy regarding the dose is needed, with limited effect on the depth of penetration.

Hydrogels present one of the most intensively used semisolid forms due to their excellent biocompatibility, solubility in water, and structural and viscoelastic resemblance to the cell membrane [27,28]. In addition, they are more cosmetically appealing, they do not cause skin maceration or folliculitis, and they can be applied to the scalp [25]. In the light of atopic dermatitis treatment, there is data suggesting that hydrogels are the most preferable dosage form [29]. The most recent studies have pointed towards the preparation of so-called novel hydrogels, which consist of novel formulations such as nanoparticles, nanoemulsions, microemulsions, liposomes, etc. [5]. These dosage forms provide the opportunity to resolve some of the issues of drug delivery to the skin as well as being capable of controlling the drug release. Various polymers could be used for hydrogel preparation, including natural ones (such as chitosan [30,31], hydroxyethylcellulose [32], and hyaluronic acid [33]) or synthetic ones (such as carbomer [34], pluronic [27], and polyvinyl alcohol [35]). The choice of gelling agent can affect the properties of the prepared hydrogel and the expected drug behavior. Even though carbomer is one of the most widely used gelling agents for semisolid formulations, its gelation is pH-dependent and occurs in a neutral medium [36]. This could be inappropriate for pH-dependent Eudragit® L100 nanoparticles. Another considerably universal gelling agent for various routes of application are Pluronic derivatives [37]. Pluronics are water-soluble non-ionic triblock copolymers (PEO-PPO-PEO) of varying numbers of polyethylene oxide (PEO) and polypropylene oxide (PPO) units. Depending on the size of the blocks and molecular weight, different grades of Pluronic copolymers exist. The PEO and PPO blocks determine their amphiphilic structure, which allows micelle formation for the solubilization of lipophilic drugs. Depending on the type and concentration, a thermo-reversible gelation can be observed [37]. The most common representative applied as gelation aid is the hydrophilic Pluronic® F127. Its ease of gelation and biocompatibility make it very suitable for semisolid topical formulations. A disadvantage is its relatively low mechanical strength. On the other hand, methylcellulose is a cellulose derivative with excellent biocompatibility properties. It is the simplest ether derivative, with methyl groups substituting the hydroxyl ones at the C-2, C-3, and/or C-6 positions. There are a lot of commercial grades of methylcellulose, varying the degree and localization of substitution. Methylcellulose hydrogels have been exploited for dermal, ocular, vaginal, rectal, and oral drug delivery [38]. Therefore, gelling agents from two different groups were chosen in the current study for comparison purposes. The consistency of the hydrogels and their spreadability are important characteristics that provide information regarding the application or delivery of a desired drug dose to the skin and the ease of gel application. These properties significantly influence the patient's preference for the respective semisolid formulation [39].

The use of corticosteroids in the pediatric population should be considered with care, as these patients possess a higher propensity to develop adverse actions due to a higher surface area-to-body weight ratio and fragile skin [40]. A formulation of budesonide in an appropriate delivery system capable of providing efficient treatment of early forms of atopic dermatitis with limited side effects is a very attractive approach. The prolonged release achieved with the help of nanocomposites can overcome issues regarding systemic absorption through topical administration. Furthermore, a semisolid formulation could be suitable for easy application, a longer stay on the affected area, and a possible hydration effect. Thus, the aim of the current study is to develop and characterize Eudragit® L100-based nanoparticles loaded with budesonide for pH-sensitive delivery of the drug. Furthermore, the drug-loaded nanoparticles were formulated into two types of semisolid hydrogels as a final dosage form for the therapy of atopic dermatitis.

2. Results and Discussion

In the present study, budesonide is encapsulated into Eudragit L100 nanoparticles that are further formulated in hydrogel dosage form. The scientific rationale is to combine the pH-dependent budesonide delivery via Eudragit L100 nanoparticles with the hydration ability of methylcellulose or F127 hydrogels as a final dosage form.

2.1. Preparation and Characterization of the Nanoparticles

Eudragit L100 nanoparticles were successfully prepared by the nanoprecipitation technique. Eudrgait L 100 and budesonide were dissolved in ethanol, and their solution was slowly precipitated via mixing with a 0.25% aqueous solution of PVA as a non-solvent. During this mixing, rapid diffusion of the ethanol occurs in the water, which is accompanied by reduced interfacial tension and the formation of small droplets of the polymer and drug. Upon ethanol evaporation, nanoprecipitation occurs [17]. It appeared that the pH of the aqueous PVA-phase was a crucial factor in the preparation of particles with a size on the nanoscale. The medium diameter of the particles obtained with PVA-phase at pH 4.0 was approximately 6268 nm, whereas those prepared at pH 5.0 had an average diameter less than 60 nm. Furthermore, the ratio between both the organic and aqueous PVA-phases also influenced the particle size. Figure 1a shows the three ratios between both phases that were evaluated (1:1, 1:6, and 1:10, v/v). The optimal ratio between the ethanol and the aqueous phase was determined to be 1:10, since only at this ratio was the size of the particles at the nanoscale. As shown, the polydispersity slightly increased at this ratio but still indicated a narrow size distribution. The results from the dynamic light scattering analysis (DLS) for the optimized batch are presented in Figure 1b. It can be seen that there is no significant difference between the size of the empty (57.2 nm) and the drug-loaded nanoparticles (55.8 nm). Both particle samples showed a narrow particle size distribution, with PDI equal to 0.309 and 0.219 for the NP and Bud-NP, respectively. Similar results were observed by other studies [41].

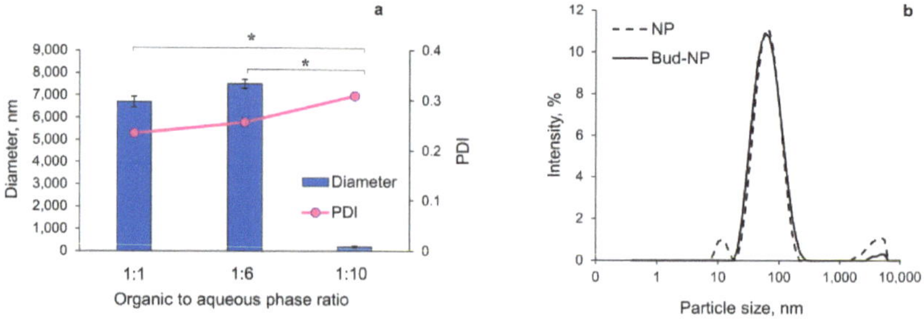

Figure 1. Mean diameter and polydispersity index with standard deviation ($n = 3$) of the particles prepared at different ratios between the ethanol and the aqueous phase (**a**) and histogram of particle size distribution by intensity of the optimized empty (NP) and budesonide-loaded nanoparticles (Bud-NP) (**b**) (significance level * $p < 0.05$).

Important information regarding nanoparticle colloidal stability can be provided by investigating their zeta potential. According to literature data, polymeric nanoparticles are considered stable if their absolute value of zeta potential is equal to or greater than 30 mV [42]. All batches of the prepared nanoparticles were characterized with similar values ranging from −30 mV to −32.7 mV. The negative zeta potential could be explained by the presence of carboxylic groups in the polymer carrier on the nanoparticle surface. Similar results regarding Eudragit L100 nanoparticles can be found in the literature [3,43,44]. The morphology of the prepared nanoparticles was characterized by TEM. The micrographs are presented in Figure 2. It can be seen that the nanoparticles were spherical in shape, and the observed diameter was correlated with that found by DLS.

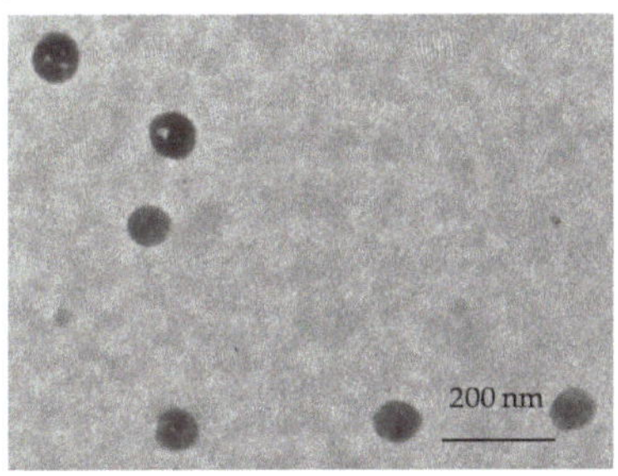

Figure 2. Transmission electron microscope image of the optimized nanoparticle batch.

The encapsulation efficiency was investigated in the case of different ratios between the drug and the polymer (correspondingly 1:5, 1:8, and 1:10, wt/wt). The results from the different batches are statistically different ($p < 0.005$). The results showed that encapsulation efficiency was paramount at a ratio of 1:8, achieving approximately 90% (Figure 3a). Similarly, the yield of the obtained lyophilized nanoparticles was highest at the same ratio (Figure 3b). The loading degree at a ratio of 1:8 was slightly lower than that obtained at a ratio of 1:5 (Figure 3a). Thus, taking into consideration the higher encapsulation efficiency and higher yield, the ratio 1:8 was selected as optimal, and all further tests were performed with these nanoparticles.

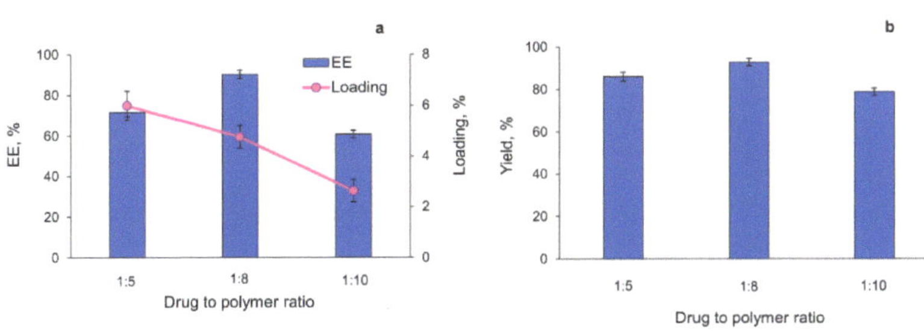

Figure 3. Influence of the ratio between the drug and the polymer on encapsulation efficiency, drug loading (**a**), and nanoparticle yield (**b**).

Figure 4 presents the FTIR spectra of budesonide, Eudragit L100, empty nanoparticles, and budesonide-loaded nanoparticles. The spectrum of budesonide consists of lots of well-resolved absorption peaks, which could be assigned as follows: a peak in the region 3600–3350 cm^{-1} due to OH-group stretching; peaks in the region 3000–2850 cm^{-1} due to the stretching of C−H bonds; peaks at 1721 cm^{-1}, 1666 cm^{-1} and 1622 cm^{-1}, attributed to stretching vibrations of C=O (carboxylic), conjugated C=O stretching, and C=C bonds, respectively. The Eudragit L100 spectrum represents a broad band in 3700–3080 cm^{-1} of stretching of the OH-group, which overlaps partially with the peaks of the C−H stretch in the 3060–2870 cm^{-1} range. A peak with high intensity at 1722 cm^{-1} is assigned to C=O ester stretching with a shoulder at 1620 cm^{-1}. The peaks of C−H bending vibrations are found in the region 1380–1470 cm^{-1}. The spectrum of the empty nanoparticles consists of the same peaks as this of Eudragit L100, with a slight difference in their intensities. The spectrum of the drug-loaded sample (Bud-NP) is characterized by the same peaks observed in the spectrum of the empty nanoparticles, accompanied by a noticeable additional peak at 1664 cm^{-1} characteristic for the budesonide, which confirms the loading of budesonide into the nanoparticles formed.

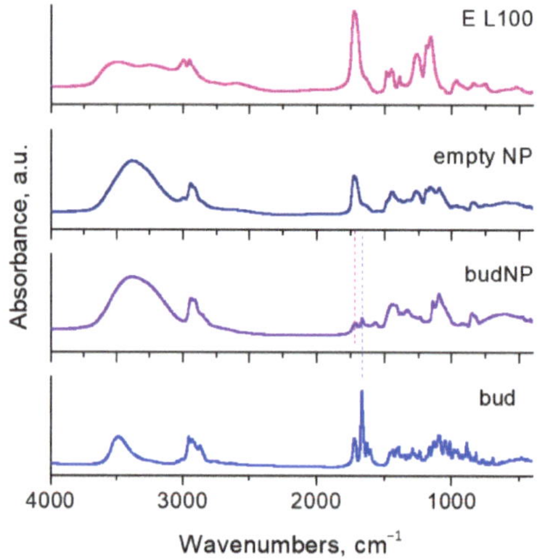

Figure 4. FTIR spectra of budesonide (Bud), budesonide-loaded nanoparticles (Bud-NP), empty nanoparticles (NP), and Eudragit L100 (E L100).

Part of the XRD patterns of budesonide, Eudragit L100, empty nanoparticles, and budesonide-loaded nanoparticles are presented in Figure 5. The pattern of budesonide represents a well-crystalline compound with two epimers (22R and 22S) [45]. Our detailed examination also revealed that it consists of patterns of the 22R and 22S epimers. The mass ratio of 3:1 (75% and 25% for 22R and 22S epimers, respectively) was calculated by comparing the total intensity of the peaks connected to both phases with the assumption of their equal density. Both epimers were found to crystallize in orthorhombic Space Group $P2_12_12_1$, and their unit cell parameters are calculated by our experimental data as follows: for 22R a = 8.516(4) Å, b = 9.185(3) Å, c = 28.87(1) Å; for 22S a = 8.449(2) Å, b = 9.127(2) Å, c = 20.099(9) Å. It is worth mentioning that they are very close to those reported in the study of Albertsson et al. [45]. The XRD pattern of Eudragit L100 shows typical amorphous humps at around $2\theta = 15°$ and 30°, as it was observed previously [46]. The well-visible shift of the maximum of the first amorphous peak from $2\theta = 15°$ to 18.5° upon the formation of empty nanoparticles can be seen. The shift indicates that, in the presence of PVA as a stabilizer, the characteristics of the encapsulated Eudragit L100 differ from those of the bulk Eudragit L100. This is a common feature of nanoparticles. In particular, the surface layer tends to have many structural defects, resulting in different types of bonding and coordination of the atoms compared to the bulk carrier [47]. The loaded sample shows the same amorphous peak as in the empty nanoparticles and some small crystalline peaks at around $2\theta = 6°$ and 10° indexed as (002) and (011) peaks of budesonide. The results confirm the successful loading of budesonide into Eudragit L100 nanoparticles.

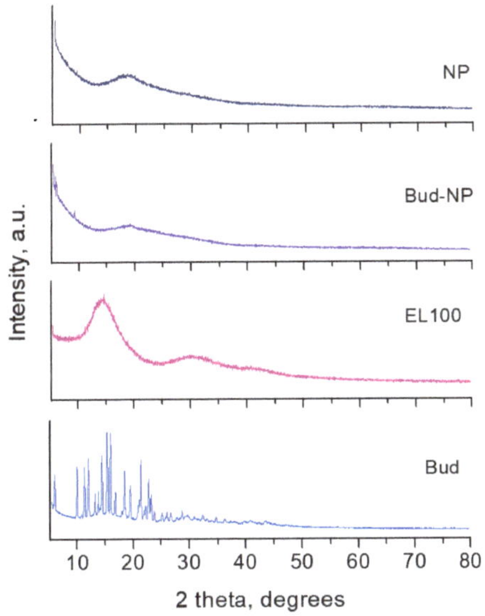

Figure 5. Powder XRD patterns of budesonide (Bud), budesonide-loaded nanoparticles (Bud-NP), empty nanoparticles (NP), and Eudragit L100 (E L100).

2.2. Cytotoxicity Evaluation of HaCaT Cells

Cell cytotoxicity assays represent one of the most frequently used in vitro bioassay methods for predicting the toxicity and irritating side effects of drugs and medical devices; thus, it is important to study thoroughly the response of cell mechanisms upon exposure to different compounds [48]. Cultured human keratinocytes offer a means to predict dermal irritancy resulting from exposure to various substances in humans [49,50]. Keratinocytes, being the first living cells that come into contact with externally applied compounds, represent a biologically relevant target for assessing skin irritants. However, primary keratinocyte cultures have inherent limitations, including limited and variable availability of source material and varying susceptibility to irritants with the number of passages. To overcome these challenges, HaCaT cells were employed as a model. These non-tumorigenic, spontaneously immortalized keratinocyte cells offer a nearly limitless supply of identical cells, thereby ensuring high levels of reproducibility within and between laboratories [50]. Furthermore, it is worth noting that in vitro cytotoxicity data obtained from the human keratinocyte line (HaCaT) closely correlates with in vivo data [51].

Therefore, we evaluated the potential cytotoxic effects of pure budesonide, budesonide loaded into the nanoparticles, and empty nanoparticles on the viability of the human keratinocyte line HaCaT. A colorimetric assay measuring the capacity for viable cells to metabolize a tetrazolium colorless salt to a blue formazan (MTT assay) was used as an indirect measurement of cell viability to predict skin irritancy. After 24 h of treatment, the empty nanoparticles (NP) at concentrations ranging from 2.65 to 85 μg/mL did not decrease cell viability and showed no toxic effects. Furthermore, both pure budesonide (ranging from 0.17 to 5.4 μg/mL) and budesonide loaded into the nanoparticles (corresponding concentrations) did not exhibit a statistically significant decrease in keratinocyte viability, as shown in Figure 6. In the tested concentrations, both samples demonstrated no in vitro toxicity and a favorable safety profile in the human keratinocyte HaCaT cell line.

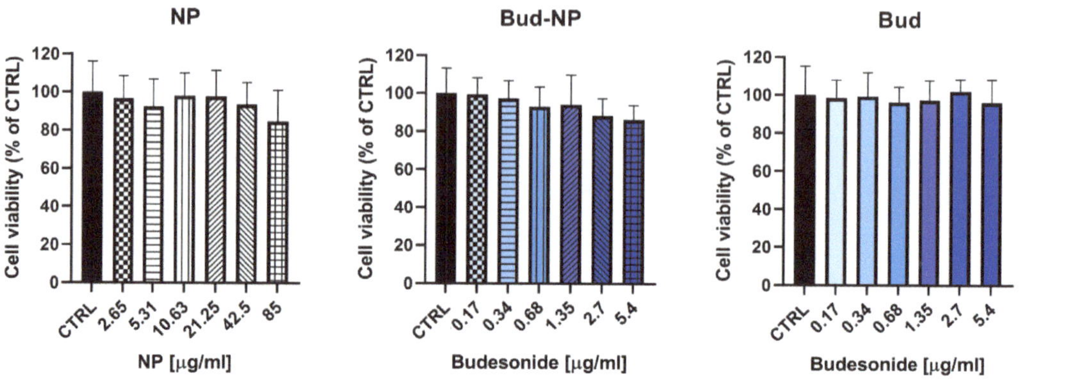

Figure 6. Cytotoxicity on HaCaT cells measured by MTT assay of pure budesonide (Bud), budesonide loaded in the nanoparticles (Bud-NP), and empty nanoparticles (NP). The results are expressed as means ± SD of triplicate assays ($n = 3$). All groups were compared statistically vs. untreated controls by one-way ANOVA with Dunnet's post hoc test.

2.3. Preparation and Characterization of the Hydrogels

The nanoparticle dispersions can be easily removed from the skin and exert limited contact with the affected area. Thus, in order to improve the retention of the drug at the site of application, the nanoparticles were incorporated into two types of hydrogels. Furthermore, the high water content of hydrogels would provide hydration for the atopic skin. The latter makes hydrogels very appropriate topical dosage forms since the disrupted barrier function of atopic skin allows transepidermal water loss. The hydrogels are prepared by simple gelling of nanoparticle dispersions with methylcellulose or Pluronic F127 (further referred to as F127). The gelling agent selection was based on their frequent use, biocompatibility, and generally regarded as safe (GRAS) status [52,53]. Light microscopic observations showed that the incorporation of the nanoparticles within both hydrogels did not lead to any changes in their appearance or stability. All gels maintained a homogeneous, transparent appearance with no visible aggregates upon nanoparticle incorporation. In addition, the pH of the prepared hydrogels was determined with or without the presence of nanoparticles. The F127 and methylcellulose plain gels had pH equal to 5.19 and 5.26, respectively. The results suggested that the semisolid vehicles are appropriate for the incorporation of the Eudragit L100 nanoparticles. The incorporation of the nanoparticles leads to a slight but insignificant decrease in the pH values (5.11 and 5.18, respectively). This can be attributed to the presence of PVA as a stabilizer for the nanoparticles. The data suggests the suitability of the proposed formulations for dermal application [54].

Preliminary dynamic rheological tests of methylcellulose- and Pluronic- based samples revealed a significant difference in the elastic properties of materials (Table 1). The hydrogel formed by F127 was much more elastic than the methylcellulose hydrogel (MC). At first glance, one of the reasons for the huge difference in the elastic modulus (G') of F127 and MC hydrogel carriers might be their different concentrations. More precisely, at the given concentration, the plain MC sample was in the form of a highly viscous solution ($G'' > G'$), which formed a soft gel upon adding the nanoparticles. On the other hand, the F127 system exhibited the typical behavior for hard gels ($G'' >> G'$) with and without NPs. It should also be noted that the gelation of the two polymers in aqueous media occurs by different mechanisms. Above certain critical concentrations and temperatures, the macromolecules of MC tend to intertwine, and some junction pints are formed to produce a weak physical hydrogel [55]. In contrast, under the reported experimental conditions, F127 macromolecules are self-assembled into nanosized micelles, which are closely packed into a three-dimensional network structure. Such material behaves as a hard gel [56].

Table 1. Elastic (G′) and loss (G″) moduli and complex dynamic viscosity (η*) of plain methylcellulose (MC) and Pluronic F127 (F127) hydrogels and the corresponding hydrogels containing empty (NP-F127 and NP-MC) and drug-loaded nanoparticles (Bud-NP-F127 and Bud-NP-MC).

Sample	F127	NP-F127	Bud-NP-F127	MC	NP-MC	Bud-NP-MC
G′ (Pa)	22,960	26,900	27,570	69	240	191
G″ (Pa)	3037	1847	1757	108	231	185
η* (Pa.s)	3686	4292	4396	20	52	49

Embedding the empty and budesonide-loaded nanoparticles into the hydrogel matrix resulted in increased elastic modulus and complex dynamic viscosity (η*) (see Table 1). The reinforcing effect of the nanoparticles can be explained by the fact that the NPs comprise a polymethacrylate derivative, which makes them more rigid than the hydrogel matrix.

The investigation of the occlusive properties of the selected gel bases was evaluated, taking into consideration their administration to atopic lesions. The occlusive properties of both hydrogels are compared to those of petrolatum, which is well known for its high occlusion [57]. The hydrogels are preferable semisolid vehicles for dermal delivery due to their more appealing properties and the reduced occlusive effect they possess [28]. Indeed, our study reveals that both hydrogels have a lower occlusion factor than petrolatum (Figure 7). The observed occlusive effect is due to the gelling agent present in the formulation, which tends to form a thin film on the surface, thus preventing water evaporation [58]. The lower occlusive effect compared to petrolatum is attributed to the hydrophilic properties of the gels. This result indicates that the hydrogels will ensure breathability during skin treatment. Thus, the combination of breathability and hydration ability of the developed hydrogels could be considered important parameters for effective healing of atopic lesions [59]. Further, the F127 gel showed more pronounced occlusion compared to the methylcellulose gel. This may be attributed to the lower concentration at which methylcellulose is used for the gelation. This is probably not the only reason since there is data in the literature that 0.5%–0.8% Carbopol-based gels showed an occlusive factor similar [60] or even higher [58] to the one of plain methylcellulose hydrogel in our study. Probably, the difference in the chemical structure of the gelling agents affects film formation and the prevention of water evaporation.

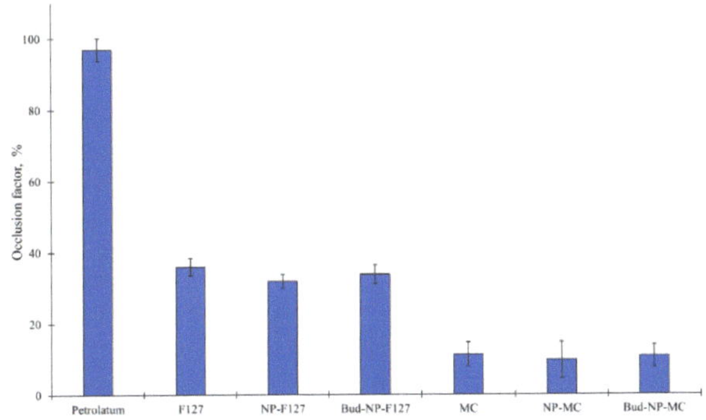

Figure 7. Occlusion factor for the plain methylcellulose (MC) and Pluronic F127 (F127) hydrogels and the same hydrogels containing the empty (NP-F127 and NP-MC) or drug-loaded nanoparticles (Bud-NP-F127 and Bud-NP-MC). Petrolatum was used as a positive control.

Further, the presence of nanoparticles was evaluated in terms of their effect on occlusive properties. It is well known that lipid nanoparticles exhibit skin occlusive effects [58,60]. There are limited data characterizing the occlusion of polymeric nanoparticles. Thus, in the present study, we investigated whether the embedment of Eudragit L100 nanoparticles within two different hydrogels affects their prevention of water evaporation.

The results showed that the incorporation of empty or drug loaded nanoparticles did not significantly alter the occlusion factor of the parent hydrogels. The retention of water could be useful in terms of the effectiveness of drug delivery as it could hydrate the stratum corneum [61] and also ameliorate the atopic skin condition [62].

The penetration and spreadability of the prepared hydrogels were also evaluated in order to provide some information regarding their ease of application. As can be seen in Figure 8, the addition of nanoparticles within the hydrogels is not related to a significant alteration of the depth of penetration. It can be seen that the F127-based gel shows a statistically lower depth of penetration compared to the methylcellulose gel (p = 0.012, one-way ANOVA). According to the rheology, it could be due to the closely packed three-dimensional network structure of this hydrogel compared to the weak physical hydrogel of methylcellulose. Similar values for the depth of penetration in F127 gels have been reported by other working groups [63].

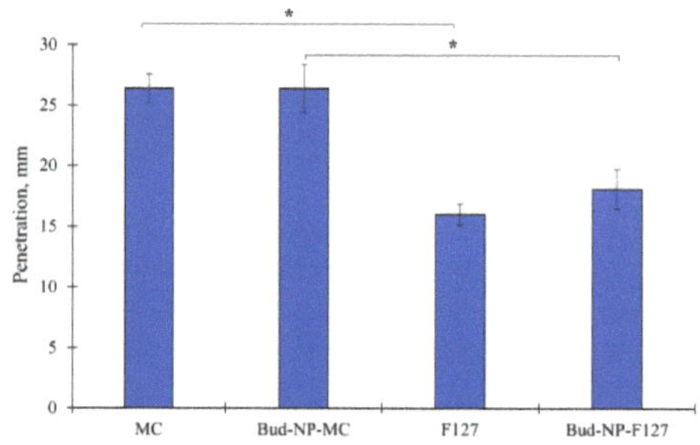

Figure 8. Penetration of plain hydrogels (MC and F127) and hydrogels containing budesonide-loaded nanoparticles (Bud-NP-MC and Bud-NP-F127); mean ± SD, n = 3; (* significant difference at level $p < 0.05$).

The results from the spreadability test for the plain and nanoparticles containing hydrogels are presented in Figure 9. It can be seen that the F127 gels are less spreadable than the ones with methylcellulose as a gelling agent. This can be attributed to the higher concentration used for gelation. Such results can be found in the literature [64]. According to the literature, methylcellulose solutions with a concentration of about 1% or less show thermogellation above 30 °C, depending on the molecular weight of the used methylcellulose [38]. Furthermore, the increase in molecular weight and concentration leads to gelation at a lower temperature [65], and typically hydrogels are formed at room temperature at a concentration of 3–6% [66]. In our study, a high-molecular weight methylcellulose was used at a concentration of 4%, leading to the formation of gel at room temperature with a spreadability factor of 5.95 mm^2/g. The incorporation of the Eudragit L100 nanoparticles in the methylcellulose hydrogels resulted in an increase in hydrogel spreadability, as shown in Figure 9a, and the corresponding spreadability factors are 9.09 mm^2/g and 8.91 mm^2/g for the NP-MC and Bud-NP-MC samples, respectively. Since Zilberman et al. reported a decrease in surface tension for the mixtures of cellulose derivatives and PVA [67], we suggest that the presence of PVA in nanoparticle dispersion may contribute to the larger spreadability of the hydrogels containing the nanoparticles.

In the case of F127-based gels, no significant difference is observed between the plain and composite gels, as shown in Figure 9b. Only a slight decrease in the spreadability factor is evidenced for the hydrogel loaded with the empty nanoparticles (4.23 ± 2.02 mm^2/g) compared to the empty hydrogel (5.49 ± 3.55 mm^2/g) and the hydrogel formulated with budesonide-loaded nanoparticles (4.96 ± 3.21 mm^2/g). Similar decrease in spreadability was observed for F127-based gels loaded with different types of nanoparticles [68].

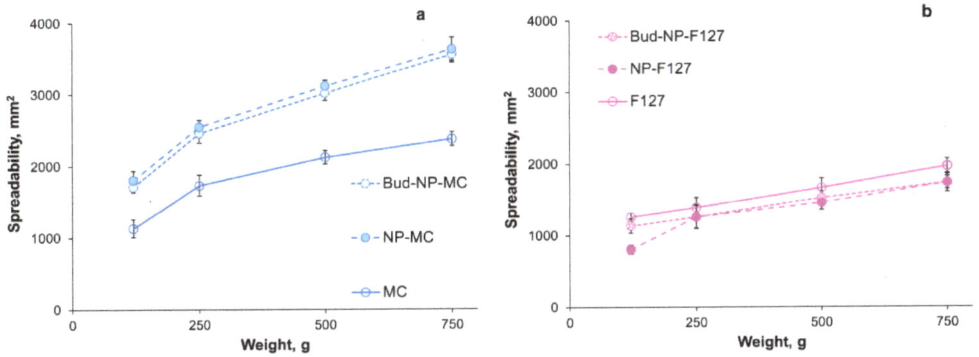

Figure 9. Extensiometric profiles of plain methylcellulose (MC) (**a**) and F127-based gels (**b**) and the same hydrogels containing empty (NP-MC and NP-F127) or drug-loaded nanoparticles (Bud-NP-MC and Bud-NP-F127). Mean ± SD, $n = 3$.

Thus, the different effects of nanoparticle incorporation in the two types of hydrogels could be explained by the different mechanisms of gelation for F127 and methylcellulose. According to most recent literature studies, methylcellulose forms gel based on the fibril theory [53,69]. The coiling of the fibrils is more pronounced at lower pH values as opposed to higher pH values, and the viscosity is correspondingly lower as there is limited possibility for polymer-polymer interaction [70]. In the present study, the incorporation of the nanoparticles leads to a slight reduction of the pH. At the same time, F127 gelation is due to the very tight packing of the formed micelles and their overlaying [37,71]. The enthropy is determining the gelation process [37], and probably the nanoparticles do not affect it.

The release profiles of the free drug from the two hydrogels as well as budesonide-loaded nanoparticles (Bud-NP) and their corresponding hydrogels are shown in Figure 10. The drug release from the nanoparticle dispersion fits the best Higuchi release kinetics (Table 2). It can therefore be expected that budesonide release is diffusion-driven through the undissolved Eudragit L100 matrix nanoparticles. Similar results for the release of Eudragit-based polymeric nanoparticles have been reported by other researchers [72,73]. Similarly, the Higuchi model best fits the release of non-encapsulated budesonide from the two hydrogels (Table 2). As shown, with values for the diffusional exponent $n > 0.5$ (Korsmeyer-Peppas model), a non-Fickian diffusion controlled the release from the hydrogels, whereas quasi-Fickian diffusion could be considered in the case of nanoparticles ($n < 0.5$).

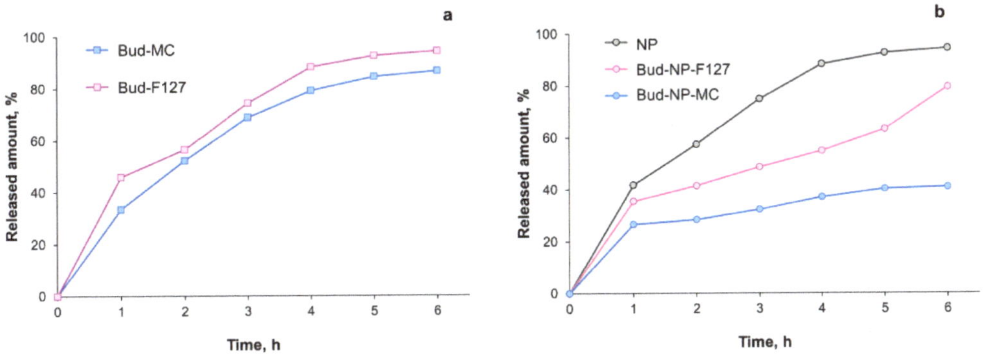

Figure 10. In vitro drug release from the hydrogels containing non-encapsulated budesonide (**a**), nanoparticles, and hydrogels with nanoparticles (**b**) in a buffer medium (pH 5.5).

Table 2. Kinetic parameters of the in vitro drug release from the nanoparticles containing methylcellulose (Bud-NP-MC) and F127 (Bud-NP-F127) hydrogels.

Formulation	Zero Order	First Order	Higuchi	Korsmeyer-Peppas
	$Q_t = Q_0 - k_0 t$	$lnQ_t = lnQ_0 - k_1 t$	$Q_t = k_H t^{1/2}$	$\frac{M_t}{M_\infty} = k.t^n$
Bud-MC	$R^2 = 0.8870$ $k = 13.895$	$R^2 = 0.9875$ $k = -0.152$	$R^2 = 0.9874$ $k = 37.678$	$R^2 = 0.8909$ $n = 0.672$
Bud-F127	$R^2 = 0.8598$ $k = 14.567$	$R^2 = 0.9848$ $k = -0.425$	$R^2 = 0.9850$ $k = 40.073$	$R^2 = 0.8344$ $n = 0.613$
Bud-NP	$R^2 = 0.871$ $k = 14.844$	$R^2 = 0.9862$ $k = -0.2176$	$R^2 = 0.9874$ $k = 40.619$	$R^2 = 0.8612$ $n = 0.413$
Bud-NP-MC	$R^2 = 0.9673$ $k = 3.175$	$R^2 = 0.9687$ $k = -0.021$	$R^2 = 0.9604$ $k = 11.007$	$R^2 = 0.9337$ $n = 0.268$
Bud-NP-F127	$R^2 = 0.9599$ $k = 8.309$	$R^2 = 0.8673$ $k = -0.090$	$R^2 = 0.9089$ $k = 28.133$	$R^2 = 0.9125$ $n = 0.418$

Q—amount of drug; *k*—rate constant; *t*—time; *n*—release exponent.

Comparing the release from both hydrogels containing non-encapsulated (Figure 10a) or encapsulated drugs (Figure 10b), it can be concluded that the methylcellulose gel is characterized by a slower and incomplete release as opposed to the F127. The drug release from the NP-loaded F127 gel follows zero-order (Table 2). Such behavior is observed by other researchers as well [74]. It is due predominantly to the F127 dissolution in the medium [75,76]. In the case of methylcellulose gel, the swelling of the polymer retards the drug release. Furthermore, the investigation of the release kinetics suggests a first-order pattern. This is in accordance with previous data suggesting that the polymer itself may retain the drug [77]. Another study comparing the release of free drugs from methylcellulose and poloxamer gel showed the slowest release from the methylcellulose gel, even though the viscosity was lower than the Pluronic F127 one [78]. The authors explain these findings due to the interaction between the polymer and drug. Another study points towards the significance of drug-polymer interactions rather than the viscosity or concentration of the polymer [79]. In the current study, probably the Eudragit nanoparticles stabilized with PVA interact with methylcellulose but not with Pluronic F127. These assumptions are also supported by the susceptibility of the methylcellulose gel's spreadability and viscosity to the incorporation of the proposed nanoparticles. The calculation of the similarity factor between the two release profiles shows that they are not similar ($f_2 = 33.1$). Possible reasons for the different effect that the gelling agent exerts on the release pattern are the surface active properties of F127 and its tendency to form micelles at concentrations above 0.725 wt% at 25 °C [80]. Therefore, in the case of F127 hydrogel, budesonide is probably solubilized, which enables the release process. Furthermore, the drug release from Pluronic F127 gel is governed by gel erosion and is not affected to a significant extent by drug diffusion [71]. Such an assumption is supported by the release kinetics findings in the current study.

3. Conclusions

Budesonide was successfully encapsulated in Eudragit nanoparticles (approximately 90% encapsulation efficiency) intended to provide local drug delivery at pH 5.5 and above, which is desired for atopic skin treatment. The nanoparticles possess appropriate physicochemical properties, particularly their small size and highly negative surface charge, that are prerequisites for improved penetration and colloid stability, respectively. Prolonged release was achieved, which could reduce the applied dose. The lack of irritancy of the prepared nanocarriers was demonstrated in vitro in the human keratinocyte cell line, HaCaT. Further, the budesonide-loaded nanoparticles were homogeneously embedded in two types of hydrogels, based on methylcellulose or Pluronic F127, able to provide ease of application and hydration ability to the topical formulation. Both hydrogels showed suitability for dermal application in terms of spreadability, penetration, pH, and occlusion properties. At the same time, the budesonide release from the F127 gel was more complete in the tested time frame, making it more practically applicable.

4. Materials and Methods

4.1. Materials

Budesonide, methylcellulose (Methocel 90HG), and ethanol (96%) were purchased from Sigma Aldrich; poly(methacrylic acid-co-methyl methacrylate) 1:1 (Eudragit® L100) from Evonik Röhm GmbH (Darmstadt, Germany); polyvinyl alcohol (PVA 22000) from Fluka Chemie AG (Germany); and Pluronic F127 from BASF (Ludwigshafen, Germany). Distilled water was prepared in the laboratory. For HPLC analysis, acetonitrile and methanol HPLC grades were used from Fisher Chemical (Thermo Fisher Scientific Inc., Waltham, MA, USA). Dulbecco's Modified Eagle's Medium, MTT (3-(4,5-dimethylthiazol-2-yl)-2,5-diphenyltetrazolium bromide), fetal bovine serum, and L-glutamine were purchased from Sigma-Aldrich (Merck KGaA, Darmstadt, Germany). The Human immortalized keratinocyte cell line HaCaT (300493) was acquired from the CLS Cell Lines Service GmbH (CLS, Eppelheim, Germany).

4.2. Preparation of the Nanoparticles

The nanoparticles were prepared by nanoprecipitation according to the procedure suggested by Sahle et al. [21], with some modifications as shown in Figure 11. First, PVA (0.25% wt/v) was dissolved in purified water, and the pH was adjusted to 5.0 by the addition of 0.1 N HCl. Eudragit® L100 and budesonide were dissolved in 95% ethanol in different concentrations, giving the following ratios in regard to Eudragit® L100: 1:5, 1:8, and 1:10 (wt/wt). Afterwards, the ethanol solution was added dropwise to the PVA while being sonicated at 80 kHz (Bandelin Sonoplus HD3100, Bandelin Electronics, Berlin, Germany) for 1 min. The sonication was continued for 1 more minute after the ethanol solution was added completely. Then, the resultant dispersion was left for 24 h under continuous stirring for ethanol evaporation. Upon the evaporation of the organic solvent, Eudragit® L100 precipitates into nanoparticles stabilized by the non-ionic surfactant PVA. The prepared dispersion was filtered (0.45 μm), and the filters were rinsed with ethanol (50%). The encapsulation efficiency was determined based on the initial amount of budesonide (Bud_{total}) and the amount found in the filter fractions (Bud_{filter}). The following equation was used for the calculation:

$$EE\% = \frac{Bud_{total} - Bud_{filter}}{Bud_{total}}.100 \quad (1)$$

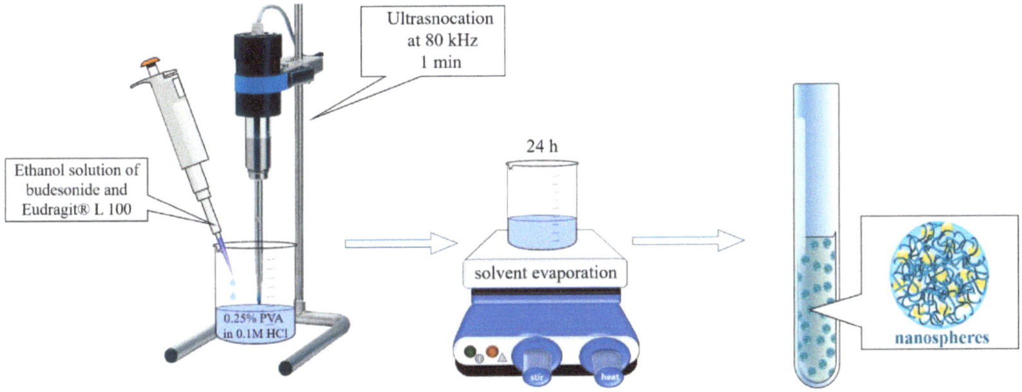

Figure 11. Schematic representation of the nanoprecipitation for nanoparticle preparation.

4.3. Determination of Nanoparticle Size, Polydispersity Index (PDI), and Zeta-Potential

Dynamic light scattering (DLS) was applied to investigate the particle size, polydispersity (PDI), and zeta-potential of the prepared empty and budesonide-loaded nanoparticles (Zeta-Master, Malvern Instruments, Worcestershire, UK). The measurements were performed in triplicate on the aqueous nanoparticle dispersions at 25 °C with a scattering angle of 90°. Transmission electron microscopy (TEM) was applied for the evaluation of nanoparticle shape and surface morphology (HR STEM JEOL JEM 2100, Tokyo, Japan).

4.4. X-ray Powder Diffraction Analysis (XRPD) and FTIR Spectrophotometry

The diffraction patterns of budesonide, budesonide-loaded nanoparticles, empty nanoparticles, and Eudragit L100 were collected from 5 to 80°2θ on a Bruker D8-Advance Diffractometer (Karlsruhe, Germany). CuKα radiation was used, and registration was performed by the LynxEye detector. The unit cell parameters were refined using the Topas 4.2 program, part of the Bruker software (Bruker AXS, Karsruhe, Germany).

The FTIR spectra of budesonide, budesonide-loaded nanoparticles, empty nanoparticles, and Eudragit L100 in KBr were recorded on a Thermo Nicolet Avatar 360 FTIR spectrometer (Thermo Fisher Scientific, Waltham, MA, USA), within the range 4000–400 cm^{-1} with a resolution of 2 cm^{-1}.

4.5. Cytotoxicity Evaluation of HaCaT Cells

The cell line was cultured in 75 cm^2 flasks in DMEM medium with glucose (4.5 g/L), to which 10% fetal bovine serum and 2 mM L-glutamine were added. The cells were maintained at a constant temperature of 37 °C within an environment comprising 5% CO_2. When the cells reached approximately 80% confluence, a series of sequential steps were performed. The cells were first harvested using a trypsin/EDTA solution, after which they were precisely seeded into the central 60 wells of 96-well plates at a density of 5×10^4 cells per milliliter. Subsequently, these plates were placed in an incubator and maintained at 37 °C with 5% CO_2 for a period of 24 h. This process was meticulously repeated three times, utilizing cells from different passages, to ensure experimental consistency and reliability.

The MTT assay was employed to assess the cytotoxicity of the tested samples according to the previously described procedure [81]. The cells were treated with a reference solution of budesonide (0.17, 0.34, 0.68, 1.35, 2.7, and 5.4 μg/mL), dispersion of budesonide loaded nanoparticles (in the same concentrations), and dispersion of empty nanoparticles (from 2.65 μg/mL to 85 μg/mL). Each plate included control wells that contained only culture medium. After 24 h of treatment, the culture medium was aspirated and replaced with 100 μL of the MTT solution (5 mg/mL in phosphate-buffered saline) in each well. Subsequently, the plates were incubated for a period of 3 h, the cell culture medium was aspirated, and 100 μL of dimethylsulfoxide (DMSO) per well was added to dissolve the purple formazan product. This was achieved by gently shaking the plates for 10 min at room temperature. The absorbance of the resulting solutions was measured at 570 nm using a multiplate reader, Synergy 2 (BioTek Instruments, Inc., Highland Park, Winooski, VT, USA).

For statistical analysis, GraphPad Prism 8 Software was utilized. The data underwent a one-way analysis of variance (ANOVA), which was followed by Dunnett's multiple comparisons post-test. This post-test was employed to assess and compare differences between the control and treatment groups. A significance level of 0.05 was selected as the threshold for determining statistical significance in all the comparisons conducted.

4.6. Hydrogel Preparation

A hydrogel was proposed as a semisolid dosage form containing budesonide loaded nanoparticles. Two types of hydrogels were formulated using methylcellulose (MC) (4 wt %) and Pluronic F127 (F127) (25 wt %) as gelling agents. Pluronic F127 was dispersed in the medium, while methylcellulose was dispersed in the hot medium, and then they were both left to dissolve completely at 4 °C overnight. The gels containing budesonide-loaded nanoparticles were prepared in a similar manner, with the liquid medium being the nanoparticle-containing dispersion. In the case of methylcellulose, the different gelling agent concentrations were compensated by the addition of distilled water with a pH of 5.0. In this way, the budesonide concentration in both hydrogels was adjusted to 0.1 mg/g gel.

4.7. Appearance and pH of Hydrogels

All formulations were visually and microscopically investigated using a Leica DM750 light microscope equipped with Air Teach software (v.1.0.9874) (Heerbrugg, Switzerland). The pH of the formulations was determined by the potentiometric method with a pH meter (Hanna HI98100, Hanna Instruments Inc., Woonsocket, RI, USA). The investigated gel samples were diluted (1:4) with distilled water, mixed vigorously for 1 min, and the pH was recorded [82,83].

4.8. In Vitro Occlusion Test

The occlusive properties of the gels were tested in vitro based on the measurement of water evaporation in controlled environmental conditions, as proposed by Caldas et al. [84]. In brief, a beaker was filled with 25 mL of distilled water, covered with a Whatman cellulose filter (0.45 μm), and tightly sealed with Teflon tape. An equal amount of the tested gels was evenly spread on the surface of the filter paper (surface area: 13.84 cm^2). The samples were accurately weighed and stored in a climate chamber (T = 32 ± 0.5 °C; RH = 50% ± 1%) in the dark. After 48 h, the samples were weighed again. The water loss of the sample (L_S) was calculated based on the change in weight. The difference with the reference sample's loss (L_R) was used for the calculation of the occlusive factor (F).

$$F\% = \frac{L_R - L_S}{L_R}.100 \quad (2)$$

The reference sample was a beaker with plain filter paper on top. An occlusive factor of 100 means a maximal occlusive effect, while an occlusive factor of 0 means no occlusion [85].

4.9. Rheology, Spreadability, and Penetrometry of Hydrogels

Dynamic rheological measurements of hydrogels were carried out with a HAAKE MARS 60 rheometer in controlled deformation mode using a parallel plate sensor system (top plate diameter = 20 mm; gap = 1 mm). The elastic (G') modulus was determined at 32 °C and constant deformation (γ = 0.01) in the 0.1–10 Hz frequency range.

The spreadability test characterizing the rheological properties of the hydrogels was performed with the parallel plate method [86,87]. A circle with a diameter of 1 cm was marked on a glass plate, and a sample of 1 g of the tested gel was placed inside. A second glass plate with a known weight was set on top. Subsequent weights are placed on top every 5 min. The diameter (d) of the spread gel was measured and recorded after each weight (W). The results were plotted to obtain the extensiometric profiles of the samples. All measurements were performed in triplicate. The spreadability (S) and spreadability factor (S_F) were calculated based on the following equations:

$$S = \frac{d^2.\pi}{4} \quad (3)$$

$$S_F = \frac{S}{W} \quad (4)$$

The consistency of the semisolid formulations was evaluated using the pharmacopoeial penetrometry test [88]. The gel samples with a sufficient amount were prepared immediately after gelation and stored in the test container for 24 h at 25 ± 0.5 °C prior to testing. The gravity-driven penetrating object was released for 5 s, and the depth of penetration was measured in millimeters.

4.10. In Vitro Dissolution Test and Release Kinetics

The dissolution test was performed in a buffer medium with a pH = 5.5, simulating the physiological acidity of the skin. A sample (corresponding to 0.65 mg budesonide) was placed in a dialysis membrane (MW 10 000 Da) and introduced into a 50 mL acceptor phase tempered at 32 ± 0.5 °C at constant shaking. Aliquot samples were withdrawn at predetermined time intervals and replaced with fresh medium. The released drug amount was evaluated using the HPLC method. The chromatographic procedure was carried out with the HPLC system UltiMate Dionex 3000 SD, Chromeleon 7.2 SR3 Systems (Thermo Fisher Scientific Inc., Waltham, MA, USA). The separation was achieved with Column Luna (Phenomenex, Torrance, CA, USA) C18, 250 × 4.60 mm, particle size 5 μm, and a Diode Array Detector. The chromatographic conditions are as follows: mobile phase acetonitrile:methanol (70:30 v/v), flow rate 1.0 mL/min, and a wavelength of 254 nm. The amount was calculated based on a standard curve prepared in the concentration range of 3.5–10 μg/mL.

The drug release mechanism of the hydrogels was investigated by fitting the release profiles according to different release kinetic equations. Further, regression analysis was performed to evaluate the best fit.

Author Contributions: Conceptualization, M.S. and K.Y.; methodology, M.S., V.T. and K.Y.; investigation, M.S., C.L., I.P.-E.T., D.S., I.S., D.K. and P.D.P.; writing—original draft preparation, M.S., C.L., I.P.-E.T., D.S., V.T., I.S. and D.K.; writing—review and editing, K.Y. and P.D.P.; visualization, M.S., C.L., D.S., I.S. and D.K.; project administration, K.Y.; funding acquisition, K.Y. All authors have read and agreed to the published version of the manuscript.

Funding: This research was funded by the European Union-NextGenerationEU through the National Recovery and Resilience Plan of the Republic of Bulgaria, project № BG-RRP-2.004-0004-C01.

Institutional Review Board Statement: Not applicable.

Informed Consent Statement: Not applicable.

Data Availability Statement: The data are contained within the article.

Conflicts of Interest: The authors declare no conflicts of interest.

References

1. Suga, H.; Sato, S. Novel Topical and Systemic Therapies in Atopic Dermatitis. *Immunol. Med.* **2019**, *42*, 84–93. [CrossRef] [PubMed]
2. Singh, S.; Behl, T.; Sharma, N.; Zahoor, I.; Chigurupati, S.; Yadav, S.; Rachamalla, M.; Sehgal, A.; Naved, T.; Pritima; et al. Targeting Therapeutic Approaches and Highlighting the Potential Role of Nanotechnology in Atopic Dermatitis. *Environ. Sci. Pollut. Res.* **2022**, *29*, 32605–32630. [CrossRef] [PubMed]
3. Dong, P.; Sahle, F.F.; Lohan, S.B.; Saeidpour, S.; Albrecht, S.; Teutloff, C.; Bodmeier, R.; Unbehauen, M.; Wolff, C.; Haag, R.; et al. pH-Sensitive Eudragit® L 100 Nanoparticles Promote Cutaneous Penetration and Drug Release on the Skin. *J. Control. Release* **2019**, *295*, 214–222. [CrossRef] [PubMed]
4. Padula, C.; Machado, I.P.; Vigato, A.A.; de Araujo, D.R. New Strategies for Improving Budesonide Skin Retention. *Pharmaceutics* **2021**, *14*, 30. [CrossRef] [PubMed]
5. Almoshari, Y. Novel Hydrogels for Topical Applications: An Updated Comprehensive Review Based on Source. *Gels* **2022**, *8*, 174. [CrossRef] [PubMed]
6. Davari, D.R.; Nieman, E.L.; McShane, D.B.; Morrell, D.S. Current Perspectives on the Management of Infantile Atopic Dermatitis. *J. Asthma Allergy* **2020**, *13*, 563–573. [CrossRef] [PubMed]
7. Shetty, K.; Sherje, A.P. Nano Intervention in Topical Delivery of Corticosteroid for Psoriasis and Atopic Dermatitis-a Systematic Review. *J. Mater. Sci. Mater. Med.* **2021**, *32*, 88. [CrossRef]
8. Kwatra, G.; Mukhopadhyay, S. Topical corticosteroids: Pharmacology. In *A Treatise on Topical Corticosteroids in Dermatology: Use, Misuse and Abuse*; Lahiri, K., Ed.; Springer: Singapore, 2018; pp. 11–22. ISBN 978-981-10-4609-4.
9. Spada, F.; Barnes, T.M.; Greive, K.A. Comparative Safety and Efficacy of Topical Mometasone Furoate with Other Topical Corticosteroids. *Australas. J. Dermatol.* **2018**, *59*, e168–e174. [CrossRef]
10. Campos, E.V.R.; Proença, P.L.F.; da Costa, T.G.; de Lima, R.; Hedtrich, S.; Fraceto, L.F.; de Araujo, D.R. Hydrogels Containing Budesonide-Loaded Nanoparticles to Facilitate Percutaneous Absorption for Atopic Dermatitis Treatment Applications. *ACS Appl. Polym. Mater.* **2021**, *3*, 4436–4449. [CrossRef]

11. Li, C.-Y.; Liu, Z. Effect of Budesonide on Hospitalization Rates among Children with Acute Asthma Attending Paediatric Emergency Department: A Systematic Review and Meta-Analysis. *World J. Pediatr. WJP* **2021**, *17*, 152–163. [CrossRef]
12. Zhou, H.; Qian, H. Preparation and Characterization of pH-Sensitive Nanoparticles of Budesonide for the Treatment of Ulcerative Colitis. *Drug Des. Devel. Ther.* **2018**, *12*, 2601–2609. [CrossRef] [PubMed]
13. Mota, F.L.; Carneiro, A.P.; Queimada, A.J.; Pinho, S.P.; Macedo, E.A. Temperature and Solvent Effects in the Solubility of Some Pharmaceutical Compounds: Measurements and Modeling. *Eur. J. Pharm. Sci. Off. J. Eur. Fed. Pharm. Sci.* **2009**, *37*, 499–507. [CrossRef] [PubMed]
14. Daley-Yates, P.T. Inhaled Corticosteroids: Potency, Dose Equivalence and Therapeutic Index. *Br. J. Clin. Pharmacol.* **2015**, *80*, 372–380. [CrossRef] [PubMed]
15. Cho, J.H.; Baek, H.H.; Lee, J.M.; Kim, J.H.; Kim, D.D.; Cho, H.K.; Cheong, I.W. Topical Delivery of Budesonide Emulsion Particles in the Presence of PEO-PCL-PEO Triblock Copolymers. *Macromol. Res.* **2009**, *17*, 969–975. [CrossRef]
16. Goyal, R.; Macri, L.K.; Kaplan, H.M.; Kohn, J. Nanoparticles and Nanofibers for Topical Drug Delivery. *J. Control. Release Off. J. Control. Release Soc.* **2016**, *240*, 77–92. [CrossRef]
17. Rao, J.P.; Geckeler, K.E. Polymer Nanoparticles: Preparation Techniques and Size-Control Parameters. *Prog. Polym. Sci.* **2011**, *36*, 887–913. [CrossRef]
18. Andrés Real, D.; Gagliano, A.; Sonsini, N.; Wicky, G.; Orzan, L.; Leonardi, D.; Salomon, C. Design and Optimization of pH-Sensitive Eudragit Nanoparticles for Improved Oral Delivery of Triclabendazole. *Int. J. Pharm.* **2022**, *617*, 121594. [CrossRef]
19. Rowe, R.C. *Handbook of Pharmaceutical Excipients*; Pharmaceutical Press: London, UK, 2020.
20. Patra, C.N.; Priya, R.; Swain, S.; Kumar Jena, G.; Panigrahi, K.C.; Ghose, D. Pharmaceutical Significance of Eudragit: A Review. *Future J. Pharm. Sci.* **2017**, *3*, 33–45. [CrossRef]
21. Sahle, F.F.; Balzus, B.; Gerecke, C.; Kleuser, B.; Bodmeier, R. Formulation and in Vitro Evaluation of Polymeric Enteric Nanoparticles as Dermal Carriers with pH-Dependent Targeting Potential. *Eur. J. Pharm. Sci.* **2016**, *92*, 98–109. [CrossRef]
22. Cardoso, A.M.L.; Oliveira, E.E.; Machado, B.A.S.; Marcelino, H.R. Eudragit®-Based Nanoparticles for Controlled Release through Topical Use. *J. Nanoparticle Res.* **2023**, *25*, 32. [CrossRef]
23. Menon, J.U.; Kona, S.; Wadajkar, A.S.; Desai, F.; Vadla, A.; Nguyen, K.T. Effects of Surfactants on the Properties of PLGA Nanoparticles. *J. Biomed. Mater. Res. A* **2012**, *100*, 1998–2005. [CrossRef] [PubMed]
24. Marwah, H.; Garg, T.; Goyal, A.K.; Rath, G. Permeation Enhancer Strategies in Transdermal Drug Delivery. *Drug Deliv.* **2016**, *23*, 564–578. [CrossRef]
25. Ference, J.D.; Last, A.R. Choosing Topical Corticosteroids. *Am. Fam. Physician* **2009**, *79*, 135–140. [PubMed]
26. Coondoo, A.; Chattopadhyay, C. Use and Abuse of Topical Corticosteroids in Children. *Indian J. Paediatr. Dermatol.* **2014**, *15*, 1. [CrossRef]
27. Cai, M.-H.; Chen, X.-Y.; Fu, L.-Q.; Du, W.-L.; Yang, X.; Mou, X.-Z.; Hu, P.-Y. Design and Development of Hybrid Hydrogels for Biomedical Applications: Recent Trends in Anticancer Drug Delivery and Tissue Engineering. *Front. Bioeng. Biotechnol.* **2021**, *9*, 630943. [CrossRef] [PubMed]
28. Slavkova, M.; Tzankov, B.; Popova, T.; Voycheva, C. Gel Formulations for Topical Treatment of Skin Cancer: A Review. *Gels* **2023**, *9*, 352. [CrossRef] [PubMed]
29. Harrison, I.P.; Spada, F. Hydrogels for Atopic Dermatitis and Wound Management: A Superior Drug Delivery Vehicle. *Pharmaceutics* **2018**, *10*, 71. [CrossRef]
30. Cardoso, A.M.; de Oliveira, E.G.; Coradini, K.; Bruinsmann, F.A.; Aguirre, T.; Lorenzoni, R.; Barcelos, R.C.S.; Roversi, K.; Rossato, D.R.; Pohlmann, A.R.; et al. Chitosan Hydrogels Containing Nanoencapsulated Phenytoin for Cutaneous Use: Skin Permeation/Penetration and Efficacy in Wound Healing. *Mater. Sci. Eng. C* **2019**, *96*, 205–217. [CrossRef]
31. Morsi, N.M.; Abdelbary, G.A.; Ahmed, M.A. Silver Sulfadiazine Based Cubosome Hydrogels for Topical Treatment of Burns: Development and in Vitro/in Vivo Characterization. *Eur. J. Pharm. Biopharm. Off. J. Arbeitsgemeinschaft Pharm. Verfahrenstechnik EV* **2014**, *86*, 178–189. [CrossRef]
32. Marafon, P.; Fachel, F.N.S.; Dal Prá, M.; Bassani, V.L.; Koester, L.S.; Henriques, A.T.; Braganhol, E.; Teixeira, H.F. Development, Physico-Chemical Characterization and in-Vitro Studies of Hydrogels Containing Rosmarinic Acid-Loaded Nanoemulsion for Topical Application. *J. Pharm. Pharmacol.* **2019**, *71*, 1199–1208. [CrossRef]
33. Wei, S.; Xie, J.; Luo, Y.; Ma, Y.; Tang, S.; Yue, P.; Yang, M. Hyaluronic Acid Based Nanocrystals Hydrogels for Enhanced Topical Delivery of Drug: A Case Study. *Carbohydr. Polym.* **2018**, *202*, 64–71. [CrossRef] [PubMed]
34. González-Delgado, J.A.; Castro, P.M.; Machado, A.; Araújo, F.; Rodrigues, F.; Korsak, B.; Ferreira, M.; Tomé, J.P.C.; Sarmento, B. Hydrogels Containing Porphyrin-Loaded Nanoparticles for Topical Photodynamic Applications. *Int. J. Pharm.* **2016**, *510*, 221–231. [CrossRef] [PubMed]
35. Yasasvini, S.; Anusa, R.S.; VedhaHari, B.N.; Prabhu, P.C.; RamyaDevi, D. Topical Hydrogel Matrix Loaded with Simvastatin Microparticles for Enhanced Wound Healing Activity. *Mater. Sci. Eng. C Mater. Biol. Appl.* **2017**, *72*, 160–167. [CrossRef] [PubMed]
36. Srivastava, P. Excipients for semisolid formulations. In *Excipient Development for Pharmaceutical, Biotechnology, and Drug Delivery Systems*; Taylor & Francis Group: Oxford, UK, 2006; p. 197.
37. Russo, E.; Villa, C. Poloxamer Hydrogels for Biomedical Applications. *Pharmaceutics* **2019**, *11*, 671. [CrossRef] [PubMed]
38. Nasatto, P.L.; Pignon, F.; Silveira, J.L.M.; Duarte, M.E.R.; Noseda, M.D.; Rinaudo, M. Influence of Molar Mass and Concentration on the Thermogelation of Methylcelluloses. *Int. J. Polym. Anal. Charact.* **2015**, *20*, 110–118. [CrossRef]

39. Mut, A.M.; Vlaia, L.; Coneac, G.; Olariu, I.; Vlaia, V.I.; Stănciulescu, C.O.; Mitu, M.A.; Szabadai, Z.; Lupuleasa, D. Chitosan/Hpmc-Based Hydrogels Containing Essential Oils for Topical Delivery of Fluconazole: Preliminary Studies. *Farmacia* **2018**, *66*, 248–256.
40. Hengge, U.R.; Ruzicka, T.; Schwartz, R.A.; Cork, M.J. Adverse Effects of Topical Glucocorticosteroids. *J. Am. Acad. Dermatol.* **2006**, *54*, 1–15. [CrossRef]
41. Saadallah, M.S.; Hamid, O.A. Eudragit® L100 Nanoparticles for Transdermal Delivery of Rosuvastatin Calcium. *J. Excip. Food Chem.* **2022**, *13*, 80–93.
42. Duman, O.; Tunç, S. Electrokinetic and Rheological Properties of Na-Bentonite in Some Electrolyte Solutions. *Microporous Mesoporous Mater.* **2009**, *117*, 331–338. [CrossRef]
43. Cetin, M.; Atila, A.; Kadioglu, Y. Formulation and in Vitro Characterization of Eudragit® L100 and Eudragit® L100-PLGA Nanoparticles Containing Diclofenac Sodium. *AAPS PharmSciTech* **2010**, *11*, 1250–1256. [CrossRef]
44. Torres-Flores, G.; Türeli Nazende, G.; Akif Emre, T. Preparation of Fenofibrate Loaded Eudragit L100 Nanoparticles by Nanoprecipitation Method. *Mater. Today Proc.* **2019**, *13*, 428–435. [CrossRef]
45. Albertsson, J.; Oskarsson, Å.; Svensson, C. X-Ray Study of Budesonide: Molecular Structures and Solid Solutions of the (22S) and (22R) Epimers of 11β, 21-Dihydroxy-16α, 17α-Propylmethylenedioxy-1, 4-Pregnadiene-3, 20-Dione. *Acta Crystallogr. B* **1978**, *34*, 3027–3036. [CrossRef]
46. Nikam, A.; Sahoo, P.R.; Musale, S.; Pagar, R.R.; Paiva-Santos, A.C.; Giram, P.S. A Systematic Overview of Eudragit® Based Copolymer for Smart Healthcare. *Pharmaceutics* **2023**, *15*, 587. [CrossRef] [PubMed]
47. Shen, X.; Yu, D.; Zhu, L.; Branford-White, C.; White, K.; Chatterton, N.P. Electrospun Diclofenac Sodium Loaded Eudragit® L 100-55 Nanofibers for Colon-Targeted Drug Delivery. *Int. J. Pharm.* **2011**, *408*, 200–207. [CrossRef] [PubMed]
48. Sanchez, L.; Mitjans, M.; Infante, M.R.; Vinardell, M.P. Potential Irritation of Lysine Derivative Surfactants by Hemolysis and HaCaT Cell Viability. *Toxicol. Lett.* **2006**, *161*, 53–60. [CrossRef] [PubMed]
49. Abruzzo, A.; Armenise, N.; Bigucci, F.; Cerchiara, T.; Gösser, M.B.; Samorì, C.; Galletti, P.; Tagliavini, E.; Brown, D.M.; Johnston, H.J.; et al. Surfactants from Itaconic Acid: Toxicity to HaCaT Keratinocytes in Vitro, Micellar Solubilization, and Skin Permeation Enhancement of Hydrocortisone. *Int. J. Pharm.* **2017**, *524*, 9–15. [CrossRef]
50. Li, H.; Toh, P.Z.; Tan, J.Y.; Zin, M.T.; Lee, C.-Y.; Li, B.; Leolukman, M.; Bao, H.; Kang, L. Selected Biomarkers Revealed Potential Skin Toxicity Caused by Certain Copper Compounds. *Sci. Rep.* **2016**, *6*, 37664. [CrossRef]
51. Wilheilm, K.P.; Samblebe, M.; Siegers, C.P. Quantitative in Vitro Assessment of N-Alkyl Sulphate-Induced Cytotoxicity in Human Keratinocytes (HaCaT). Comparison with in Vivo Human Irritation Tests. *Br. J. Dermatol.* **1994**, *130*, 18–23. [CrossRef]
52. Giuliano, E.; Paolino, D.; Fresta, M.; Cosco, D. Mucosal Applications of Poloxamer 407-Based Hydrogels: An Overview. *Pharmaceutics* **2018**, *10*, 159. [CrossRef]
53. Coughlin, M.L.; Liberman, L.; Ertem, S.P.; Edmund, J.; Bates, F.S.; Lodge, T.P. Methyl Cellulose Solutions and Gels: Fibril Formation and Gelation Properties. *Prog. Polym. Sci.* **2021**, *112*, 101324. [CrossRef]
54. Nair, A.; Jacob, S.; Al-Dhubiab, B.; Attimarad, M.; Harsha, S. Basic Considerations in the Dermatokinetics of Topical Formulations. *Braz. J. Pharm. Sci.* **2013**, *49*, 423–434. [CrossRef]
55. Li, L.; Thangamathesvaran, P.M.; Yue, C.Y.; Tam, K.C.; Hu, X.; Lam, Y.C. Gel Network Structure of Methylcellulose in Water. *Langmuir* **2001**, *17*, 8062–8068. [CrossRef]
56. Hyun, K.; Nam, J.G.; Wilhellm, M.; Ahn, K.H.; Lee, S.J. Large Amplitude Oscillatory Shear Behavior of PEO-PPO-PEO Triblock Copolymer Solutions. *Rheol. Acta* **2006**, *45*, 239–249. [CrossRef]
57. Pinto, J.R.; Monteiro e Silva, S.A.; de Holsback, V.S.S.; Leonardi, G.R. Skin Occlusive Performance: Sustainable Alternatives for Petrolatum in Skincare Formulations. *J. Cosmet. Dermatol.* **2022**, *21*, 4775–4780. [CrossRef] [PubMed]
58. Montenegro, L.; Parenti, C.; Turnaturi, R.; Pasquinucci, L. Resveratrol-Loaded Lipid Nanocarriers: Correlation between In Vitro Occlusion Factor and In Vivo Skin Hydrating Effect. *Pharmaceutics* **2017**, *9*, 58. [CrossRef] [PubMed]
59. Op't Veld, R.C.; Walboomers, X.F.; Jansen, J.A.; Wagener, F.A.D.T.G. Design Considerations for Hydrogel Wound Dressings: Strategic and Molecular Advances. *Tissue Eng. Part B Rev.* **2020**, *26*, 230–248. [CrossRef] [PubMed]
60. Choi, W.-S.; Cho, H.-I.; Lee, H.-Y.; Lee, S.-H.; Choi, Y.-W. Enhanced Occlusiveness of Nanostructured Lipid Carrier (NLC)-Based Carbogel as a Skin Moisturizing Vehicle. *J. Pharm. Investig.* **2010**, *40*, 373–378. [CrossRef]
61. Noval, N.; Rosyifa, R.; Annisa, A. Effect of HPMC concentration variation as gelling agent on physical stability of formulation gel ethanol extract bundung plants (Actinuscirpus Grossus). In Proceedings of the First National Seminar Universitas Sari Mulia (NS-UNISM 2019), Banjarmasin, Indonesia, 23 November 2019. [CrossRef]
62. Barbosa, A.I.; Torres, T.; Lima, S.A.C.; Reis, S. Hydrogels: A Promising Vehicle for the Topical Management of Atopic Dermatitis. *Adv. Ther.* **2021**, *4*, 2100028. [CrossRef]
63. Sipos, E.; Szász, N.; Vancea, S.; Ciurba, A. Evaluation and Selection of Gel Base for the Formulation of Dexpanthenol Products. *Trop. J. Pharm. Res.* **2015**, *13*, 1987. [CrossRef]
64. Helal, D.A.; El-Rhman, D.A.; Abdel-Halim, S.A.; El-Nabarawi, M.A. Formulation and Evaluation of Fluconazole Topical Gel. *Int J Pharm Pharm Sci* **2012**, *4*, 176–183.
65. Levy, G.; Schwarz, T.W. The Effect of Certain Additives on the Gel Point of Methylcellulose. *J. Am. Pharm. Assoc. Sci. Ed* **1958**, *47*, 44–46. [CrossRef] [PubMed]
66. Vlaia, L.; Coneac, G.; Olariu, I.; Vlaia, V.; Lupuleasa, D. Cellulose-Derivatives-Based Hydrogels as Vehicles for Dermal and Transdermal Drug Delivery. *Emerg. Concepts Anal. Appl. Hydrogels* **2016**, *2*, 64.

67. Zilberman, E.N.; Lerner, F.; Joseph, H.M.; Alon, M. Properties of Hydroxypropyl Methylcellulose–Polyvinyl Alcohol Water Systems, Dispersants in Vinyl Chloride Suspension Polymerization. *J. Appl. Polym. Sci.* **1993**, *48*, 435–442. [CrossRef]
68. Sguizzato, M.; Mariani, P.; Ferrara, F.; Drechsler, M.; Hallan, S.S.; Huang, N.; Simelière, F.; Khunti, N.; Cortesi, R.; Marchetti, N.; et al. Nanoparticulate Gels for Cutaneous Administration of Caffeic Acid. *Nanomaterials* **2020**, *10*, 961. [CrossRef] [PubMed]
69. Niemczyk-Soczynska, B.; Sajkiewicz, P.; Gradys, A. Toward a Better Understanding of the Gelation Mechanism of Methylcellulose via Systematic DSC Studies. *Polymers* **2022**, *14*, 1810. [CrossRef]
70. Punitha, S.; Uvarani, R.; Panneerselvam, A. Effect of pH in Aqueous (Hydroxy Propyl Methyl Cellulose) Polymer Solution. *Results Mater.* **2020**, *7*, 100120. [CrossRef]
71. Abdeltawab, H.; Svirskis, D.; Sharma, M. Formulation Strategies to Modulate Drug Release from Poloxamer Based in Situ Gelling Systems. *Expert Opin. Drug Deliv.* **2020**, *17*, 495–509. [CrossRef]
72. Gandhi, A.; Jana, S.; Sen, K.K. In-Vitro Release of Acyclovir Loaded Eudragit RLPO®Nanoparticles for Sustained Drug Delivery. *Int. J. Biol. Macromol.* **2014**, *67*, 478–482. [CrossRef]
73. Das, S.; Suresh, P.K.; Desmukh, R. Design of Eudragit RL 100 Nanoparticles by Nanoprecipitation Method for Ocular Drug Delivery. *Nanomedicine Nanotechnol. Biol. Med.* **2010**, *6*, 318–323. [CrossRef]
74. Inal, O.; Yapar, E.A. Effect of Mechanical Properties on the Release of Meloxicam from Poloxamer Gel Bases. *Indian J. Pharm. Sci.* **2013**, *75*, 700–706.
75. Zhang, L.; Parsons, D.L.; Navarre, C.; Kompella, U.B. Development and In-Vitro Evaluation of Sustained Release Poloxamer 407 (P407) Gel Formulations of Ceftiofur. *J. Control. Release* **2002**, *85*, 73–81. [CrossRef] [PubMed]
76. Dumortier, G.; Grossiord, J.L.; Agnely, F.; Chaumeil, J.C. A Review of Poloxamer 407 Pharmaceutical and Pharmacological Characteristics. *Pharm. Res.* **2006**, *23*, 2709–2728. [CrossRef] [PubMed]
77. Musial, W. The Effect of Methylcellulose on Metronidazole Release from Polyacrylic Acid Hydrogels. *Chem. Pharm. Bull.* **2007**, *55*, 1141–1147. [CrossRef] [PubMed]
78. Bonina, F.P.; Montenegro, L. Vehicle Effects on in Vitro Heparin Release and Skin Penetration from Different Gels. *Int. J. Pharm.* **1994**, *102*, 19–24. [CrossRef]
79. Tas, Ç.; Özkan, Y.; Savaser, A.; Baykara, T. In Vitro Release Studies of Chlorpheniramine Maleate from Gels Prepared by Different Cellulose Derivatives. *Il Farm.* **2003**, *58*, 605–611. [CrossRef] [PubMed]
80. Basak, R.; Bandyopadhyay, R. Encapsulation of Hydrophobic Drugs in Pluronic F127 Micelles: Effects of Drug Hydrophobicity, Solution Temperature, and pH. *Langmuir* **2013**, *29*, 4350–4356. [CrossRef] [PubMed]
81. Mosmann, T. Rapid Colorimetric Assay for Cellular Growth and Survival: Application to Proliferation and Cytotoxicity Assays. *J. Immunol. Methods* **1983**, *65*, 55–63. [CrossRef]
82. Joshi, V.; Yashaswini, G.; Acharya, A.; Bheemachari; Annegowda, H.V.; Niraula, B. Formulation and Evaluation of Semisolid Dosage Forms of an Anti-Inflammatory Drug. *3 Biotech* **2019**, *9*, 248. [CrossRef]
83. Batista, C.M.; de Queiroz, L.A.; Alves, Â.V.F.; Reis, E.C.A.; Santos, F.A.; Castro, T.N.; Lima, B.S.; Araújo, A.A.S.; Godoy, C.A.P.; Severino, P.; et al. Photoprotection and Skin Irritation Effect of Hydrogels Containing Hydroalcoholic Extract of Red Propolis: A Natural Pathway against Skin Cancer. *Heliyon* **2022**, *8*, e08893. [CrossRef]
84. Caldas, A.R.; Catita, J.; Machado, R.; Ribeiro, A.; Cerqueira, F.; Horta, B.; Medeiros, R.; Lúcio, M.; Lopes, C.M. Omega-3- and Resveratrol-Loaded Lipid Nanosystems for Potential Use as Topical Formulations in Autoimmune, Inflammatory, and Cancerous Skin Diseases. *Pharmaceutics* **2021**, *13*, 1202. [CrossRef]
85. Stamatas, G.N.; de Sterke, J.; Hauser, M.; von Stetten, O.; van der Pol, A. Lipid Uptake and Skin Occlusion Following Topical Application of Oils on Adult and Infant Skin. *J. Dermatol. Sci.* **2008**, *50*, 135–142. [CrossRef] [PubMed]
86. Mendes Aciole, I.H.; de Andrade Júnior, F.P.; Vilar Cordeiro, L.; Pereira de Souza, J.B. Aloe Gel: Manipulation and Characterization of Physical-Chemical Quality Adjustment. *Rev. Colomb. Cienc. Quím.-Farm.* **2020**, *49*, 790–805. [CrossRef]
87. Borghetti, G.S.; Knorst, M.T. Desenvolvimento e avaliação da estabilidade física de loções O/A contendo filtros solares. *Rev. Bras. Ciênc. Farm.* **2006**, *42*, 531–537. [CrossRef]
88. Zaharieva, M.M.; Kaleva, M.; Kroumov, A.; Slavkova, M.; Benbassat, N.; Yoncheva, K.; Najdenski, H. Advantageous Combinations of Nanoencapsulated Oregano Oil with Selected Antibiotics for Skin Treatment. *Pharmaceutics* **2022**, *14*, 2773. [CrossRef] [PubMed]

gels

MDPI

Article

Electrosprayed Stearic-Acid-Coated Ethylcellulose Microparticles for an Improved Sustained Release of Anticancer Drug

Yuexin Ji [1,†], Hua Zhao [2,*,†], Hui Liu [1], Ping Zhao [1] and Deng-Guang Yu [1,*]

1 School of Materials and Chemistry, University of Shanghai for Science and Technology, No. 516 Jungong Road, Shanghai 200093, China; 213353172@st.usst.edu.cn (Y.J.); huiliu@usst.edu.cn (H.L.); 203613027@st.usst.edu.cn (P.Z.)
2 Medical School, Quzhou College of Technology, No. 18 Jiangyuan Road, Quzhou 324000, China
* Correspondence: zhaohua1215@163.com (H.Z.); ydg017@usst.edu.cn (D.-G.Y.)
† These authors contributed equally to this work.

Abstract: Sustained release is highly desired for "efficacious, safe and convenient" drug delivery, particularly for those anticancer drug molecules with toxicity. In this study, a modified coaxial electrospraying process was developed to coat a hydrophobic lipid, i.e., stearic acid (SA), on composites composed of the anticancer drug tamoxifen citrate (TC) and insoluble polymeric matrix ethylcellulose (EC). Compared with the electrosprayed TC-EC composite microparticles M1, the electrosprayed SA-coated hybrid microparticles M2 were able to provide an improved TC sustained-release profile. The 30% and 90% loaded drug sustained-release time periods were extended to 3.21 h and 19.43 h for M2, respectively, which were significantly longer than those provided by M1 (0.88 h and 9.98 h, respectively). The morphology, inner structure, physical state, and compatibility of the components of the particles M1 and M2 were disclosed through SEM, TEM, XRD, and FTIR. Based on the analyses, the drug sustained-release mechanism of multiple factors co-acting for microparticles M2 is suggested, which include the reasonable selections and organizations of lipid and polymeric excipient, the blank SA shell drug loading, the regularly round shape, and also the high density. The reported protocols pioneered a brand-new manner for developing sustained drug delivery hybrids through a combination of insoluble cellulose gels and lipid using modified coaxial electrospraying.

Keywords: sustained release; anticancer drug; ethylcellulose; coaxial electrospraying; stearic acid; microparticle; anticancer; insoluble gels

Citation: Ji, Y.; Zhao, H.; Liu, H.; Zhao, P.; Yu, D.-G. Electrosprayed Stearic-Acid-Coated Ethylcellulose Microparticles for an Improved Sustained Release of Anticancer Drug. *Gels* **2023**, *9*, 700. https://doi.org/10.3390/gels9090700

Academic Editor: David Díaz Díaz

Received: 2 August 2023
Revised: 21 August 2023
Accepted: 26 August 2023
Published: 29 August 2023

1. Introduction

Most of the anticancer drugs, regardless of the active biomolecules (such as curcumin, quercetin, silybum marianum, and paclitaxel) or the synthetic therapeutic molecules, on the one hand, are poorly water soluble or even insoluble [1–4]. On the other hand, they are toxic due to a high blood drug concentration after oral administration resulting from the initial burst release, which is particularly a negative case for numerous nano/micro drug delivery systems [5–8]. Thus, better drug dissolution and sustained release of these drugs after oral delivery is highly desired for a "safe, efficacious, and convenient" delivery to the patients [9–12].

For the sustained release of a drug, the common strategies that can be relied upon can be divided into two approaches. One is to encapsulate the drug molecules into an insoluble inert matrix [13,14]. Some examples are phospholipid, insoluble polymers, biodegradable polymers, and also many inorganic materials, such as silicon, carbon nano tubes, and graphene [15–17]. The other is to treat the drug and the excipients through advanced pharmaceutical techniques, which are frequently introduced into the field of pharmaceuticals from other material conversion methods [18–21]. Certainly, these two approaches are frequently integrated together to develop a wide variety of drug sustained-release DDSs.

Cellulose and its derivatives are highly popular as polymeric matrices for providing all types of drug controlled-release profiles in many fields, such as drug delivery, tissue engineering, and food packaging engineering [22–27]. In general, soluble cellulose derivatives can be exploited for promoting the fast release of the poorly water-soluble drugs, such as a series of hydroxypropyl methylcellulose [28]. Insoluble cellulose derivatives are often exploited for drug extended- or sustained-release profiles, such as ethylcellulose (EC) and cellulose acetate (CA) [29,30]. These cellulose derivatives have fine film-forming, filament-forming, and other processing properties [31]. Thus, in the literature, they are frequently converted with a guest drug to form new kinds of DDSs. Particularly in this nano era, they are often transferred into nanoparticles, nanofibers, and beads-on-a-string products through a certain pharmaceutical nanotechnology [32–34].

One most recent pharmaceutical nanotechnology is electrospinning, which belongs to an electrohydrodynamic atomization (EHDA) process [35–39]. In comparison, electrospraying, as a sister EHDA method of electrospinning, receives less attention for drug delivery applications [40,41]. In the past several years, single-fluid electrospinning has been quickly moving to coaxial, tri-axial, side-by-side, and tri-layer side-by-side processes [42–47]. Furthermore, many fluids without electrospinnability have also taken part in the multiple-fluid electrospinning processes because only one of the fluids must be electrospinnable [48]. However, the electrospraying process is still mainly the single-fluid process and also the traditional coaxial process. Inspired by modified coaxial electrospinning, in which unspinnable fluids can be explored as the sheath working fluids for creating core–sheath nanofibers [49], unsolidable fluids may also be utilized as the shell working fluids for generating core–shell particles.

Tamoxifen citrate (TC) is a non-steroidal anti-estrogen drug with a structure similar to estrogen. It mainly competes with estrogen for estrogen receptors so as to prevent estrogen from entering tumor cells, prevent estrogen from playing its role, and thus inhibit the proliferation of breast cancer cells. Its most important role is to treat recurrent and metastatic breast cancer in women and to serve as an adjuvant treatment for postoperative metastasis of breast cancer to prevent recurrence [50,51]. In addition, its anti-estrogen effect can also be used to improve breast hyperplasia, pain, and discomfort symptoms. However, there are a series of possible common adverse reactions after the oral administration of TC, which are included as follows: (1) Gastrointestinal reactions, mainly manifested as nausea, vomiting, abdominal pain, and diarrhea; (2) The main side effects of the reproductive system are menstrual disorder, menopause, and vaginal bleeding; (3) Skin side effects, manifested as facial flushing and rash; (4) Symptoms of the mental nerves, mainly manifested as headaches and dizziness; (5) Changes in blood routine can lead to a decrease in white blood cells and platelets in patients; and (6) Some patients may also experience abnormal liver function. Thus, sustained release of TC after oral administration is highly desired for the patient's compliance and a better therapeutic effect [5,6,50,51].

In this study, we hypothesized that unsolidable, i.e., dilute stearic acid (SA) solution, can be utilized as a shell liquid to conduct a modified coaxial electrospraying process, in which the core fluid was a solidable drug polymer co-dissolved solution composed of EC and TC. We further hypothesized that the coating SA layer, as a hydrophobic shell, was able to remarkably modify the sustained-release profile of TC from the insoluble EC matrix. A series of characterizations were carried out to disclose the products' morphologies and inner structures, the components' physical states and their compatibility, and the targeted sustained release performances.

2. Results and Discussion

2.1. The Coaxial Electrospraying and the Core–Shell Structures in Drug Delivery

Just as coaxial electrospinning is used for creating core–sheath nanofibers [52–55], coaxial electrospraying is useful for generating core–shell micro- or nanoparticles [40,41]. A diagram shows the main components of a coaxial electrospraying apparatus, and the key elements in a spraying process are included in Figure 1. A concentric spraying head is the convergent place for the working fluids and the electrostatic energy. Two syringe pumps are exploited to quantitatively send the core and shell working fluids to the spraying head. A power supply is utilized to generate the applied high voltage, and a homemade collector is utilized to collect the deposited particles under the spraying head. Besides these four essential parts, often, a camera can be added to monitor the working process. During the working process, the first key is to form a Taylor cone, which is followed by the Coulomb explosion [40,41]. The second key is to achieve the solid particles, i.e., the designed products, after the effective removal of the solvents. A failed electrospraying process often results in a wet film. By the way, all the components of the electrospraying apparatus should be grounded for the safe operation of an electrospinning system [56,57].

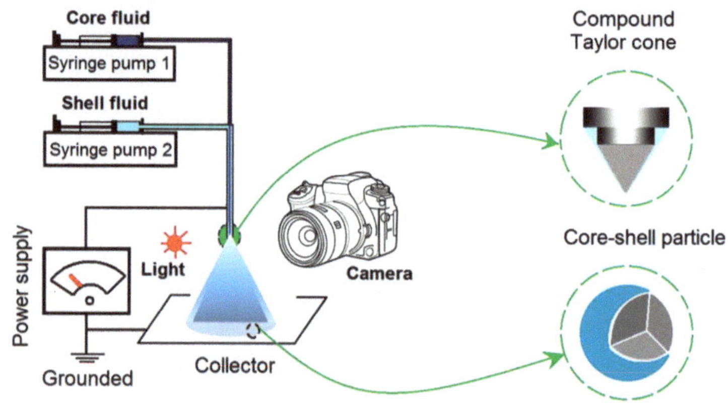

Figure 1. A diagram showing the modified coaxial electrospraying system, its main components, and key information about the working process.

In the traditional coaxial electrospraying process, the shell fluid should be solidable when it experiences an electrospraying process for ensuring the formation of solid core–shell particles. However, Li et al. reported that pure organic solvent (a non-solidable property) can be explored for creating high-quality nanoparticles [40]. Along this direction, the present work developed a modified coaxial electrospraying method, in which the dilute unsolidable SA solution was explored as the shell liquid to coat the core solidable EC-TC solution. The parameters for the electrospraying processes are included in Table 1.

Table 1. Fabrication parameters of the microparticles.

No.	Electrospraying	Applied Voltage (kV)	Fluid Flow Rate (mL/h)		Products
			Core [a]	Shell [b]	
M1	Single fluid	21	--	1.0	Monolithic microparticles
M2	Coaxial	18	0.4	1.0	Core–shell microparticles
M3	Single fluid	12	0.4	--	--

[a] The core fluid consisted of 2.0 g of TC and 10.0 g of EC in 100 mL of solvent mixture of anhydrous ethanol and DCM with a volume ratio of 6:4. [b] The shell solution was composed of 1.0 g of stearic acid in 50 mL of DCM.

The key points about the implementation of electrospraying are recorded in Figure 2. A whole image of the modified coaxial electrospraying system is shown in Figure 2a. The homemade spraying head is exhibited in the upper left inset, which can also be explored in coaxial electrospinning and modified coaxial electrospinning. A coaxial outlet of the nozzle was utilized to guide the core and shell working fluids into the electrical field. The homemade collector was prepared by placing aluminium foil on a cardboard box. The working distance between the collector and the nozzle of the spraying head was fixed at

20 cm. The transportation of electrostatic energy and working fluids is given in Figure 2b. The copper line could be directly wrapped around the stainless steel capillary guiding the core fluid to transfer the high voltage. The core fluid was delivered to the spraying head directly by its syringe fixed on a pump. The shell fluid was sent to the inlet of the spraying head by the polymeric tubes, which were composed of a hard Teflon capillary and a section of highly elastic silicon tube as a connection between it and the metal inlet of the spraying head. This arrangement was exploited to avoid the absorbance of dichloromethane (DCM) and swelling of the silicon tube during the working process.

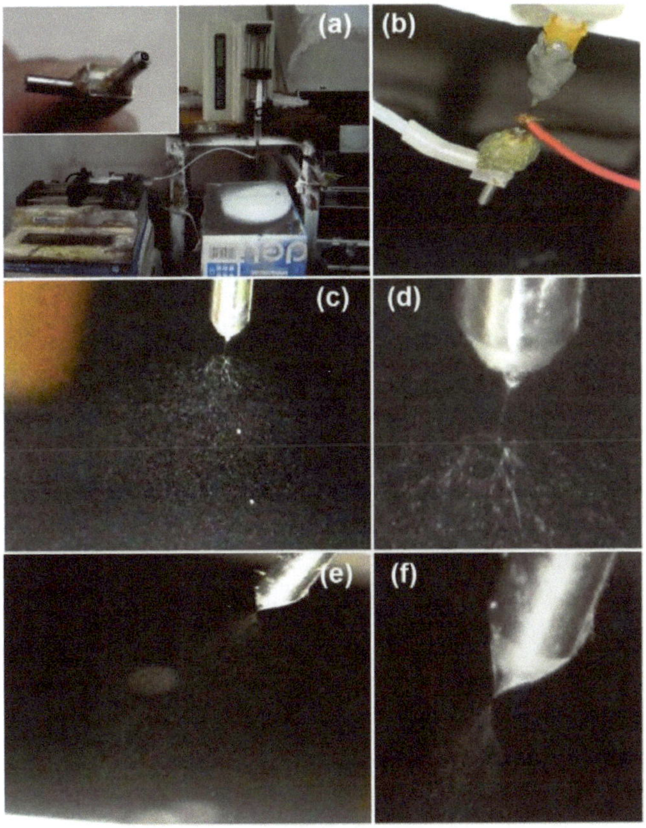

Figure 2. The coaxial electrospraying apparatus and observations of the working processes: (**a**) a digital picture of the electrospraying apparatus, with the upper left inset showing the concentric spraying head; (**b**) a digital picture showing the connection of the working fluids and the transferring of electrostatic energy; (**c**,**d**) a single-fluid electrospraying process experienced by the core solidable TC-EC solution under different magnifications for preparing the microparticles M1; (**e**,**f**) the digital photos of modified coaxial electrospraying processes for observing the whole process and the compound Taylor cone for producing the microparticles M2, respectively.

The applied voltage was adjusted to a suitable value based on a continuous and robust electrospraying process. When one of the two working fluids was switched off, then the coaxial electrospraying process downgraded into a single-fluid electrospraying process. As indicated by Figure 2c,d, for creating microparticles M1, a trio emerged between the Taylor cone, the straight fluid jet, and the Coulomb explosion, and the Taylor cone from the inner metal capillary was stable and in a typical cone shape. A high voltage of 21 kV is needed due to the higher surface tension of the core TC-EC fluid for a fluid rate of 1.0 mL/h.

After some optimization, the shell and core fluid flow rates were selected as 0.4 and 1.0 mL/h, respectively. The applied voltage was 18 kV. Although there was a generally larger fluid flow rate from the concentric nozzle (1.0 + 0.4 = 1.4 mL/h) for the modified coaxial process compared to the single-fluid electrospraying of the core fluid, the applied voltage was reduced to a smaller value of 18 kV compared to 21 kV. This strange phenomenon occurred because the Coulomb explosion is always initiated at the surface of the working liquid, and the electrostatic charges tend to gather on the surface of a working liquid. Thus, the smaller applied voltage was a direct result of the smaller surface tension of the shell SA solution. The working processes for generating microparticles M2 are recorded in Figure 2e (a typical whole coaxial electrospraying process) and Figure 2f (the compound core–shell Taylor cone). The single-fluid electrospraying of the shell SA solution (switching off the core fluid) resulted in only wet films due to a diluted SA concentration. The exploration of unsolidable SA solution as the shell fluid had the following advantages, besides providing a hydrophobic coating on the TC-EC medicated composites: (1) a good encapsulation of the core fluid at the nozzle of the spray head; (2) a facile initiation of the electrospraying process; (3) an easy fission of the sprayed droplets during the Coulomb explosion procedure; and (4) a stabler drying process of the core TC-EC solution, which would be demonstrated by the resultant products.

2.2. The Morphology and Structure of the Microparticles

The SEM images of the morphologies and the diameter distributions of microparticles from the single-fluid and modified coaxial processes are included in Figure 3. The differences between microparticles M1 and M2 can be concluded as follows: (1) particles M1 are dented from different directions and thus in an irregular shape (Figure 3a,b), whereas particles are mainly in a round shape (Figure 3d,e); (2) although both have satellites, particles M1 are more severe than M2; (3) particles M1 had a larger average diameter value (1.31 $\pm$ 0.29 μm, Figure 3c) than particles M2 (1.13 $\pm$ 0.34 μm, Figure 3f), although the applied voltage in producing M1 was bigger than that in producing M2.

The TEM images of particles M1 and M2 are included in Figure 4. In Figure 4a, it is obvious that microparticles M1 are irregular in their shape, with recessed sections at their surface. This indicates that the thicknesses of particles M1 varied, with no rules, which is reflected by the various gray levels in one particle, as the darker places mean a thicker region than the lighter gray places. In sharp contrast, microparticles M2 are mainly in a round shape, with a regular gray level change trend, i.e., the shell gray levels are always smaller than the core gray levels. By estimation, the thicknesses of these SA shell layers are between 10 and 30 nm.

A diagram showing the microformation mechanism of the single-fluid electrospraying process is exhibited in Figure 5a. The interaction between the electric energy, the surface tension, and the viscosity of the working fluid results in the Taylor cone, which is followed by a straight fluid jet and the Coulomb explosion. The most fundamental rule is that the "same charge repels each other," by which the droplets are continuously splitting, reducing in volume, and being solidified to form the microparticles. During the process and at the late stage of the Coulomb explosion, the split droplets may have the semi-solid state of their surface but maintain a fluid state in the inner part. The formation of a solid film on the droplets not only retards further splitting but would also trap some solvent. Later, when the trapped solvent escapes to the environment, the barometric pressure would deform the round "droplet" shape to the irregular, dented shape in Figure 3b. This phenomenon has been reported in some other investigations about the preparations of electrosprayed CA particles [40] and zein nanoparticles [41].

When a diluted SA solution was exploited as a shell fluid to implement the modified coaxial electrospraying process, the shell solution, on the one hand, would dominate the Coulomb splitting process because of the surface distribution property of charges and its small surface tension (Figure 5b). On the other hand, the shell solvent would act as a bridge for the core solvents to move from the core sections to the atmosphere, and a more continuous and robust drying process could be ensured. Meanwhile, the shell solution may help to resist the outer disturbances for a stabler and more robust microparticles generation process. These positive factors should be attributed to the formation of the round shape of microparticles M2, and they would make the density of microparticles M2 larger than that of microparticles M1.

Figure 3. The morphologies and diameters of the electrosprayed microparticles: (**a**,**b**) SEM images of microparticles M1 at different magnifications; (**c**) the diameters of microparticles M1 and their size distributions; (**d**,**e**) SEM images of microparticles M2 at different magnifications; (**f**) the diameters of microparticles M2 and their size distributions.

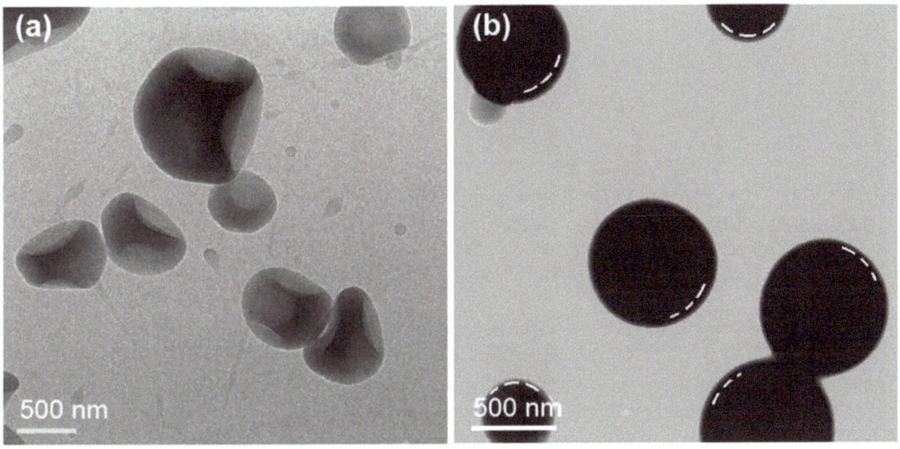

Figure 4. Inner structures of the electrosprayed microparticles: (**a**) TEM images of microparticles M1; and (**b**) TEM images of the microparticles M2.

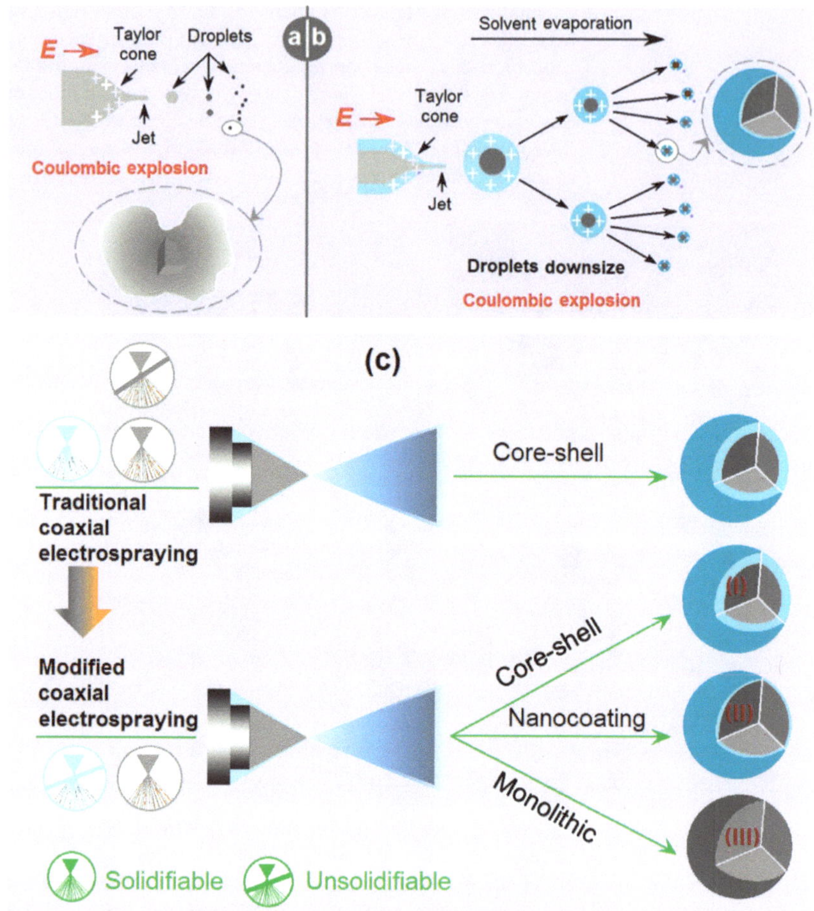

Figure 5. The microformation mechanisms: (**a**) the single-fluid electrospraying for producing microparticles M1; (**b**) the modified coaxial electrospraying for preparing microparticles M2; and (**c**) the different materials processing capabilities of traditional coaxial electrospraying and modified coaxial electrospraying processes; the case (II) represents the SA-coated core–shell microparticles M2 in this work.

Compared with traditional coaxial electrospraying, the modified coaxial process has a stronger capability to create different kinds of particulate micro products. Shown in Figure 5c, the main difference in the implementation is that the shell fluid must be solidable in the traditional coaxial electrospraying, whereas the shell fluid in a modified coaxial process is unsolidable. From a standpoint of created products, traditional coaxial electrospraying can only create core–shell particles. In contrast, modified coaxial electrospraying can create core–shell particles (Case I), coat the particles' surfaces in a continuous manner (Case II, i.e., the microparticles M2 in this work) or in a discontinuous way, and be exploited to create monolithic nanoparticles (Case III), such as by using only solvent as a shell fluid. This last case has been demonstrated by some previous investigations [40,41].

2.3. The Physical State and Compatibility

The ATR-FTIR spectra of the initial materials, i.e., EC, TC, and SA, and their electrosprayed microparticles M1 and M2, are included in the left section of Figure 6. The molecular formulas of EC, TC, and SA are given in the right side of Figure 6. TC spectra have a series of characteristic peaks, such as at 1724, 1581, and 1508 cm^{-1}. SA has characteristic peaks at 1704, 2917, and 2850 cm^{-1}, and EC's characteristic peaks are at 1104 and 1062 cm^{-1}. In the spectra of microparticles M1 from the single-fluid electrospraying process, the EC's characteristic peaks are still clearly there. However, the characteristic peaks of TC are greatly reduced or even disappear, with only one peak distinguishable at 1727 cm^{-1}. The reasons should be the formation of composites between EC and TC in the electrosprayed microparticles M1. In the spectra of microparticles M2, the peaks of SA are very obvious, which should be the reason that the detection depth of ATR-FTIR is about 10 nm. However, the information from the core section can still be discerned, such as the shoulder of the peak at 1704 cm^{-1} (as indicated by the "A" red arrow) and also the peaks at 1104 and 1062 cm^{-1}. These results suggest that the shell SA and the core TC-EC composites co-existed in the core–shell particles in a hybrid manner, concurring with the TEM observations in Figure 4b. From the molecular formula of EC, SA, and TC, it can be anticipated that these components are highly compatible due to the secondary interactions between their molecules, such as hydrogen bonds, hydrophobic interactions, electrostatic interactions, and the van der Waals interaction [58,59].

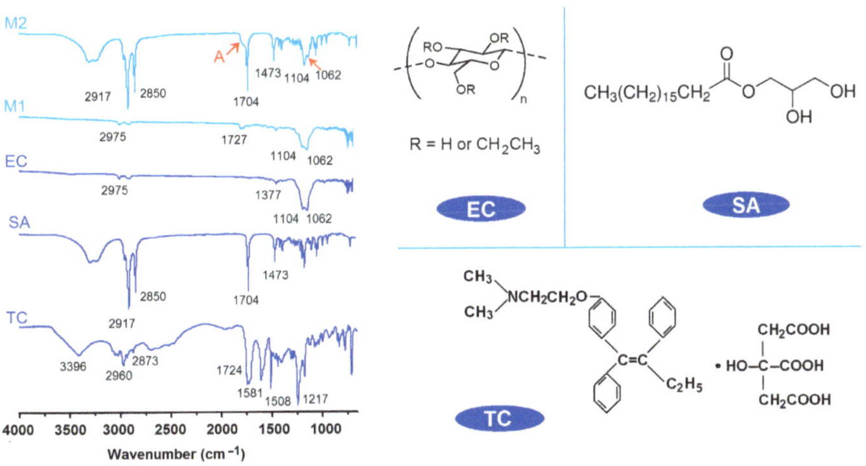

Figure 6. The ATR-FTIR spectra of the raw materials (TC, EC, and SA) and their electrosprayed microparticles (M1 and M2); the molecular formulas of EC, SA, and TC.

The XRD patterns of the raw materials of EC, TC, and SA, and their electrosprayed microparticles M1 and M2, are included in the left section of Figure 7. All the samples except the raw TC powders present in an amorphous state. TC patterns have some sharp Bragg peaks, suggesting their raw crystalline state. However, when TC was experiencing the electrospraying processes, regardless of whether the single-fluid monoaxial or the modified coaxial process was used, TC was converted into an amorphous state within its carrier EC in both monolithic composite microparticles M1 and core–shell hybrid microparticles M2. The electrospraying process is a very rapid fluid drying process, essentially. The short drying time period leaves almost no time for the TC molecules in the working fluids to recrystallize into new particles. The homogeneous state of the solutions was sufficiently maintained after the removal of organic solvents. Because there are abundant favorable secondary interactions between the components, the homogeneous state can be stably maintained. Shown in the right section of Figure 7 are the possible hydrogen bonds between EC and TC in the core section, between EC and SA, and between TC and SA in the core–shell interfaces.

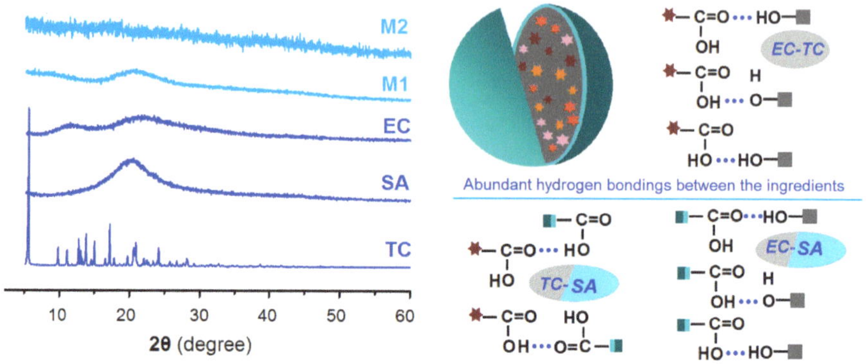

Figure 7. XRD patterns of the raw materials (TC, EC, and SA) and their electrosprayed microparticles (M1 and M2); the potential hydrogen bonds among the three components in the core–shell microparticles M2.

2.4. The Drug Encapsulation Ratio of the Dual-Stage Drug Controlled-Release Profile

The measured entrapment efficiency (*EE*%) of TC was 99.4 ± 3.5% and 100.1 ± 3.3% for microparticles M1 and M2, respectively. The results indicate that all of the drug TC was successfully encapsulated into the microparticles through the electrospraying processes, regardless of whether the single-fluid process for creating particles M1 or the double-fluid modified coaxial process for generating the particles M2 was used. During the rapid electrospraying process, the TC solutions were converted into solid microparticles because of the evaporation of volatile solvents of ethanol and DCM to the environment. The solutes TC, EC, and also SA were solidified together, with few chances to escape from the working processes to the environment. The fate of a drug molecule from the preparation of its dosage form to its final clinical application is influenced by many factors [60–64]. Compared with many "bottom–up" nanofabrication methods, EHDA has its advantages to create drug-loaded micro/nano products with a higher *EE*% value.

The in vitro drug release profiles of microparticles M1 and M2 are exhibited in Figure 8a–c. Figure 8a shows the full time period experimental results of two kinds of microparticles. The comparisons between the two kinds of particles' release profiles can help discern that the microparticles M2 from the modified coaxial process showed an obvious improved effect of the TC sustained release performances. In Figure 8a, it is clear that microparticles M2 showed a smaller tailing-off release phenomenon at the end of in vitro release, which is a negative phenomenon in drugs' sustained-release profile. The initial burst release is another negative phenomenon in drug sustained release. Figure 2b shows that the initial burst release from the microparticles M1 is obvious. The microparticles M2 had no initial burst release. It is after almost 4 h that microparticles M2 released a drug release amount that matched what was released at the first hour from the microparticles M1. A further regressed treatment of the experimental data can calculate the time needed to release a certain percentage of drug from the microparticles. The results are included in Figure 8c. The release times for core–shell microparticles M2 for releasing 30%, 50%, and 90% of the loaded drugs were 3.21, 7.33, and 19.43 h, respectively. These results are better than those from the monolithic microparticles M1, whose corresponding values were 0.88, 1.81, and 93.98 h, respectively.

To further determine the drug release mechanisms from the two sorts of microparticles, the Peppas Equation ($Q = kt^n$; Q, k, t, and n represent the accumulative drug release amount, a constant, the sampling time point, and the exponent [65]) was exploited to regress the in vitro drug release data. The results are shown in Figure 8d for microparticles M1 and in Figure 8e,f for microparticles M2. Just as expected, TC released from the microparticles M1 was manipulated by a typical Fickian diffusion mechanism. The regressed equation is LogQ1 = 1.61 + 0.30 Logt (R = 0.9545), in which the exponent value is 0.30, which is smaller than the judge standard of 0.45. Unexpectedly, the full time period regressions of microparticles M2 indicated a combination of diffusion and erosion mechanisms of TC release. The corresponding equation is LogQ2 = 0.89 + 0.84 Logt (R = 0.9293), which is shown in Figure 8e. The exponent is 0.84, between 0.45 and 0.90. However, when the data in the treatment are started from the fourth hour, a new regressed equation is achieved as LogQ2′ = 1.38 + 0.43 Logt (R = 0.9927). These results suggest that the shell SA had exerted a remarkable influence on the TC released from the core–shell microparticles M2, particularly during the first several hours. When the shell SA was removed at about 4 h, the drugs released from the core TC-EC composites were still manipulated by the typical Fickian diffusion mechanism, an anticipated result of drug molecules released from their insoluble matrices [66–69].

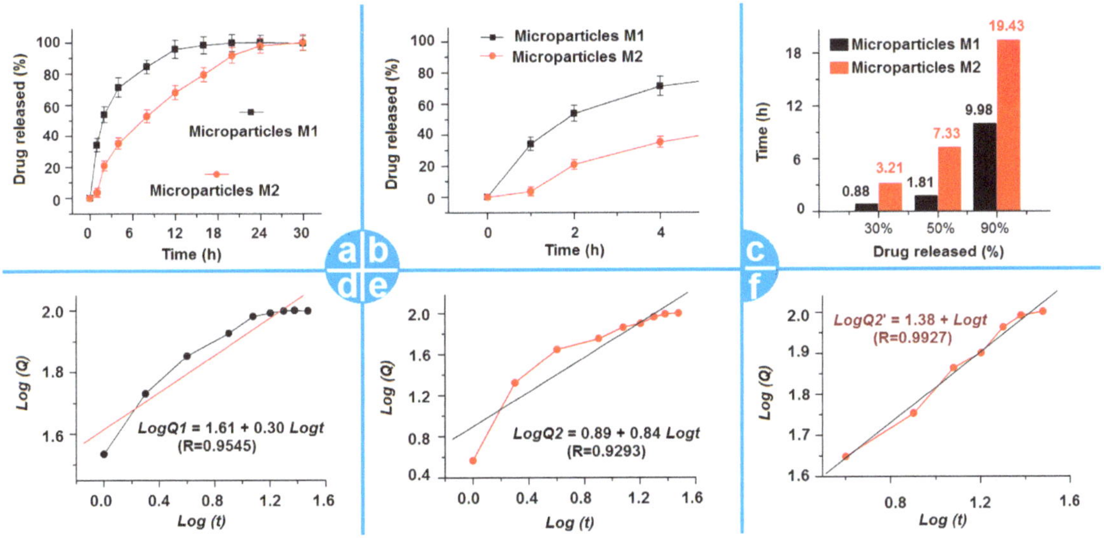

Figure 8. The drug sustained-release functional performances and the related drug controlled-release mechanisms: (**a**) The full time period drug sustained-release profiles of microparticles M1 and M2; (**b**) The first 4 h drug release profiles of the particles M1 and M2; (**c**) The regressed drug release time periods for releasing a certain percentage (30%, 50%, and 90%) of TC from the microparticles M1 and M2; (**d**) The regressed drug release mechanism for microparticles M1; (**e**,**f**) The regressed drug release mechanisms for microparticles M2 for the whole time period and from the 4th hour to the final time point, respectively.

2.5. The Proposed Drug Controlled-Release Mechanism Based on the Insoluble Gel Forming

Initially, polymer properties are the most important factor for providing a wide variety of drug controlled-release profiles, including drug sustained release [70–74]. Numerous reports in the literature have demonstrated that insoluble and biodegradable polymers and their composites and hybrids can be exploited as carriers to manipulate the gradual release of the loaded drug molecules [75–78]. Meanwhile, lipid materials are also frequently explored for the extended release of drugs in the formation of microparticles, micelles, composites, solid nanoparticles, emulsions, and liposome [79–83]. Thus, firstly, in this study, both insoluble polymer EC gels and the lipid SA were selected as the TC carrier, which are materials selected from experiences with the traditional pharmaceutics. Secondly, SA was exploited as a shell coating material to cover the polymer EC and the loaded drug molecules. This core–shell organization format at the micro scale is difficult to realize through the traditional chemical and physical methods. The coaxial electrospraying process proceeds in a one-step, straightforward manner. Meanwhile, modified coaxial electrospraying can organize the core–shell particles in a more regular, round shape. In contrast, the EC-TC particles M1 from the single-fluid electrospraying process have an irregular shape, which means an even more enlarged surface area for aggravating the initial burst release (Figure 9). Thirdly, the blank SA coating can make it so that there are no drug molecules on the surface of the electrosprayed microparticles, and thus completely eliminates the initial burst release phenomenon. Fourthly, the regular shape of microparticles M2 not only determines a relatively small surface for drug distribution and the amount of the drug initially released, but also determines a regular long diffusion distance for the penetration of both water molecules and drug molecules (e.g., routes R1 and R2 in Figure 9). In the irregular microparticles M1, the drug molecules would always diffuse to the dissolution bulk solution through the thinnest places (e.g., Route R3 is more possible than route R4), which is negative for the drug's sustained release. Fifthly, although the fluid processing capacity per unit time in the modified coaxial electrospraying process is larger than that of the single-fluid electrospraying, the microparticles M2 are smaller than M1. This means that the particles M2 have a greater density than M1. The drug release route comprises water molecules penetrating into the EC-TC composites; absorbance of water, gelling, and swelling of EC molecules; dissolution of TC from the EC-TC composites; and diffusion of TC molecules from the inside of the particles to the bulk dissolution solutions. The greater density thus means a slower process of the above-mentioned procedure, which is favorable for the sustained release of TC. Thus, the fine sustained-release effect of microparticles is the result of multiple factors co-acting. One is the reasonable selection of the starting materials. The other four reasons come from the materials conversions through modified coaxial electrospraying. EHDA processes, as the popular material conversion method currently, hold many new approaches for creating novel, functional materials by updating the working strategies, such as the design of a biomimetic spraying head [84], exploring the alternating current [85,86], and integrating the chemical reaction into the physical drying procedure [87]. Meanwhile, the microformation mechanisms of these new EHDA processes deserve further investigations, which are completely different with those "bottom–up" fabrication processes, such as assemblies [88,89].

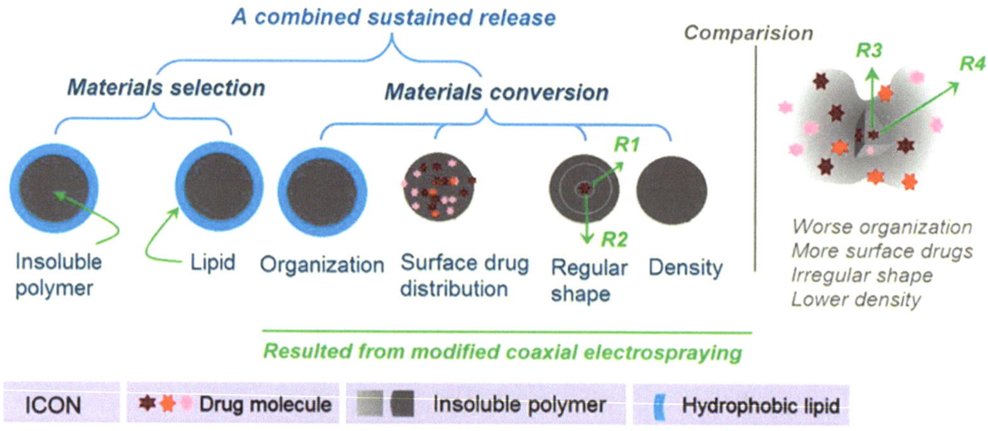

Figure 9. A sketch showing those factors that have co-acted to promote an improved sustained release of TC from the electrosprayed microparticles M2, in which a core insoluble gel composite is coated by a lipid shell layer.

3. Conclusions

In this study, a modified coaxial electrospraying process was developed to encapsulate the anticancer drug TC in the core section of a new type of core–shell microparticle, in which the shell sections are the lipid SA coatings. Although the shell SA solution had no solidable property alone, it could ensure a robust and continuous modified coaxial electrospraying process to fabricate high-quality core–shell microparticles M2. SEM and TEM results demonstrated that M2 had a round shape, an obvious core–shell structure, and an estimated diameter of 1.13 ± 0.34 μm. XRD and FTIR verified that the drug TC presented in the electrosprayed products in an amorphous state, and TC had fine compatibility with EC and also SA. Compared with the electrosprayed TC-EC microparticles M1, the electrosprayed SA-coated microparticles M2 were able to provide an improved TC sustained-release profile. In total, 30% and 90% of the loaded drug sustained-release time periods were extended to 3.21 h and 19.43 h for M2, respectively, which was significantly longer than those provided by M1 (0.88 h and 9.98 h, respectively). Both the microformation mechanism of the modified coaxial electrospraying and the drug sustained-release mechanism from the core–shell microparticles M2 are suggested. The present study pioneered a brand-new manner for developing sustained drug delivery hybrids through a combination of insoluble cellulose gels and lipid using a modified coaxial electrospraying process.

4. Materials and Methods

4.1. Materials

TC with a purity greater than 99% was purchased from Shanghai Haosheng Bioengineering Company (Shanghai, China). The polymer EC and lipid SA were bought from Shanghai Huashi Big Pharmacy (Shanghai, China). The organic solvents DCM and anhydrous ethanol were analytical grade and obtained from Shanghai First Reagent Factory (Shanghai, China). Water was double distilled just before use.

4.2. Electrospraying

The homemade electrospraying apparatus comprised two syringe pumps (KDS100, USA) and a high voltage generator (ZGF2000/6 mA, Wuhan Huatian, Wuhan, China). The concentric spray head and the collector were homemade. The collecting distance from the nozzle of the spray head to the collector was fixed at 20 cm. The collected powders were kept in a desiccator until the characterizations.

4.3. Characterization

4.3.1. Morphology and Inner Structure

A filed-emission scanning electron microscope (Quanter 450, FEI, Hillsboro, OR, USA) was used to evaluate the morphologies of microparticles M1 and M2. Before the assessments, the collected samples were placed on the conductive adhesives and sputtered with a thin layer of Pt. The applied voltage was fixed at 10 keV. The average diameters of the microparticles were evaluated using ImageJ software V1.8.0 (National Institutes of Health, Bethesda, MD, USA) by randomly measuring 100 places in the SEM images.

A transmission electron microscope (TEM, JEM2100F, JEOL, Tokyo, Japan) was used to evaluate the inner structures of microparticles M1 and M2. The samples were prepared by placing a carbon film supported by 200 × 200 Cu Mesh on the collector to collect microparticles for 1 min. The operational voltage was 300 keV.

4.3.2. Physical State and Compatibility

A Spectrum 100 FTIR Spectrometer (Perkin-Elmer, Billerica, MA, USA) was used to conduct the ATR-FTIR detection. The samples included the raw TC, EC, and SA powders, and their electrosprayed products, microparticles M1 and M2. A Bruker X-ray Diffractometer (Karlsruhu, Germany) was utilized to achieve the XRD patterns of raw TC, EC, and SA powders and the electrosprayed microparticles M1 and M2. The X-rays were emitted at 40 kV and 30 mA. The recorded range of 2θ was between 5° and 60°.

4.4. Functional Performances

4.4.1. Entrapment Efficiency

TC has a maximum absorbance at λ_{max} = 278 nm, which was exploited for its quantitative measurements. The *EE*% was measured by extracting the TC from the prepared microparticles M1 and M2 as follows: an accurately weighted product of microparticles was dissolved into a mixture of ethanol and DCM (6:4 in volume); then, 1 mL of the solution was dripped into 1000 mL of distilled water; after being centrifuged at 5000 rpm for 8 min at room temperature, the supernatant was measured using a UV-vis Spectrophotometer (UV-2102PC, Unico Instrument Co. Ltd., Shanghai, China). The encapsulated TC in the electrosprayed products could be calculated through the predetermined calibration standard equation. The value of *EE*% could be achieved through the following Equation (1):

$$EE(\%) = W_m/W_p \times 100\% \quad (1)$$

where *EE*% is the entrapment efficiency, W_m is TC measured in the microparticles, and W_p represents the TC added in the working fluids. All measurements were conducted in triplicate.

4.4.2. In Vitro Dissolution Tests

The paddle method in the Chinese Pharmacopoeia (2020 Ed.) was explored to measure the in vitro release profiles of microparticles M1 and M2. The phosphate buffer solution (PBS, 0.01 M, pH 7.0) was utilized as the dissolution bulk media. The experimental conditions included 600 mL of PBS, 37 °C, and a rotation rate of 50 rpm. At predetermined time points, a 5.0 mL aliquot was withdrawn, and 5.0 mL of fresh PBS solution was added. All in vitro experiments were repeated six times.

Author Contributions: Conceptualization, Y.J., H.Z. and D.-G.Y.; methodology, Y.J., H.L., P.Z. and D.-G.Y.; writing—original draft preparation, Y.J. and D.-G.Y.; writing—review and editing, H.L. and D.-G.Y.; visualization, Y.J. and P.Z.; supervision, D.-G.Y.; project administration, D.-G.Y.; funding acquisition, H.Z. and H.L. All authors have read and agreed to the published version of the manuscript.

Funding: The financial support from the Quzhou Science and Technology Bureau (No. 2021K32 to H.Z.), the Natural Science Foundation of China (No. 52203062), and the Shanghai Sailing Program (No. 22YF1430300) are appreciated.

Institutional Review Board Statement: Not applicable.

Informed Consent Statement: Not applicable.

Data Availability Statement: The data supporting the findings of this manuscript are available from the corresponding authors upon reasonable request.

Conflicts of Interest: The authors declare no conflict of interest.

References

1. Zahra, N.; Seyed, A.S.; Sepideh, H. A Novel Self-Assembled Micelles Based on Stearic Acid Modified Schizophyllan for Efficient Delivery of Paclitaxel. *Colloids Surf. B Biointerfaces* **2021**, *199*, 111524. [CrossRef]
2. Gupta, N.; Yadav, V.; Patel, R. A Brief Review of the Essential Role of Nanovehicles for Improving the Therapeutic Efficacy of Pharmacological Agents Against Tumours. *Curr. Drug Deliv.* **2022**, *19*, 301–316. [CrossRef] [PubMed]
3. Murugesan, R.; Raman, S. Recent Trends in Carbon Nanotubes Based Prostate Cancer Therapy: A Biomedical Hybrid for Diagnosis and Treatment. *Curr. Drug Deliv.* **2022**, *19*, 229–237. [CrossRef] [PubMed]
4. Gupta, C.; Naik, I.; Menon, M.; Ambre, P.; Coutinho, E. A Review on Exploring the Opportunities of Polymer Drug Conjugated Systems for Targeted Cancer Treatment. *Curr. Drug Deliv.* **2023**, *20*, 8–30. [CrossRef]
5. Dalia, S.S.; Mohamed, A.S.; Mohmoud, S.H. Cellular Uptake, Cytotoxicity and in-Vivo Evaluation of Tamoxifen Citrate Loaded Niosomes. *Int. J. Pharm.* **2015**, *493*, 285–294. [CrossRef]
6. Fahima, M.H.; Mohamed, N.; Ahmed, K. In Vitro Cytotoxicity and Bioavailability of Solid Lipid Nanoparticles Containing Tamoxifen Citrate. *Pharm. Dev. Technol.* **2014**, *19*, 824–832. [CrossRef]

7. Assi, S.; El Hajj, H.; Hayar, B.; Pisano, C.; Saad, W.; Darwiche, N. Development and Challenges of Synthetic Retinoid Formulations in Cancer. *Curr. Drug Deliv.* **2022**, *20*, 1314–1326. [CrossRef]
8. Ajalli, N.; Pourmadadi, M.; Yazdian, F.; Abdouss, M.; Rashedi, H.; Rahdar, A. PVA Based Nanofiber Containing GO Modified with Cu Nanoparticles and Loaded Curcumin; High Antibacterial Activity with Acceleration Wound Healing. *Curr. Drug Deliv.* **2023**, *20*, 1569–1583. [CrossRef]
9. Kolisnyk, O.; Vashchenko, O.; Ruban, O.; Fil, N.; Slipchenko, G. Assessing Compatibility of Excipients Selected for A Sustained Release Formulation of Bilberry Leaf Extract. *Braz. J. Pharm. Sci.* **2022**, *58*, e19753. [CrossRef]
10. Zhang, Y.; Liu, X.; Geng, C.; Shen, H.; Zhang, Q.; Miao, Y.; Wu, J.; Ouyang, R.; Zhou, S. Two Hawks with One Arrow: A Review on Bifunctional Scaffolds for Photothermal Therapy and Bone Regeneration. *Nanomaterials* **2023**, *13*, 551. [CrossRef]
11. Qi, Q.; Wang, Q.; Li, Y.; Silva, D.Z.; Ruiz, M.E.L.; Ouyang, R.; Liu, B.; Miao, Y. Recent Development of Rhenium-Based Materials in the Application of Diagnosis and Tumor Therapy. *Molecules* **2023**, *28*, 2733. [CrossRef] [PubMed]
12. Thang, N.H.; Chien, T.B.; Cuong, D.X. Polymer-Based Hydrogels Applied in Drug Delivery: An Overview. *Gels* **2023**, *9*, 523. [CrossRef] [PubMed]
13. Xie, D.; Zhou, X.; Xiao, B.; Duan, L.; Zhu, Z. Mucus-Penetrating Silk Fibroin-Based Nanotherapeutics for Efficient Treatment of Ulcerative Colitis. *Biomolecules* **2022**, *12*, 1263. [CrossRef]
14. Shen, Y.; Yu, X.; Cui, J.; Yu, F.; Liu, M.; Chen, Y.; Wu, J.; Sun, B.; Mo, X. Development of Biodegradable Polymeric Stents for the Treatment of Cardiovascular Diseases. *Biomolecules* **2022**, *12*, 1245. [CrossRef]
15. Huang, H.; Wu, Z.; Zhou, Z.; Xu, Q.; Yan, J.; Li, Q. Study on Sustained-Release Pesticides Blended with Fosthiazate-Stearic Acid/Expanded Perlite. *J. Renew Mater.* **2023**, *11*, 257–272. [CrossRef]
16. Blinov, A.V.; Kachanov, M.D.; Gvozdenko, A.A.; Nagdalian, A.A.; Blinova, A.A.; Rekhman, Z.A.; Golik, A.B.; Vakalov, D.S.; Maglakelidze, D.G.; Nagapetova, A.G.; et al. Synthesis and Characterization of Zinc Oxide Nanoparticles Stabilized with Biopolymers for Application in Wound-Healing Mixed Gels. *Gels* **2023**, *9*, 57. [CrossRef]
17. Zhu, L.F.; Zheng, Y.; Fan, J.; Yao, Y.; Ahmad, Z.; Chang, M.W. A Novel Core-Shell Nanofiber Drug Delivery System Intended for the Synergistic Treatment of Melanoma. *Eur. J. Pharm. Sci.* **2019**, *137*, 105002. [CrossRef]
18. Bita, B.; Stancu, E.; Stroe, D.; Dumitrache, M.; Ciobanu, S.C.; Iconaru, S.L.; Predoi, D.; Groza, A. The Effects of Electron Beam Irradiation on the Morphological and Physicochemical Properties of Magnesium-Doped Hydroxyapatite/Chitosan Composite Coatings. *Polymers* **2022**, *14*, 582. [CrossRef]
19. Vidakis, N.; Petousis, M.; Velidakis, E.; Korlos, A.; Kechagias, J.D.; Tsikritzis, D.; Mountakis, N. Medical-Grade Polyamide 12 Nanocomposite Materials for Enhanced Mechanical and Antibacterial Performance in 3D Printing Applications. *Polymers* **2022**, *14*, 440. [CrossRef]
20. Du, X.-Y.; Li, Q.; Wu, G.; Chen, S. Multifunctional Micro/Nanoscale Fibers Based on Microfluidic Spinning Technology. *Adv. Mater.* **2019**, *31*, 1903733. [CrossRef]
21. Windbergs, M.; Zhao, Y.; Heyman, J.; Weitz, D.A. Biodegradable Core-Shell Carriers for Simultaneous Encapsulation of Synergistic Actives. *J. Am. Chem. Soc.* **2013**, *135*, 7933–7937. [CrossRef] [PubMed]
22. Jiang, L.; Huang, X.; Tian, C.; Zhong, Y.; Yan, M.; Miao, C.; Wu, T.; Zhou, X. Preparation and Characterization of Porous Cellulose Acetate Nanofiber Hydrogels. *Gels* **2023**, *9*, 484. [CrossRef] [PubMed]
23. Zhang, S.; Yang, Z.; Huang, X.; Wang, J.; Xiao, Y.; He, J.; Feng, J.; Xiong, S.; Li, Z. Hydrophobic Cellulose Acetate Aerogels for Thermal Insulation. *Gels* **2022**, *8*, 671. [CrossRef] [PubMed]
24. Riva, L.; Lotito, A.D.; Punta, C.; Sacchetti, A. Zinc- and Copper-Loaded Nanosponges from Cellulose Nanofibers Hydrogels: New Heterogeneous Catalysts for the Synthesis of Aromatic Acetals. *Gels* **2022**, *8*, 54. [CrossRef]
25. Huang, H.; Song, Y.; Zhang, Y.; Li, Y.; Li, J.; Lu, X.; Wang, C. Electrospun Nanofibers: Current Progress and Applications in Food Systems. *J. Agric. Food. Chem.* **2022**, *70*, 1391–1409. [CrossRef] [PubMed]
26. Huang, X.; Jiang, W.; Zhou, J.; Yu, D.-G.; Liu, H. The Applications of Ferulic-Acid-Loaded Fibrous Films for Fruit Preservation. *Polymers* **2022**, *14*, 4947. [CrossRef]
27. Narayanan, K.B.; Bhaskar, R.; Sudhakar, K.; Nam, D.H.; Han, S.S. Polydopamine-Functionalized Bacterial Cellulose as Hydrogel Scaffolds for Skin Tissue Engineering. *Gels* **2023**, *9*, 656. [CrossRef]
28. Tabakoglu, S.; Kolbuk, D.; Sajkiewicz, P. Multifluid Electrospinning for Multi-drug Delivery Systems: Pros and Cons, Challenges, and Future Directions. *Biomater. Sci.* **2022**, *11*, 37–61. [CrossRef]
29. Weon, S.H.; Han, J.; Choi, Y.-K.; Park, S.; Lee, S.H. Development of Blended Biopolymer-Based Photocatalytic Hydrogel Beads for Adsorption and Photodegradation of Dyes. *Gels* **2023**, *9*, 630. [CrossRef]
30. Wang, M.; Hou, J.; Yu, D.-G.; Li, S.; Zhu, J.; Chen, Z. Electrospun Tri-layer Nanodepots for Sustained Release of Acyclovir. *J. Alloys Compd.* **2020**, *846*, 156471. [CrossRef]
31. Wang, H.; Lu, Y.; Yang, H.; Yu, D.-G.; Lu, X. The Influence of the Ultrasonic Treatment of Working Fluids on Electrospun Amorphous Solid Dispersions. *Front. Mol. Biosci.* **2023**, *10*, 1184767. [CrossRef] [PubMed]
32. Zhou, J.; Dai, Y.; Fu, J.; Yan, C.; Yu, D.-G.; Yi, T. Dual-Step Controlled Release of Berberine Hydrochloride from the Trans-Scale Hybrids of Nanofibers and Microparticles. *Biomolecules* **2023**, *13*, 1011. [CrossRef] [PubMed]
33. He, H.; Wu, M.; Zhu, J.; Yang, Y.; Ge, R.; Yu, D.G. Engineered Spindles of Little Molecules Around Electrospun Nanofibers for Biphasic Drug Release. *Adv. Fiber Mater.* **2022**, *4*, 305–317. [CrossRef]

34. Vanheusden, C.; Vanminsel, J.; Reddy, N.; Samyn, P.; D'Haen, J.; Peeters, R.; Ethirajan, A.; Buntinx, M. Fabrication of poly(3-hydroxybutyrate-co-3-hydroxyhexanoate) Fibers Using Centrifugal Fiber Spinning: Structure, Properties and Application Potential. *Polymers* **2023**, *15*, 1181. [CrossRef]
35. Mouro, C.; Martins, R.; Gomes, A.P.; Gouveia, I.C. Upcycling Wool Waste into Keratin Gel-Based Nanofibers Using Deep Eutectic Solvents. *Gels* **2023**, *9*, 661. [CrossRef]
36. Yu, D.-G.; Xu, L. Impact Evaluations of Articles in Current Drug Delivery Based on Web of Science. *Curr. Drug Deliv.* **2023**, *20*, 1–8. [CrossRef]
37. Xia, M.; Ji, S.; Fu, Y.; Dai, J.; Zhang, J.; Ma, X.; Liu, R. Alumina Ceramic Nanofibers: An Overview of the Spinning Gel Preparation, Manufacturing Process, and Application. *Gels* **2023**, *9*, 599. [CrossRef]
38. Chuang, Y.-C.; Chang, Y.-C.; Tsai, M.-T.; Yang, T.-W.; Huang, M.-T.; Wu, S.-H.; Wang, C. Electrospinning of Aqueous Solutions of Atactic Poly(*N*-isopropylacrylamide) with Physical Gelation. *Gels* **2022**, *8*, 716. [CrossRef]
39. Liu, H.; Dai, Y.; Li, J.; Liu, P.; Zhou, W.; Yu, D.-G.; Ge, R. Fast and Convenient Delivery of Fluidextracts Liquorice Through Electrospun Core-Shell Nanohybrids. *Front. Bioeng. Biotechnol.* **2023**, *11*, 1172133. [CrossRef]
40. Li, X.-Y.; Zheng, Z.-B.; Yu, D.-G.; Liu, X.-K.; Qu, Y.-L.; Li, H.-L. Electrosprayed Sperical Ethylcellulose Nanoparticles for An Improved Sustained Release Profile of Anticancer Drug. *Cellulose* **2017**, *24*, 5551–5564. [CrossRef]
41. Liu, Z.-P.; Zhang, Y.-Y.; Yu, D.-G.; Wu, D.; Li, H.-L. Fabrication of Sustained-Release Zein Nanoparticles via Modified Coaxial Electrospraying. *Chem. Eng. J.* **2018**, *334*, 807–816. [CrossRef]
42. Zhu, M.; Yu, J.; Li, Z.; Ding, B. Self-Healing Fibrous Membranes. *Angew. Chem. Int. Ed.* **2022**, *61*, e202208949. [CrossRef]
43. Lv, H.; Liu, Y.; Zhao, P.; Bai, Y.; Cui, W.; Shen, S.; Liu, Y.; Wang, Z.; Yu, D.-G. Insight into the Superior Piezophotocatalytic Performance of $BaTiO_3$//ZnO Janus Nanofibrous Heterostructures in the Treatment of Multi-pollutants from Water. *Appl. Catal. B* **2023**, *330*, 122623. [CrossRef]
44. Du, Y.; Yang, Z.; Kang, S.; Yu, D.-G.; Chen, X.; Shao, J. A Sequential Electrospinning of a Coaxial and Blending Process for Creating Double-Layer Hybrid Films to Sense Glucose. *Sensors* **2023**, *23*, 3685. [CrossRef] [PubMed]
45. Lv, H.; Liu, Y.; Bai, Y.; Shi, H.; Zhou, W.; Chen, Y.; Liu, Y.; Yu, D.-G. Recent Combinations of Electrospinning with Photocatalytic Technology for Treating Polluted Water. *Catalysts* **2023**, *13*, 758. [CrossRef]
46. Du, Y.; Yu, D.-G.; Yi, T. Electrospun Nanofibers as Chemosensors for Detecting Environmental Pollutants: A Review. *Chemosensors* **2023**, *11*, 208. [CrossRef]
47. Nguyen, D.-N.; Moon, W. Significant Electromechanical Characteristic Enhancement of Coaxial Electrospinning Core–Shell Fibers. *Polymers* **2022**, *14*, 1739. [CrossRef]
48. Kan, Y.; Bondareva, J.V.; Statnik, E.S.; Cvjetinovic, J.; Lipovskikh, S.; Abdurashitov, A.S.; Kirsanova, M.A.; Sukhorukhov, G.B.; Evlashin, S.A.; Salimon, A.I.; et al. Effect of Graphene Oxide and Nanosilica Modifications on Electrospun Core-Shell PVA–PEG–SiO_2@PVA–GO Fiber Mats. *Nanomaterials* **2022**, *12*, 998. [CrossRef]
49. Yao, L.; Sun, C.; Lin, H.; Li, G.; Lian, Z.; Song, R.; Zhuang, S.; Zhang, D. Electrospun Bi-decorated BixBiyOz/TiO_2 Flexible Carbon Nanofibers and Their Applications on Degradating of Organic Pollutants under Solar Radiation. *J. Mater. Sci. Technol.* **2023**, *150*, 114–123. [CrossRef]
50. Pankaj, K.K.; Nivedita, G.; Karthikeyan, K. Tamoxifen Citrate Loaded Chitosan-Gellan Nanocapsules for Breast Cancer Therapy: Development, Characterisan and in-Vitro Cell Viability Study. *J. Microencapsul.* **2018**, *35*, 292–300. [CrossRef]
51. Hiremath, J.G.; Rudani, C.G.; Domb, A.J.; Suthar, R.V.; Khamar, N.S. Preparation and in Vitro Characterization of Poly(Sebacic Acid-co-Ricinoleic Acid)-Based Tamoxifen Citrate-Loaded Microparticles for Breast Cancer. *J. Appl. Polym. Sci.* **2011**, *124*, 4747–4754. [CrossRef]
52. Qian, C.; Liu, Y.; Chen, S.; Zhang, C.; Chen, X.; Liu, Y.; Liu, P. Electrospun Core–Sheath PCL Nanofibers Loaded with nHA and Simvastatin and Their Potential Bone Regeneration Applications. *Front. Bioeng. Biotechnol.* **2023**, *11*, 1205252. [CrossRef] [PubMed]
53. Cheng, G.; Ma, X.; Li, J.; Cheng, Y.; Cao, Y.; Wang, Z.; Shi, X.; Du, Y.; Deng, H.; Li, Z. Incorporating Platelet-Rich Plasma into Coaxial Electrospun Nanofibers for Bone Tissue Engineering. *Int. J. Pharm.* **2018**, *547*, 656–666. [CrossRef] [PubMed]
54. Brimo, N.; Serdaroglu, D.C.; Uysal, B. Comparing Antibiotic Pastes with Electrospun Nanofibers as Modern Drug Delivery Systems for Regenerative Endodontics. *Curr. Drug Deliv.* **2022**, *19*, 904–917. [CrossRef]
55. Yang, S.; Li, X.; Liu, P.; Zhang, M.; Wang, C.; Zhang, B. Multifunctional Chitosan/Polycaprolactone Nanofiber Scaffolds with Varied Dual-Drug Release for Wound-Healing Applications. *ACS Biomater. Sci. Eng.* **2020**, *6*, 4666–4676. [CrossRef] [PubMed]
56. Yu, D.-G.; Du, Y.; Chen, J.; Song, W.; Zhou, T. A Correlation Analysis between Undergraduate Students' Safety Behaviors in the Laboratory and Their Learning Efficiencies. *Behav. Sci.* **2023**, *13*, 127. [CrossRef]
57. Kang, S.; Hou, S.; Chen, X.; Yu, D.-G.; Wang, L.; Li, X.; Williams, R.G. Energy-Saving Electrospinning with a Concentric Teflon-Core Rod Spinneret to Create Medicated Nanofibers. *Polymers* **2020**, *12*, 2421. [CrossRef]
58. Liu, Z.-H.; Huang, Y.-C.; Kuo, C.-Y.; Chuang, C.-C.; Chen, C.-C.; Chen, N.-Y.; Yip, P.K.; Chen, J.-P. Co-Delivery of Docosahexaenoic Acid and Brain-Derived Neurotropic Factor from Electrospun Aligned Core–Shell Fibrous Membranes in Treatment of Spinal Cord Injury. *Pharmaceutics* **2022**, *14*, 321. [CrossRef]
59. Huang, J.; Feng, C. Aniline Dimers Serving as Stable and Efficient Transfer Units for Intermolecular Charge-Carrier Transmission. *iScience* **2023**, *26*, 105762. [CrossRef]

60. Wu, W.; Li, T. Deepening the Understanding of the in Vivo and Cellular Fate of Nanocarriers. *Adv. Drug Deliv. Rev.* **2022**, *189*, 114529. [CrossRef]
61. Zhang, T.; Li, L.; Chunta, S.; Wu, W.; Chen, Z.; Lu, Y. Enhanced Oral Bioavailability from Food Protein Nanoparticles: A Mini Review. *J. Control. Release* **2023**, *354*, 146–154. [CrossRef] [PubMed]
62. Man, F.; Yang, Y.; He, H.; Qi, J.; Wu, W.; Lu, Y. Establishment of In Vitro Dissolution Based on Similarity with In Vivo Dissolution: A Case Study on Aripiprazole. *Mol. Pharm.* **2023**, *20*, 2579–2588. [CrossRef]
63. Cai, Y.; Qi, J.; Lu, Y.; He, H.; Wu, W. The in Vivo Fate of Polymeric Micelles. *Adv. Drug Deliv. Rev.* **2022**, *188*, 114463. [CrossRef] [PubMed]
64. Tan, P.K.; Kuppusamy, U.R.; Chua, K.H.; Arumugam, B. Emerging Strategies to Improve the Stability and Bioavailability of Insulin: An Update on Formulations and Delivery Approaches. *Curr. Drug Deliv.* **2022**, *20*, 1141–1162. [CrossRef] [PubMed]
65. Peppas, N. Analysis of Fickian and Non-Fickian Drug Release from Polymers. *Pharm. Acta Helv.* **1985**, *60*, 110–111.
66. Bobokalonov, J.; Muhidinov, Z.; Nasriddinov, A.; Jomnurodov, A.; Khojaeva, F.; Komilova, G.; Yusufi, S.; Liu, L. Evaluation of Extended-Release of Piroxicam-Loaded Pectin-Zein Hydrogel Microspheres: In Vitro, Ex Vivo, and In Vivo Studies. *Curr. Drug Deliv.* **2022**, *19*, 1093–1101. [CrossRef] [PubMed]
67. Liao, Q.; Kim, E.J.; Tang, Y.; Xu, H.; Yu, D.-G.; Song, W.; Kim, B.J. Rational Design of Hyper-Crosslinked Polymers for Biomedical Applications. *J. Polym. Sci.* **2023**. [CrossRef]
68. Fan, W.; Peng, H.; Yu, Z.; Wang, L.; He, H.; Ma, Y.; Qi, J.; Lu, Y.; Wu, W. The Long-Circulating Effect of Pegylated Nanoparticles Revisited via Simultaneous Monitoring of Both the Drug Payloads and Nanocarriers. *Acta Pharm. Sin. B* **2022**, *12*, 2479–2493. [CrossRef]
69. Yu, D.G.; Zhou, J. How Can Electrospinning Further Service Well for Pharmaceutical Researches? *J. Pharm. Sci.* 2023. [CrossRef]
70. Guler, E.; Nur Hazar-Yavuz, A.; Tatar, E.; Morid Haidari, M.; Sinemcan Ozcan, G.; Duruksu, G.; Graça, M.P.F.; Kalaskar, D.M.; Gunduz, O.; Emin Cam, M. Oral Empagliflozin-Loaded Tri-layer Core-Sheath Fibers Fabricated Using Tri-axial Electrospinning: Enhanced in Vitro and in Vivo Antidiabetic Performance. *Int. J. Pharm.* **2023**, *635*, 122716. [CrossRef]
71. Zhou, J.; Wang, P.; Yu, D.-G.; Zhu, Y. Biphasic Drug Release from Electrospun Structures. *Exp. Opin. Drug Deliv.* **2023**, *20*, 621–640. [CrossRef] [PubMed]
72. Ilić-Stojanović, S.; Nikolić, L.; Cakić, S. A Review of Patents and Innovative Biopolymer-Based Hydrogels. *Gels* **2023**, *9*, 556. [CrossRef] [PubMed]
73. Li, J.; Song, W.; Li, F. Polymeric DNA Hydrogels and Their Applications in Drug Delivery for Cancer Therapy. *Gels* **2023**, *9*, 239. [CrossRef] [PubMed]
74. Zhou, J.; Wang, L.; Gong, W.; Wang, B.; Yu, D.-G.; Zhu, Y. Integrating Chinese Herbs and Western Medicine for New Wound Dressings through Handheld Electrospinning. *Biomedicines* **2023**, *11*, 2146. [CrossRef] [PubMed]
75. Han, W.; Wang, L.; Li, Q.; Ma, B.; He, C.; Guo, X.; Nie, J.; Ma, G. A Review: Current Status and Emerging Developments on Natural Polymer-Based Electrospun Fibers. *Macromol. Rapid Commun.* **2022**, *43*, 2200456. [CrossRef]
76. Dias, J.R.; Sousa, A.; Augusto, A.; Bártolo, P.J.; Granja, P.L. Electrospun Polycaprolactone (PCL) Degradation: An In Vitro and In Vivo Study. *Polymers* **2022**, *14*, 3397. [CrossRef]
77. Liu, H.; Wang, H.; Lu, X.; Murugadoss, V.; Huang, M.; Yang, H.; Wan, F.; Yu, D.G.; Guo, Z. Electrospun structural nanohybrids combining three composites for fast helicide delivery. *Adv. Compos.Hybrid Mater.* **2022**, *5*, 1017–1029. [CrossRef]
78. Yu, D.G.; Huang, C. Electrospun Biomolecule-Based Drug Delivery Systems. *Biomolecules* **2023**, *13*, 1152. [CrossRef]
79. Pooja, C.; Bright, B.; Siew, H.N.; Aaron, W.; Lynn, P.W.; Heather, L.W. Solidified Saturated Fats Coating Subunit Vaccines Greatly Extended Vaccine Booster Release and Contributed to a Th1/Th2 Mixed Immune Response in Mice. *Vaccine* **2023**, *41*, 3989–4001. [CrossRef]
80. Agrawal, A.; Joshi, A.; Bhattacharya, S. Recent Excavation of Nanoethosomes in Current Drug Delivery. *Curr. Drug Deliv.* **2022**, *21*, 168–183. [CrossRef]
81. Sare, F.; Mohammad, R.A.; Seyyedeh, E.M.; Seyyed, M.R.S.; Mandana, K.; Shirzad, A.; Solmaz, G. Diazepam Loaded Solid Lipid Nanoparticles: In Vitro and In Vivo Evaluations. *Adv. Pharm. Bull.* **2022**, *12*, 86–92. [CrossRef]
82. Alam, M.; Rizwanullah, M.; Mir, S.R.; Amin, S. Statistically Optimized Tacrolimus and Thymoquinone Co-Loaded Nanostructured Lipid Carriers Gel for Improved Topical Treatment of Psoriasis. *Gels* **2023**, *9*, 515. [CrossRef] [PubMed]
83. Ahmadi, N.; Rincón, M.; Silva-Abreu, M.; Sosa, L.; Pesantez-Narvaez, J.; Calpena, A.C.; Rodríguez-Lagunas, M.J.; Mallandrich, M. Semi-Solid Dosage Forms Containing Pranoprofen-Loaded NLC as Topical Therapy for Local Inflammation: In Vitro, Ex Vivo and In Vivo Evaluation. *Gels* **2023**, *9*, 448. [CrossRef] [PubMed]
84. Song, W.; Tang, Y.; Qian, C.; Kim, B.J.; Liao, Y.; Yu, D.-G. Electrospinning Spinneret: A Bridge between the Visible World and the Invisible Nanostructures. *Innovation* **2023**, *4*, 100381. [CrossRef]
85. Sivan, M.; Madheswaran, D.; Hauzerova, S.; Novotny, V.; Hedvicakova, V.; Jencova, V.; Kostakova, E.K.; Schindler, M.; Lukas, D. AC Electrospinning: Impact of High Voltage and Solvent on the Electrospinnability and Productivity of Polycaprolactone Electrospun Nanofibrous Scaffolds. *Mater. Today Chem.* **2022**, *26*, 101025. [CrossRef]
86. Sivan, M.; Madheswaran, D.; Valtera, J.; Kostakova, E.K.; Lukas, D. Alternating Current Electrospinning: The Impacts of Various High-Voltage Signal Shapes and Frequencies on the Spinnability and Productivity of Polycaprolactone Nanofibers. *Mater. Des.* **2022**, *213*, 110308. [CrossRef]

87. Xu, J.; Zhong, M.; Song, N.; Wang, C.; Lu, X. General Synthesis of Pt and Ni Co-Doped Porous Carbon Nanofibers to Boost HER Performance in Both Acidic and Alkaline Solutions. *Chin. Chem. Lett.* **2023**, *34*, 107359. [CrossRef]
88. Eslami, H.; Gharibi, A.; Muller-Plathe, F. Mechanisms of Nucleation and Solid−Solid-Phase Transitions in Triblock Janus Assemblies. *J. Chem. Theory Comput.* **2021**, *17*, 1742–1754. [CrossRef]
89. Bahri, K.; Eslami, H.; Muller-Plathe, F. Self-Assembly of Model Triblock Janus Colloidal Particles in Two Dimensions. *J. Chem. Theory Comput.* **2022**, *18*, 1870–1882. [CrossRef]

gels

MDPI

Article

Green Synthesized Silver Nanoparticles Loaded in Polysaccharide Hydrogel Applied to Chronic Wound Healing in Mice Models

Fahad M. Aldakheel [1], Dalia Mohsen [2,3,*], Marwa M. El Sayed [4], Mohammed H. Fagir [2] and Dalia K. El Dein [2]

[1] Department of Clinical Laboratory Sciences, College of Applied Medical Sciences, King Saud University, Riyadh 11433, Saudi Arabia; faldakheel@ksu.edu.sa
[2] Clinical Laboratory Sciences Program, Inaya Medical College, Riyadh 12211, Saudi Arabia; husseinfagir@inaya.edu.sa (M.H.F.); dkmohammed@inaya.edu.sa (D.K.E.D.)
[3] Microbiology Department, National Research Centre, Giza 12622, Egypt
[4] Chemical Engineering and Pilot Plant Department, National Research Centre, Giza 12622, Egypt; dr.marwameid@gmail.com
* Correspondence: dr.dalia@inaya.edu.sa

Abstract: The prevalence of chronic wounds is increasing owing to the expanding population and the growing number of individuals suffering from diabetes. Such a chronic wound continues to be a significant healthcare burden for diabetic patients because it frequently carries a high chance of limb loss due to amputation and reduces survival as a result. Development of innovative wound dressing materials with the potential to stop bacterial infections and accelerate the process of tissue regeneration is needed to increase the effectiveness of diabetic wound healing. In the current study, a co-polymerization process based on a free radical reaction was used to create a hydrogel of polysaccharides blend graft acrylamide (PsB-g-Am). Starch, chitosan, and alginate make up the polysaccharides blend (PsB). The produced hydrogel's structure was characterized using FTIR spectroscopy. The antibacterial activities of silver nanoparticles synthesized through the green method using garlic bulb (Allium sativum) is reported. The silver nanoparticles' physical characteristics were examined using scanning electron microscopy, transmission electron microscopy analysis, and UV-visible spectroscopy and they were found to range in size from 50 to 100 nm. The agar well diffusion technique is used to investigate the antibacterial characteristics. Inclusion of silver nanoparticles in the hydrogels demonstrated concentration-dependent antibacterial behavior against Gram-negative *Klebsiella pneumoniae* and Gram-positive *Staphylococcus aureus* during antimicrobial testing of the hydrogels. When hydrogels were applied to diabetic mice, the system was examined for its healing abilities, and positive therapeutic results were obtained in as little as 14 days. Thus, it can be inferred that graft copolymer of chitosan-AgNPs hydrogels can promote healing in chronic wounds over time and can be utilized as an alternative to conventional therapies for chronic wounds (such as those brought on by diabetes) in mouse models.

Keywords: chronic wound; hydrogel; graft copolymer of chitosan; antibacterial; mice induced diabetes

check for updates

Citation: Aldakheel, F.M.; Mohsen, D.; El Sayed, M.M.; Fagir, M.H.; El Dein, D.K. Green Synthesized Silver Nanoparticles Loaded in Polysaccharide Hydrogel Applied to Chronic Wound Healing in Mice Models. *Gels* **2023**, *9*, 646. https://doi.org/10.3390/gels9080646

Academic Editors: Adina Magdalena Musuc, Magdalena Mititelu and Mariana Chelu

Received: 16 July 2023
Revised: 5 August 2023
Accepted: 8 August 2023
Published: 11 August 2023

1. Introduction

Diabetic foot ulcer (DFU) is a common and serious complication of diabetes; characterized by slow-healing wounds on the skin of diabetic patients [1]. DFUs pose significant challenges due to compromised physiological conditions and weakened immune responses in individuals with diabetes [2,3]. Despite advancements in medical technology, DFU continues to burden patients' access to healthcare. Consequently, there is an urgent need to develop effective therapeutic approaches to enhance DFU [4].

The skin is the most crucial organ in the body because it protects the body from harm from the environment, is vulnerable to traumas and wounds, and can effectively heal damaged tissues [5]. A chronic wound develops when tissue does not heal within the predicted time frame, which is why wound healing is a specialized biological process connected to the general phenomena of tissue development and regeneration. Hemostasis, inflammation, migration, proliferation, and maturation are the five discrete steps that make up this process. These processes result in a complex web of interactions between various cell types, mediator chemicals, and extracellular matrix components [6].

The wounds may contain a variety of bacteria including multidrug-resistant forms of *Staphylococcus aureus*; *Pseudomonas aeruginosa*; and *Klebsiella pneumonia* organisms. A better method of treating chronic wounds is therefore desperately needed. Antibacterial medications are used in traditional therapies. However; this strategy adds to the problem of bacterial resistance development, which is still a problem today [7]. More targeted approaches to build a more powerful solution are needed. Although antibacterial applications may aid the wound more successfully; it still struggles with being quickly broken down by the body. To avoid this, a specially designed carrier that delivers the anticipated medicine while providing protection will maximize the healing process and allow for longer-lasting treatment [8].

There are various kinds of wound care dressings. These dressings, however, are developed from substances like semi-permeable gels or foams that can either be combined with antibacterial drugs or naturally possess microbiological qualities. A good wound dressing should, in general, be simple to remove, not adhere to the wound, keep the area moist permitting air permeation, and keep external detritus like bacteria out. The patient should have the least amount of discomfort and anxiety possible [9]. According to the published data in the same field, increasing the porosity can have a beneficial effect on the diffusion of nutrients and oxygen, especially in the absence of a functional vascular system [10,11]. As a result, there are many medications available for the treatment of chronic injuries. However, using these products has drawbacks, including the need for repeated applications owing to their short-term effects and high-investment costs. In order to achieve appropriate healing, new techniques were developed. One such technique is the use of hydrogels, which may absorb part of the wound's exudate and provide moisture to tissue that has lost any.

The capacity of vinyl monomers to create hydrogels by grafting copolymerization onto polysaccharides, such as starch, chitosan, sodium alginate, and carrageenan, are extensively described [12]. Additionally, hydrogels are frequently sensitive to the circumstances of the surrounding environment and are referred to as "intelligent materials" or "smart materials" due to the presence of various functional groups along the polymer chains [13,14]. It was claimed that by employing hydrogels made from natural polymers mixed with nanostructures as an innovation for controlled drug release, taking into consideration its structure, permanence in the wound, stimulus sensitivity, and ultimately duration and temperature of breakdown, chitosan was suggested as an alternative to create hydrogels because research showed that it has curative properties, including the ability to regenerate skin tissue and control bleeding by working with inflammatory cells (leukocytes, macrophages, and fibroblasts) [15]. As a result, it has an antibacterial effect; this feature is a result of its positive charge and chelating ability [16].

The incorporation of silver nanoparticles (AgNPs) as a bactericidal and bacteriostatic agent to chitosan-made hydrogels was proposed; for example, their main advantage is to cover a greater surface area with a smaller amount of material when compared to their macroscopic structures, which inhibit different bacterial concentrations.

The unique characteristics of nanomaterials contribute to the rapid advancement of nanotechnology. Silver nanoparticles in particular have attracted significant interest from scientists due to their ability to exhibit various distinctive properties that can be adjusted based on their size. These properties include remarkable chemical stability a wide range of radiation absorption easy accessibility and non-toxicity [17,18]. Among the frequently employed nanoparticles, silver nanoparticles are known for their potent antimicrobial properties [19]. Various techniques, including electrochemical reduction, photochemical reduction, heat evaporation and biological methods, are employed to produce silver nanoparticles. However, these methods are often costly and involve the use of hazardous chemicals, posing risks to both biological systems and the environment. Recently, there is a growing interest in utilizing plant extracts for the synthesis of silver nanoparticles primarily due to their environmentally friendly nature [20]. Plant extracts serve as both reducing agents and capping agents during the nanoparticle synthesis process. Several plant extracts such as garlic (Allium sativum) [21], Z. officinale (ginger), Aloe vera and coffee [22] can be utilized.

Garlic, which is widely consumed as a spice food additive and medicinal herb, is known to have gastric stimulant properties. It contains a range of organosulfur compounds that contribute to its various biological activities. These compounds include allyl sulfide allicin allyl cysteine, ajoene and alliin. Additionally, the presence of phenols, terpenoids, ketones aldehydes and amides in plants plays a role in the synthesis of metal nanoparticles [23].

According to [24], hydrogels coated with chitosan-AgNPs showed higher antibacterial activity than hydrogels without the coating. Evaluations of in vitro antibacterial activity were conducted against wound infections brought on by the presence of methicillin-resistant *S. aureus* and *P. aeruginosa*. Hydrogels made of chitosan and AgNPs showed significant antibacterial activity. In addition, [16,25] produced chitosan and chitosan-PVP-silver nano-oxide (CPS) films having antibacterial and therapeutic capabilities for wound healing. A greater level of antibacterial activity was seen in the CPS film. There are many reports on the use of dressings with particular materials that encourage tissue development and favor cellular recruitment in the early phases of cauterization, despite the fact that wound healing is a natural mechanism in the regeneration of injured tissues inside the human body. As cauterization dressings for skin wounds, chitosan hydrogels developed with Gaps were used in the current study. An in vivo case study using mice that were previously generated with diabetes and other diseases that interfered with the tissues' normal cauterization process was carried out.

2. Results and Discussion

2.1. *Ultraviolet–Visible Spectroscopy (UV–Vis)*

The UV-vis spectra of the AgNPs synthesized at three different powers (50, 60, and 70%) are shown in Figure 1. Because the temperature developed in the solution was insufficient to carry out an effective reduction, the AgNPs synthesized at 50% power displayed the lowest intensity in the peak of 420 nm.

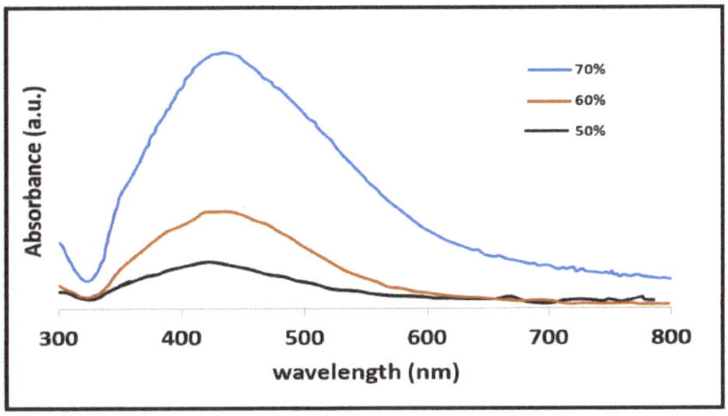

Figure 1. Spectra obtained by UV–Vis the black line is for 50% power the brown for 60% and finally the blue line for 70%.

The maximal absorbance peak was seen when the synthesis process was run at a greater power (70%). This change could be the result of using more energy per unit of time, which makes the reaction more intense and permits the production of bigger particles [26]. Furthermore, it can be deduced that as the power of the microwave oven grows, the density of the particles in the solution rises and the distance between them decreases, allowing the van der Waals forces to dominate and cluster formation to occur [27]. The sample of 70% powder will be used for the hydrogel synthesis in regard to the UV-Vis spectrum analysis.

2.2. Characterization of the Synthesied Hydrogel

FTIR

By contrasting the FTIR spectra of the raw PsB mixture with the grafted PsB, as shown in Figure 2, the grafting evidence of acrylamide onto polysaccharides blend (PsB) was validated. The characteristic peaks of chitosan are 1600 cm^{-1} (N-H bend), 1327 cm^{-1} (C-N stretch), 1155 cm^{-1} (bridge O stretch), a peak at 1440 cm^{-1} (-COO- stretching), while the characteristic peak of alginate appeared at 619 cm^{-1} (Na-O). Furthermore, for starch, the broad band was 3459 cm^{-1} due to the stretching mode of O-H groups. The adsorption band at 1648 cm^{-1} is attributed to an intermolecular H-bond involving the carboxyl group. The band at 2931 cm^{-1} is assigned to C-H stretching. For the grafted PsB, in addition to the peaks related to the three used poly saccharides, adding Am shows a broad band located at 3428 cm^{-1}, which was attributed to the N-H vibrations and a smaller peak at 2936 cm^{-1}, corresponding to the C-H stretching vibrations of the methylene group. The bands at 1667 cm^{-1} are assigned to C=O moiety of the -$CONH_2$ group (amide-I) [28,29]. Furthermore, the absorption band at 1450 cm^{-1} was related to the vibrations of C-N bond. Furthermore, there are peaks at 1157 and 1080 cm^{-1} due to the C=O stretching. Hence, the newly appeared peaks found in PsB-g-Am hydrogel support the grafting findings.

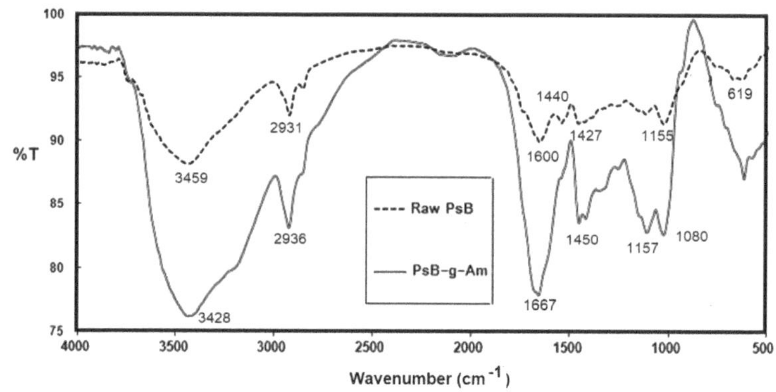

Figure 2. FTIR for PsB raw and PsB-g-Am.

2.3. SEM Analysis of AgNPs

At magnifications of 2.9 and 11, Figure 3 shows a wide range of AgNP sizes, which are consistent with the UV–Vs spectra, where the breadth of the absorption curve indicated the potential development of NP aggregates brought on by the high concentration of AgNPs.

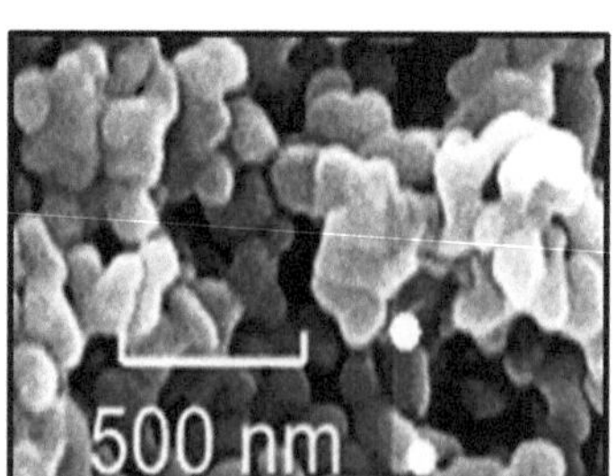

Figure 3. Images obtained by SEM of AgNPs.

2.4. TEM Analysis of AgNPs

Figure 4 demonstrates the tendency of AgNPs to cluster. Images produced by TEM of NPs morphology shows a zoom of the sampled NPs (20 nm). Similar outcomes were reported by different authors [30–33] as their generated NPs had a 100 nm size. Additionally, a significant proportion of AgNPs were found to have faceted morphologies (wireframe structures, nanorods, and truncated prisms).

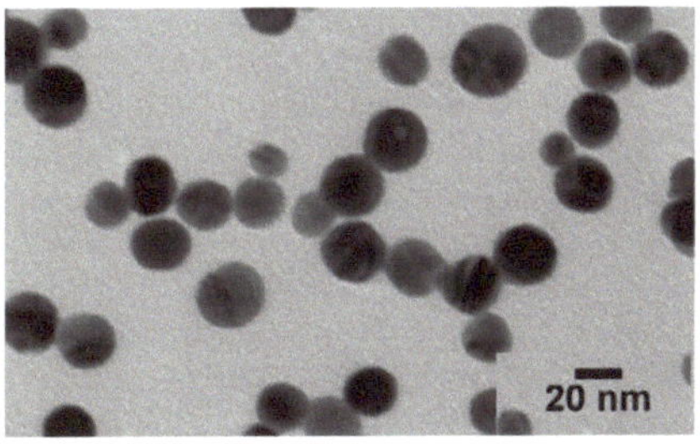

Figure 4. TEM of the synthesized AgNPs.

2.5. Swelling Water Ratio (SWR)

SWR for the two synthesized samples (PsB-g-Am and PsB-g-Am-AgNPs) are shown in Figure 5. For all grafted samples, the maximum SWR values after swelling time of 24 h was 50 g/g. Furthermore, the maximum obtained SWR for the hydrogel loaded with silver nano-particles was 74 g/g. This result means that adding the silver ion to the hydrogel main substrate improves the swelling properties by 24%, which matches with the results obtained by previously published work [29].

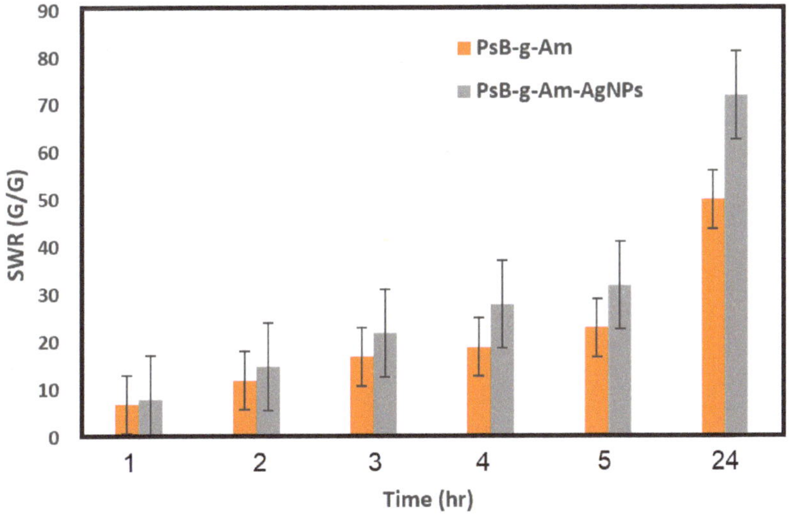

Figure 5. SWR of PsB-g-Am and PsB-g-Am-AgNPs.

2.6. In Vitro Test Examination

The study demonstrated that hydrogels made from chitosan and containing AgNPs exhibited the highest effectiveness in combating bacteria that are resistant to treatment. These bacteria typically emerge in the wounds of diabetic mice and have the ability to isolate themselves, leading to the development of secondary infections. The hydrogels not only prevented these secondary infections caused by the bacteria's resistance, but also promoted the faster wound healing.

In Figure 6, it is illustrated that the assessment focused on detecting the efficacy of PsB-g-Am- and PsB-g-Am-loaded sliver nanoparticles for the chronic wound healing group. When evaluating the impact on *Klebsiella pneumoniae*, it was observed that neither of the fixed films with the lowest silver concentration (0.5 mL of AgNPs) displayed any inhibition zone. However, when using the highest concentration of silver nanoparticles (AgNPs) at 2 mL, it was noted that the zone of inhibition indicating the effectiveness against bacterial growth increased to approximately 0.4 mm in size. When testing against *Staphylococcus aureus*, it was observed that the hydrogel exhibited a similar inhibition zone size of 0.4 mm to what was observed with *Klebsiella pneumoniae*. However, even at the lowest concentration of AgNPs, an inhibition zone of 0.26 mm was formed, while *Klebsiella pneumoniae* was 0.28 mm the size of the inhibition zones that increased as the concentration of AgNPs increased. The largest observed inhibition zone in this case was 0.4 mm. The results of both analyses align with the findings of Reiad [34], as the chitosan films without AgNPs exhibit a similar outcome. The gradual release of NPs leads to bacterial inhibition, as confirmed by the technique.

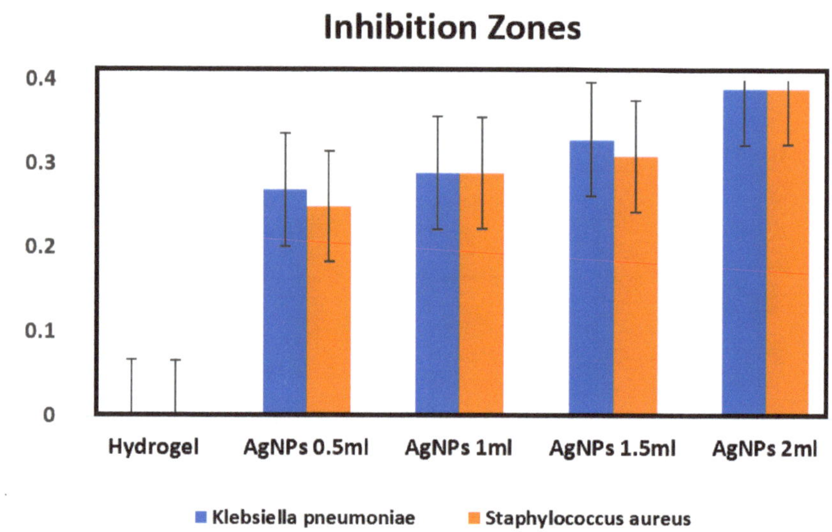

Figure 6. The Efficacy of PsB-g-Am and PsB-g-Am loaded sliver nanoparticles for infected chronic wound by Gram-negative *Klebsiella pneumoniae* and Gram-positive *Staphylococcus aureus*.

2.7. In Vivo Study

The previously induced diabetic mice were divided into three groups; first, not treated (control); and the second and third treated by (PsB-g-Am) and (PsB-g-Am-AgNPs), respectively, by applying of 50-microliter volume of (PsB-g-Am) and (PsB-g-Am-AgNPs) solutions (10 ppm) to the wound. Each group of mice was accommodated in separate housing. The duration of the experiment spanned 14 days.

When (PsB-g-Am) and (PsB-g-Am-AgNPs) are present on the surface of the wound area PsB-g-Am and silver nanoparticles could accelerate chronic wound healing as compared with the control group. As shown in Figure 7, both the two groups (2 and 3) lines of healing indicate the (second and third) groups of treated mice in days (2, 7, and 14) when the mice were sacrificed.

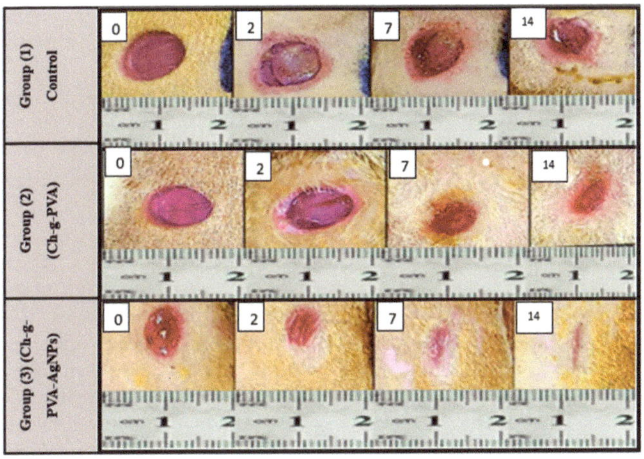

Figure 7. Wound healing efficacy in vivo using diabetic mice in days (0; 2; 7 & 14) Group 1 (control) Group 2 of diabetic mice treated by (PsB-g-Am) and Group 3 diabetic mice treated by (PsB-g-Am-AgNPs).

As inflammation is a normal part of the wound healing process, the results suggest that (PsB-g-Am) and (PsB-g-Am-AgNPs) could reduce chronic wound size and enhance skin wound healing in the mice treated with (PsB-g-Am) alone or (PsB-g-Am-AgNPs) as compared with the control group. However, a significant difference was found between the three groups at day 14 due to the effect of (PsB-g-Am) only or when PsB-g-Am-AgNPs were used. Moreover, since the formation of scars plays a crucial role in the process of wound healing [35], the extent of scarring on the skin's surface was measured.

It was observed that there was a noticeable contrast in the overall visible characteristics of the healed wounds after a period of 14 days, following the initial injury (Figure 7). The findings of the current research demonstrate that the utilization of silver nanoparticles (PsB-g-Am-AgNPs) has the potential to enhance the process of healing for skin wounds while minimizing the visibility of scars [36]. Silver nanoparticles (Ag-NPs) loaded in hydrogel display strong antimicrobial properties that effectively combat infections followed by when using (PsB-g-Am) without silver nanoparticles as hydrogels have emerged as a promising substitute for treating various challenging wounds that struggle to heal. Certain severe skin injuries often suffer from bacterial contamination, leading to delayed healing due to the presence of necrotic tissue that creates an ideal environment for bacterial growth [37]. However, by employing silver nanoparticles (AgNPs) as an antimicrobial agent, the use of hydrogels effectively diminishes the bacterial burden, thereby facilitating a proper healing process. Hydrogel and silver nanoparticles exhibit high toxicity towards microorganisms, making them effective in eliminating bacteria responsible for diseases transmitted through food water and wounds [38]. Although the exact mechanism by which (PsB-g-Am-AgNPs) affect microorganisms is not fully understood, they interact with various molecular processes within microorganisms leading to growth inhibition loss of infectivity and cell death [39]. The antimicrobial activity of Ag-NPs is attributed to the generation of free radicals on their surface [40]. The healing of chronic wounds is a dynamic process involving the coordinated interplay of blood cells proteins proteases growth factors and extracellular matrix components. This physiological process is vital for the regeneration and reorganization of damaged tissue, ultimately restoring its normal structure [2].

Furthermore, the mice treated with (PsB-g-Am) and (PsB-g-Am-AgNPs) showed reduced scar visibility and inflammation levels at the site of the wound. As a result, the size of the wound area was smaller and the healing process was shorter compared to the control group. Similarly, in a study conducted by Nadworny [41] that examined the effects of (PsB-g-Am) only or when loaded with Ag-NPs on wound healing, it was observed that the healing process was expedited and the cosmetic appearance of the wound improved in an animal model.

Based on the aforementioned explanations, it was suggested to use chitosan hydrogels containing silver nanoparticles as the best results than when using the chitosan g PVA alone. This choice was made due to the combined benefits offered by each component of the dressing, which promote the healing and regeneration of injured tissue through granulation and epithelialization processes.

3. Conclusions

In conclusion, the extract derived from garlic (alum sativum) extract effectively generates AgNPs that exhibit excellent stability in solution. These synthesized silver nanoparticles demonstrate their effectiveness as an active agent against both Gram-negative and Gram-positive bacteria. Ultimately the biosynthesis of silver nanoparticles shows great promise as a potential solution for medical applications where antimicrobial activity is crucial. The utilization of graft copolymer of chitosan hydrogels alone and graft copolymer of chitosan hydrogels containing silver nanoparticles (PsB-g-Am-AgNPs) in diabetic mice offers beneficial effects in the recovery of chronic wounds. This is due to the hydrogel only creating a moist environment around the wound facilitating the required physiological and environmental conditions for healing in the underlying tissues of ulcers or chronic wounds. Moreover, research successfully developed the (PsB-g-Am-AgNPs) as a versatile composite hydrogel that was identified as a highly effective material for treating long-term wounds. The (PsB-g-Am-AgNPs) exhibited strong antimicrobial properties as it displayed a significant ability to kill bacteria, specifically *K. pneumoniae* and *S. aureus* due to the presence of Ag^+ in its composition. Both graft copolymer of chitosan hydrogel alone and graft copolymer of chitosan hydrogel developed with AgNPs when applied to the diabetic mice wounds, led to decreased scarring and enhancement of the healing process, diminishing the likelihood of infection.

4. Materials and Methods

4.1. Materials

Corn starch (Sigma-Aldrich, Hamburg, Germany) chitosan medium molecular weight (Sigma Aldrich, Hamburg, Germany) alginic acid sodium salt from brown algae "alginate" (Routh, Hamburg, Germany) chitosan (Ch) with medium molecular weight and deacetylation $\geq$ 75% (Sigma Aldrich, Hamburg, Germany) and acrylamide (Am), MW 71.08 (Baker Chemical Co., Phillipsburg, NJ, USA) were the basic raw materials used for the hydrogel preparation. Methylene bisacryl-amide (MBA) MW 154.2 (Fluka, Buchs, Germany) and potassium persulfate (KPS) MW 270.322 (Merck, Darmstadt, Germany) were used as the crosslinker and initiator, respectively. Other chemicals include acetone acetic acid and ethanol (El Nasr Pharmaceutical Chemicals Co., Cairo, Egypt) and sodium hydroxide pellets (Laboratory chemicals Modern Lab. Egypt). The applied experiments were conducted using double distilled water (DDW). In addition, drying was conducted via an oven hot-plate magnetic stirrer; Microwave normal saline and Ultraviolet apparatus (ES-13080UV2A).

Garlic bulb (Allium sativum), a beaker, magnetic stirrer, hot plate, incubator, power supply, thermometer, Whatman No. 1 filter papers, digital electronic analytical balance (Model FA2104, Shanghai Selon Scientific Instrument Co., Ltd., Shanghai, China), furnace (Model BK-5-12GJ), ceramic crucible cups, drying oven (Model 101-0 Biobased Biodustry Shandong Co., Ltd.), China cylinders, and a centrifuge (Model AVI-558 max RPM: 5000 rpm) were used.

For the production of silver nanoparticles using garlic bulb extract, silver nitrate ($AgNO_3$) was utilized as the precursor. A 1 mM solution of silver nitrate was prepared by dissolving it in double distilled water and subsequently stored at a temperature of 4 °C in a refrigerator.

4.2. Ethical Considerations

Research protocols for animal injection were approved by the National research Centre (Animal Facility Unit). Animal testing was performed with compliance of the local ethics committee and Biosafety Committee under number (RSP 2023R506) from KSU. Mice were sacrificed through anesthetic overdose from Dimethyl ether.

4.3. Preparation of the Polysaccharides Blend (PsB)

Three grammes of starch were added to 70 mL of DW, and the mixture was stirred for 30 min at 80 °C. One gram of chitosan was added, and the mixture was stirred for five hours at room temperature. Finally, one gram of alginate was added, and the mixture was stirred for four hours at room temperature. The pre-prepared starch colloid was then added to the chitosan and alginate solutions, and the mixture was stirred for 10 min to produce the PsB solution.

4.3.1. Grafting of Acrylamide onto PsB

The pre-made PsB solution was mixed with 0.6 g of KPS and five different weights of Am, resulting in Am/PsB weight ratios of 0.6, 0.69, 0.78, 0.87, and 0.96 "g/g". Then, to each of the five produced combinations, 0.1 g of MBA was added. The next phase, grafting, was completed using a traditional technique using a three-necked round bottom flask. Nitrogen inlets and thermometers were installed in the right and left necks, respectively. The intermediate neck had a mechanical stirrer, and was condensed for an hour at 60 °C, and three neck quick-fit adapter reactants were inputted. The reaction product was cooled to room temperature.

4.3.2. Post Treatment

After the grafted hydrogels were brought to a pH of 8 using 1 N NaOH, a solution of 70% ethanol was added, and the gel product was agitated for 150 min (five times) to dissolve the homopolymer that had formed. Final steps included filtering, two new ethanol washes, and drying the product at 70 °C until a consistent weight was achieved [39,40].

4.4. Hydrogel Characterization and Analysis

4.4.1. FTIR

A FT/IR-6100 type A Jasco Japan TGS detector with the absorbance technique ranging from 500 to 4000 cm^{-1} with scanning speed of 2 mm/s was used.

4.4.2. SEM Analysis of AgNPs

The synthesized solution at 70% power was chosen for the remaining experiments after the samples' spectroscopy analysis was completed because it has a higher concentration of particles and a lower concentration of silver ions (which could produce a secondary effect in its application). Using a Tescan model MIRA LMU scanning electron microscope, SEM analysis was produced.

4.4.3. TEM Analysis of AgNPs

It was required to do a transmission electron microscopy analysis on the particles in order to precisely identify their size and form. Equipment from the TEM Jeol JSM-1010 was used for the analysis. A power of 90 keV was used to analyze the sample.

4.5. Swelling Water Ratio (SWR)

The dried hydrogel samples were immersed in RO water with different pH values (3, 5, 7, 9 and 11) and with different ionic strengths (0.1%, 2% and 3%). Samples were taken, and then weighed after indicated time intervals, where the excess water on their surface was gently removed by filter paper.

Swelling ratio (SR) was calculated by the following Equation (1) [40,41]:

$$SR(\%) = \left[\frac{W_s - W_d}{W_d}\right] \times 100 \tag{1}$$

where, W_s and W_d are the weight of the swollen and dry samples, respectively. SWR was calculated as grams of absorbed water per grams of dry hydrogel (g/g).

4.6. Ultraviolet–Visible Spectroscopy (UV–Vis)

A UV–Vis examination was carried out in a 10 S spectrometer to confirm the existence of these NPs, and spectrum scanning was performed in the 300–800 nm wavelength range. The development of a maximal absorbance peak at about 420 nm was used to assess the existence of AgNPs [30,31]. The UV-vis spectra of the AgNPs was synthesized at three different powers (50, 60, and 70%).

4.7. Synthesis of AgNPs

4.7.1. Extraction Alum Sativum

Fresh roots of garlic were obtained from a local market in Egypt. The outer skin of the garlic was removed and then washed with distilled water. Roots were dried completely to remove any moisture content. The dried roots were crushed using a mortar and pestle. Subsequently, 10 g of the powder was boiled in 100 mL of deionized water (DW) for 30 min. Finally, the extracts were filtered using Whatman No. 1 filter paper and stored at 4 °C for further use.

4.7.2. Silver Nanoparticles Biosynthesis

The synthesis of silver nanoparticles was conducted using the green synthesis method. To reduce Ag^+ ions, 1 mL of garlic extract solution was added drop by drop into a 100 mL aqueous solution containing 1 mM of $AgNO_3$. The mixture was heated at a temperature ranging from 60 to 80 °C for a duration of 1 h. During this process, a noticeable change in color was observed as the dark brown solution transformed into a reddish-brown, thus indicating the successful formation of silver nanoparticles, as shown in Figure 8.

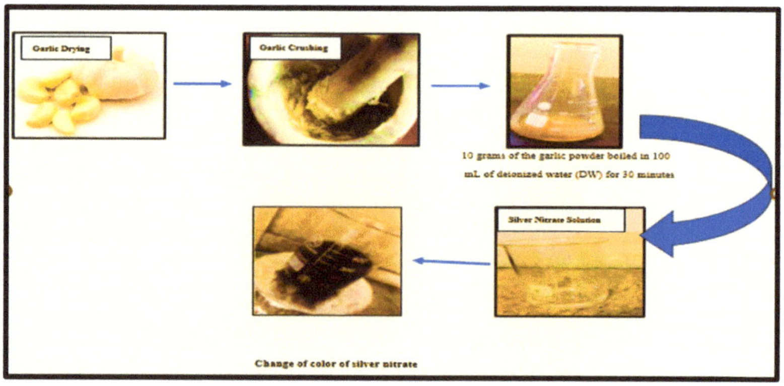

Figure 8. Silver Nanoparticles Biosynthesis from garlic extract.

4.8. Synthesis of PsB-g-Am Loaded with AgNPs

As mentioned in Section 2.3, the hydrogel was prepared. Then, this step was followed by dividing the resulting sample into five parts and the silver nanoparticles were added to each sample, as illustrated in Table 1. The reaction mixture was mixed using mild stirring for 30 min in 70 °C under nitrogen conditions.

Table 1. Sample Coding with the NPs concentration by mL.

	1	2	3	4	5
NPs (mL)	0	0.5	1	1.5	2

Then, the five samples were stirred for 10 min then poured into a Petri dish and subjected to UV irradiation for 10 min. Finally, for both in vitro and in vivo tests, the samples produced were sterilized using UV irradiation. The hydrogel was removed and stored at 4 °C.

4.9. Laboratory Animals and Housing Conditions

Laboratory animals and the care conditions were carried out in agreement with the international guidelines governing animal care. As a result, 30 BALB-c mice (weighing 35–40 g) were attainted from a laboratory animal facility at the National Research Center, Egypt. The mice were housed in cages with temperature controls between (20–22 °C), and 50–70% humidity light/dark cycles for 12 h. Mice were free of infections when they arrived at the workplace.

After a period of two weeks to adapt, the mice were divided into three groups randomly: The first group of (10) mice was untreated (control), the second group of (10) mice was treated by hydrogel only (PsB-g-Am), and the third group was treated by (PsB-g-Am with AgNPs). The animals were deprived of food overnight prior to the treatment. The mice were assessed regularly for any signs of infections. The mice that ate less through the provided commercial pellet diet were given deionized water.

4.10. In Vitro Study

The current study examined bacterial inhibition using the agar well diffusion method (at concentration about 10−5 CFU/mL), which provides qualitative insights into antimicrobial activity. Each test was conducted in triplicate to ensure accuracy. Two types of bacteria were tested against standard pathogens, namely *Klebsiella pneumoniae* and *Staphylococcus aureus*, which were selected for analysis. The assessment focused on detection of the efficacy of (PsB-g-Am) and (PsB-g-Am loaded sliver nanoparticles) with different concentrations for chronic wound healing group.

The tryptic soy broth (TSB) was used to cultivate *Klebsiella pneumoniae* (*K. pneumoniae* ATCC 9637) and *Staphylococcus aureus* (*S. aureus*-ATCC 6538) at 37 °C in an aerobic environment. Prior to the experiment, the standard curve of the absorbance (optical density OD600) vs. colony forming units (CFUs mL5) for each microbial species was developed. Both types of bacteria were quantified by spectrophotometry at abs = 600 nm. The overnight cultures were diluted 1:100 and continued to be grown until the OD600 value was 1.0.

The animal protocols used in this study followed the guidelines set by the laboratory animal facility at the National Research Center, Egypt for care and operation of laboratory animals. A total of 30 C-BALB mice (8–10 weeks), weighing between 35–40 g were employed in the wound healing experiments. Diabetes mellitus of mice were induced by single intravenous injection of 45 mg/kg streptozotocin (STZ). Blood glucose of each mouse was measured every week and the one with >300 mg /dL of blood glucose level on the 21st day was considered diabetic [41].

4.11. Surgical Wound Creation Model

Anesthesia for experimentation was achieved with dimethyl ether. After administering anesthesia, a full-thickness excisional wound measuring 2.0 cm^2 was surgically developed. The shaved dorsal area of each mouse was meticulously cleaned with iodine to disinfect the skin. The injury was developed by administering anesthesia and causing wounds with diameters of up to 10 mm. In the second and the third groups treated by (Ch-g-PVA) and (Ch-g-PVA-AgNPs), respectively, a 50-microliter volume of (Ch-g-PVA) and (Ch-g-PVA-AgNPs) solutions (10 ppm) were applied to the wound once a day at a specific time. In the first untreated (control) group, no (Ch-g-PVA nor (Ch-g-PVA-AgNPs) solution were used, but the wound area was cleansed with a normal saline. Each group of mice were accommodated in separate housing. The duration of the experiment spanned 14 days. Sampling occurred on days 0, 2, 7 and 14 days when the animals were sacrificed to determine the wound healing efficacy of the (Ch-g-PVA) and (Ch-g-PVA-AgNPs) in the treated groups and non-treated control group.

Author Contributions: Conceptualization; F.M.A. and D.M.; methodology; D.M., M.M.E.S. and D.K.E.D.; formal analysis; D.M. and M.M.E.S., D.K.E.D. and M.H.F.; resources; M.H.F.; writing—original draft preparation; D.M.; writing—review and editing; D.M. and M.M.E.S.; supervision; F.M.A.; funding acquisition; F.M.A. All authors have read and agreed to the published version of the manuscript.

Funding: This work was supported by King Saud University, Riyadh, Saudi Arabia (RSP2023R506).

Institutional Review Board Statement: Not applicable.

Informed Consent Statement: Not applicable.

Data Availability Statement: The data presented in this study are available on request from the corresponding author.

Acknowledgments: The authors extend their appreciation to the Researchers Supporting Project number (RSP2023R506) at King Saud University, Riyadh, Saudi Arabia.

Conflicts of Interest: The authors declare no conflict of interest.

References

1. Boateng, J.S.; Matthews, K.H.; Stevens, H.N.; Eccleston, G.M. Wound healing dressings and drug delivery systems: A review. *J. Pharm. Sci.* **2008**, *97*, 2892–2923. [CrossRef]
2. Pickwell, K.M.; Siersma, V.D.; Kars, M.; Holstein, P.E.; Schaper, N.C.; Eurodiale consortium. Diabetic foot disease: Impact of ulcer location on ulcer healing. *Diabetes Metab. Res. Rev.* **2013**, *29*, 377–383. [CrossRef] [PubMed]
3. Morbach, S.; Furchert, H.; Gröblinghoff, U.; Hoffmeier, H.; Kersten, K.; Klauke, G.T.; Klemp, U.; Roden, T.; Icks, A.; Haastert, B.; et al. Long-termprognosis of diabetic foot patients and their limbs: Amputation and death over the course of a decade. *Diabetes Care* **2012**, *35*, 2021–2027. [CrossRef] [PubMed]
4. Brennan, M.B.; Hess, T.M.; Bartle, B.; Cooper, J.M.; Kang, J.; Huang, E.S.; Smith, M.; Sohn, M.W.; Crnich, C. Diabetic foot ulcer severity predicts mortality among veterans with type 2 diabetes. *Diabetes Complicat.* **2017**, *31*, 556–561. [CrossRef] [PubMed]
5. Liang, Y.; He, J.; Guo, B. Functional hydrogels as wound dressing to enhance wound healing. *ACS Nano* **2021**, *15*, 12687–12722. [CrossRef]
6. Wenjun, M.; Sida, L.; Yingzhe, L.; Zhuo, C.; Jianhong, X. Bio-Inspired Low-Cost Fabrication of Stretchable, Adhesive, Transparent, and Multi-Functionalized Joint Wound Dressings. *ACS Appl. Mater. Interfaces* **2023**, *15*, 22915–22928.
7. Flemming, H.C.; Neu, T.R.; Wozniak, D.J. The EPS matrix: The "house of biofilm cells". *J. Bacteriol* **2007**, *189*, 7945–7947. [CrossRef]
8. Zhang, W.; Sun, Y.; Zhang, L. In situ synthesis of monodisperse silver nanoparticles on sulfhydryl-functionalized poly(glycidyl methacrylate) micro-spheres for catalytic reduction of 4-nitrophenol. *Ind. Eng. Chem. Res.* **2015**, *54*, 6480–6488. [CrossRef]
9. Lionelli, G.T.; Lawrence, W.T. Wound dressings. *Surg. Clin. N. Am.* **2003**, *83*, 617–638. [CrossRef]
10. Annabi, N.; Nichol, J.W.; Zhong, X.; Ji, C.; Koshy, S.; Khademhosseini, A.; Dehghani, F. Controlling the porosity and microarchitecture of hydrogels for tissue engineering. *Tissue Eng. Part B Rev.* **2010**, *16*, 371–383. [CrossRef]
11. Ma, Y.; Wang, X.; Su, T.; Lu, F.; Chang, Q.; Gao, J. Recent advances in macroporous hydrogels for cell behavior and tissue engineering. *Gels* **2022**, *8*, 606. [CrossRef] [PubMed]
12. Hong, T.T.; Okabe, H.; Hidaka, Y.; Hara, K. Radiation synthesis and characterization of super-absorbing hydrogel from natural polymers and vinyl monomer. *Environ. Pollut.* **2018**, *242*, 1458–1466. [CrossRef]
13. Alexander, S.; Holger, S.; Axel, H. Smart hydrogels based on responsive star-block copolymers. *Soft Matter* **2012**, *8*, 9436–9445.
14. El Sayed, M.M. Production of Polymer Hydrogel Composites and Their Applications. *J. Polym. Environ.* **2023**, *31*, 2855–2879. [CrossRef]
15. Li, S.; Dong, S.; Xu, W.; Tu, S.; Yan, L.; Zhao, C.; Ding, J.; Chen, X. Antibacterial hydrogels. *Adv. Sci.* **2018**, *5*, 1700527. [CrossRef] [PubMed]
16. Aldakheel, F.M.; Mohsen, D.; El Sayed, M.M.; Alawam, K.A.; Binshaya, A.S.; Alduraywish, S.A. Silver Nanoparticles Loaded on PsB -g- Am Hydrogel for the Wound-Healing Applications. *Molecules* **2023**, *28*, 3241. [CrossRef]
17. Haider, A.; Ijaz, M.; Ali, S.; Haider, J.; Imran, M.; Majeed, H.; Shahzadi, I.; Ali, M.M.; Khan, J.A.; Ikram, M. Green synthesized phytochemically (*Zingiber officinale* and *Allium sativum*) reduced nickel oxide nanoparticles confirmed bactericidal and catalytic potential. *Nanoscale Res. Lett.* **2020**, *15*, 50. [CrossRef]

18. Chavali, M.S.; Nikolova, M.P. Metal oxide nanoparticles and their applications in nanotechnology. *SN Appl. Sci.* **2017**, *1*, 607. [CrossRef]
19. Lei, L.; Wang, X.; Zhu, Y.; Su, W.; Lv, Q.; Li, D. Antimicrobial hydrogel microspheres for protein capture and wound healing. *Mater. Des.* **2022**, *215*, 110478. [CrossRef]
20. Verma, A.; Mehata, M.S. Controllable synthesis of silver nanoparticles using neem leaves and their antimicrobial activity. *J. Radiat. Res. Appl. Sci.* **2016**, *9*, 109–115. [CrossRef]
21. Von White, G.; Kerscher, P.; Brown, R.M.; Morella, J.D.; McAllister, W.; Dean, D.; Kitchens, C.L. Green synthesis of robust, biocompatible silver nanoparticles using garlic extract. *J. Nanomater.* **2012**, *12*, 730–746. [CrossRef]
22. Mofid, H.; Sadjadi, M.S.; Sadr, M.H.; Banaei, A.; Farhadyar, N. Green synthesis of zinc oxide nanoparticles using Aloe vera plant for investigation of antibacterial properties. *Adv. Nanoche.* **2020**, *2*, 32–35.
23. Dulta, K.; Ağçeli, G.K.; Chauhan, P.; Jasrotia, R.; Chauhan, P.K. A novel approach of synthesis zinc oxide nanoparticles by *Bergenia ciliata* rhizome extract: Antibacterial and anticancer potential. *J. Inorg. Organomet. Polym. Mater.* **2021**, *31*, 180–190. [CrossRef]
24. Vijayakumar, S.; Malaikozhundan, B.; Parthasarathy, A.; Saravanakumar, K.; Wang, M.H.; Vaseeharan, B. Nano biomedical potential of biopolymer chitosan-capped silver nanoparticles with special reference to antibacterial, antibiofilm, anticoagulant and wound dressing material. *J. Clust. Sci.* **2020**, *31*, 355–366. [CrossRef]
25. Archana, D.; Singh, B.K.; Dutta, J.; Dutta, P.K. Chitosan-PVP-nano silver oxide wound dressing: In vitro and in vivo evaluation. *Int. J. Biol. Macromol.* **2015**, *73*, 49–57. [CrossRef] [PubMed]
26. Rodríguez-Acosta, H.; Tapia-Rivera, J.M.; Guerrero-Guzmán, A.; Hernández-Elizarraráz, E.; Hernández-Díaz, J.A.; Garza-García, J.J.; Pérez-Ramírez, P.E.; Velasco-Ramírez, S.F.; Ramírez-Anguiano, A.C.; Velázquez-Juárez, G.; et al. Chronic wound healing by controlled release of chitosan hydrogels loaded with silver nanoparticles and calendula extract. *J. Tissue Viability* **2022**, *31*, 173–179. [CrossRef] [PubMed]
27. Masood, N.; Ahmed, R.; Tariq, M.; Ahmed, Z.; Masoud, M.S.; Ali, I.; Asghar, R.; Andleeb, A.; Hasan, A. Silver nanoparticle impregnated chitosan-PEG hydrogel enhances wound healing in diabetes induced rabbits. *Int. J. Pharm.* **2019**, *559*, 23–36. [CrossRef]
28. Torres-Figueroa, A.V.; Pérez-Martínez, C.J.; Castillo-Castro, T.D.; Bolado-Martínez, E.; Corella-Madueño, M.A.; García-Alegría, A.M.; Armenta-Villegas, L. Composite hydrogel of poly (acrylamide) and starch as potential system for controlled release of amoxicillin and inhibition of bacterial growth. *J. Chem.* **2020**, *2020*, 5860487. [CrossRef]
29. Sorour, M.H.; Hani, H.A.; Shaalan, H.F.; El Sayed, M.M.; El-Sayed, M.M. Softening of seawater and desalination brines using grafted polysaccharide hydrogels. *Desalination Water Treat.* **2015**, *55*, 2389–2397. [CrossRef]
30. Varaprasad, K.; Mohan, Y.M.; Ravindra, S.; Reddy, N.N.; Vimala, K.; Monika, K.; Sreedhar, B.; Raju, K.M. Hydrogel–silver nanoparticle composites: A new generation of antimicrobials. *J. Appl. Polym. Sci.* **2010**, *115*, 1199–1207. [CrossRef]
31. Peng, T.; Shi, Q.; Chen, M.; Yu, W.; Yang, T. Antibacterial-Based Hydrogel Coatings and Their Application in the Biomedical Field—A Review. *J. Funct. Biomater.* **2023**, *14*, 243. [CrossRef]
32. Dankovich, T.A. Microwave-assisted incorporation of silver nanoparticles in paper for point-of-use water purification. *Environ. Sci. Nano* **2014**, *1*, 367–378. [CrossRef] [PubMed]
33. Diniz, F.R.; Maia, R.C.A.; de Andrade, L.R.M.; Andrade, L.N.; Vinicius Chaud, M.; da Silva, C.F.; Severino, P. Silver nanoparticles-composing alginate/gelatine hydrogel improves wound healing in vivo. *Nanomaterials* **2020**, *10*, 390. [CrossRef]
34. Reiad, N.A.; Abdel Salam, O.E.; Abadir, E.F.; Harraz, F.A. Green synthesis of antibacterial chitosan films loaded with silver nanoparticles. *Chin. J. Polym. Sci.* **2013**, *31*, 984–993. [CrossRef]
35. Goodman, G. Postacne Scarring: A Review of its Pathophysiology and Treatment. *Dermatol. Surg.* **2000**, *26*, 857–871. [CrossRef] [PubMed]
36. Zolnik, B.S.; Gonzalez-Fernandez, A.; Sadrieh, N.; Dobrovolskaia, M.A. Mini review: Nanoparticles and the Immune System. *Endocrinology* **2010**, *151*, 458–465. [CrossRef] [PubMed]
37. Verma, J.; Kanoujia, J.; Parashar, P.; Tripathi, C.B.; Saraf, S.A. Wound healing applications of sericin/chitosan-capped silver nanoparticles incorporated hydrogel. Drug Deliv. *Translat. Res.* **2017**, *7*, 77–88.
38. Faunce, T.; Watal, A. Anosilver and global public health: International regulatory issues. *Nanomedicine* **2010**, *5*, 617–632. [CrossRef]
39. Lara, H.; Garza-Treviño, N.; IxtepanTurrent, L.; Singh, D.K. Silver nanoparticles are broad-spectrum bactericidal and virucidal compounds. *J. Nanobiotechnol.* **2011**, *9*, 30. [CrossRef]
40. Danilczuk, M.; Lund, A.; Sadlo, J.; Yamada, H.; Michalik, J. Conduction electron spin resonance of small silver particles. *Spectrochim. Acta Part A Mol. Biomol. Spectrosc.* **2006**, *63*, 189–191. [CrossRef]
41. Alaraby, R.; El Sayed, M.M. Methylene Blue Cationic Dye Removal using AA-Am Hydrogel as An Efficient Adsorbent. *Egypt. J. Chem.* **2022**, *65*, 1–10.

gels

MDPI

Article

GastroPlus- and HSPiP-Oriented Predictive Parameters as the Basis of Valproic Acid-Loaded Mucoadhesive Cationic Nanoemulsion Gel for Improved Nose-to-Brain Delivery to Control Convulsion in Humans

Afzal Hussain [1,*], Mohammad A. Altamimi [1], Mohhammad Ramzan [2], Mohd Aamir Mirza [3] and Tahir Khuroo [4]

1 Department of Pharmaceutics, College of Pharmacy, King Saud University, Riyadh 11451, Saudi Arabia; maltamimi@ksu.edu.sa
2 School of Pharmaceutical Sciences, Lovely Professional University, Phagwara 144411, India; mohhammad.26652@lpu.co.in
3 Department of Pharmaceutics, School of Pharmaceutical Education and Research, Jamia Hamdard, New Delhi 110062, India; aamir.mirza@jamiahamdard.ac.in
4 PGx Global Foundation, Houston, TX 77035, USA; tahir@pgxglobal.org
* Correspondence: amohammed2@ksu.edu.sa

check for updates

Citation: Hussain, A.; Altamimi, M.A.; Ramzan, M.; Mirza, M.A.; Khuroo, T. GastroPlus- and HSPiP-Oriented Predictive Parameters as the Basis of Valproic Acid-Loaded Mucoadhesive Cationic Nanoemulsion Gel for Improved Nose-to-Brain Delivery to Control Convulsion in Humans. *Gels* **2023**, *9*, 603. https://doi.org/10.3390/gels9080603

Academic Editors: Adina Magdalena Musuc, Magdalena Mititelu and Mariana Chelu

Received: 7 June 2023
Revised: 10 July 2023
Accepted: 12 July 2023
Published: 26 July 2023

Abstract: Oral and parenteral delivery routes of valproic acid (VA) are associated with serious adverse effects, high hepatic metabolism, high clearance, and low bioavailability in the brain. A GastroPlus program was used to predict in vivo performance of immediate (IR) and sustained release (SR) products in humans. HSPiP software 5.4.08 predicted excipients with maximum possible miscibility of the drug. Based on the GastroPlus and HSPiP program, various excipients were screened for experimental solubility, nanoemulsions, and respective gel studies intended for nasal-to-brain delivery. These were characterized by size, size distribution, polydispersity index, zeta potential, morphology, pH, % transmittance, drug content, and viscosity. In vitro drug release, ex vivo permeation profile (goat nasal mucosa), and penetration studies were conducted. Results showed that in vivo oral drug dissolution and absorption were predicted as 98.6 mg and 18.8 mg, respectively, from both tablets (IR and SR) at 8 h using GastroPlus. The predicted drug access to the portal vein was substantially higher in IR (115 mg) compared to SR (82.6 mg). The plasma drug concentration–time profile predicted was in good agreement with published reports. The program predicted duodenum and jejunum as the prime sites of the drug absorption and no effect of nanonization on T_{max} for sustained release formulation. Hansen parameters suggested a suitable selection of excipients. The program recommended nasal-to-brain delivery of the drug using a cationic mucoadhesive nanoemulsion. The optimized CVE6 was associated with the optimal size (113 nm), low PDI (polydispersity index) (0.26), high zeta potential (+34.7 mV), high transmittance (97.8%), and high strength (0.7% *w*/*w*). In vitro release and ex vivo permeation of CVE6 were found to be substantially high as compared to anionic AVE6 and respective gels. A penetration study using confocal laser scanning microscopy (CLSM) executed high fluorescence intensity with CVE6 and CVE6-gel as compared to suspension and ANE6. This might be attributed to the electrostatic interaction existing between the mucosal membrane and nanoglobules. Thus, cationic nanoemulsions and respective mucoadhesive gels are promising strategies for the delivery of VA to the brain through intransal administration for the treatment of seizures and convulsions.

Keywords: valproic acid; GastroPlus-based prediction; cationic nanoemulsion; gels; in vitro–ex vivo permeation profile; CLSM study

1. Introduction

Epilepsy is defined as a group of neurological issues of the central nervous system and is characterized as a predisposition to epileptic seizures due to the complexity of its characteristics. The World Health Organization (WHO) estimated that 50 million people are affected annually around the world [1]. In the USA, 2.3 million adults and 500,000 children are affected by varied forms of epilepsy due to unknown and known possible reasons (genetics, trauma, stroke, brain tumor, and any factors responsible for disturbing the normal pattern of the brain circuit) [2]. In Saudi Arabia, the reported prevalence of cases of epilepsy is 6.45 per 1000 people, which is responsible for affecting children's mental health, behavior, and academic performance [3].

Valproic acid (VA) is the most effective first-line anticonvulsant to control grand mal epilepsy and tonic–clonic fits (seizure), various seizures, and idiopathic generalized seizures. Several characteristics, including low molecular weight (144 g/mol), hydrophobic nature (log P = 2.54), high oral dose (not more than 600 mg/kg/day), high first-pass metabolism (methylation, sulfation, and glucuronidation), and poor brain bioavailability after oral administration, are possible reasons for nasal delivery of the drug to control seizure [4]. Commercial products (oral and parenteral) showed high plasma levels of active metabolites (90% such as 4-ene-VA and undergoes beta oxidation of fatty acid) due to hepatic metabolism (causing hepatotoxicity) and rapid clearance due to efflux (P-gp pump of microvessel endothelial cell in the blood–brain barrier). Parenteral delivery causes serious side effects, possibly due to the reticuloendothelial system (Kuffer cells)-based metabolism and low bioavailability to the brain. Hammond et al. investigated the pharmacokinetics profiles of the drug in a cat model (six adult cats) after rapid intravenous infusion (60 mg/kg within 3 min of infusion using saline) wherein the maximum level of the drug was obtained at 1 min (brain distribution half-life as 6 min estimated from α-phase) followed by the Vd rapid clearance (mean elimination half-life of 41 min), and the volume of distribution (Vd) as 0.125 L/kg. Low in vivo uptake (low brain–plasma ratio), low Vd, and rapid clearance (brain elimination half-life of 41 min estimated from the β-phase) from the brain indicated poor binding of the drug to the cerebral cortex [5]. As per the US FDA label of DEPAKENE, oral absorption was dependent on age and dosage forms (tablet versus capsule). In adult patients, the absorption rate on monotherapy (250 mg of oral delivery) is nonlinear, whereas the kinetics of the unbound drug is linear. Notably, the drug is primarily metabolized through the liver (30–50% as glucuronide conjugate) and mitochondrial β-oxidation (>40%) for excretion through urine (3% as unchanged). In humans, the mean plasma clearance and Vd values were reported as 0.56 L/Kg and 11 L, respectively, following 250 mg of oral administration in adults (70 kg or 1.73 m^2 as body surface area) [6].

Several drugs (35–40 molecules) have been exploited for brain delivery using the nasal route of administration. The route is the most preferred one to circumvent the aforementioned issues of oral and parenteral delivery in conventional dosage forms. Various nanocarriers have been reported for drug delivery through the nasal route. These are lipidic nanocarriers (lipid nanoparticles, nanoemulsion, and liposomes), nanotubes, and dendrimers [7–11]. Tan et al. tailored stable nanoemulsion comprised of safflower (70–80% linolenic acid-rich natural oil capable of drug delivery across the blood–brain barrier and cerebrospinal fluid barrier) for delivery of the drug to the brain, and brain bioavailability was improved [10,12].

Nanoemulsion is a well-explored nanocarrier system for drug delivery due to desired innate features such as nanoscale globular size, suitability to load small molecules, and thermodynamically stable isotropic mixture. An imposed cationic charge on the nanoglobule further improves its pharmaceutical utility for the facilitated permeation across the biological membrane for the extended residence time. Nasal drug delivery is usually challenged by its short residence time and high washout after nasal administration. Cationic-charged nanocarriers interact with the biological membrane for maximized internalization and increased passive permeation and drug deposition (enhanced drug access across the biological membrane) [13,14]. The nasal route of administration offers several advantages over oral and parenteral routes such as (a) high patient compliance, (b) avoiding hepatic metabolism and related drug degradation, (c) direct drug access from the olfactory region to the cranial cavity of the brain, (d) avoiding unnecessary administration of excipients, (e) dose mitigation and reduction in dose-related side effects, (f) low therapy cost, (g) ease in regulatory constraints for approval, and (h) safety and biocompatibility [14].

We predicted the in vivo performance of the drug using GastroPlus (predictive and simulation program) using the literature, default values, and experimental data. The program assisted in predicting the dose-dependent pharmacokinetic parameters (time required to reach C_{max} as T_{max}, area under the curve as AUC, and maximum drug concentration reached in the blood as C_{max}) considering an oral commercial dose (250 mg) and the dosage form (tablet) in healthy adults. Moreover, cationic nanoemulsions were prepared, optimized, and characterized for in vitro (size, size distribution, zeta potential, morphology, thermodynamic stability, and release profile at pH 5.5 and 6.8) and ex vivo performance (permeation flux, drug deposition, and enhancement ratio) (goat nasal tissue).

2. Results and Discussion

2.1. Prediction and Simulation Study Using GastroPlus

VA is orally administered in different dosage forms such as IR/SR tablets, an oral solution, and an oral capsule. Considering 200 mg as an adult dose in an IR tablet and SR tablet, the program was run for a simulation of 24 h (Table 1). The program was used to predict PK parameters (C_{max}, AUC, and T_{max}) for both of them in adults. Limited data are available for comparative PK studies of VA using IR and SR tablets in humans. No data are available for predicting PK parameters using the GastroPlus program and comparing an IR tablet and an SR tablet at a fixed dose and dosing frequency. Teixeira-da-Silva et al. predicted population pharmacokinetics of VA monotherapy considering different doses, body weights, and age groups. The regimen depicted was designed to achieve a VA concentration within the acceptable therapeutic range. The steady-state plasma concentrations were predicted to be >120 mg/L for age groups of 15 (1000 mg in tablet) and 35 years (1200 mg in tablet), whereas this value was predicted as <100 mg/L for children aged 1 (dose of 100 mg in solution) and 6 years (dose of 200 mg in solution) [15]. Thus, the authors found that there was no significant difference in the plasma drug concentration from the tablet with 1000 mg or 1200 mg in adults of different ages (15 versus 35 years old) and body weights (56 versus 70 kg) [16]. In the present study, we used a 200 mg dose for an adult weighing 70 kg to predict in vivo dissolution and in vivo absorption of the IR tablet and SR tablet of VA. The result is illustrated in Figure 1A,B. It is clear that the predicted pattern of in vivo dissolution of the IR VA tablet and the SR VA tablet are closely related without a substantial difference in a fast-state adult. Interestingly, the amount of drug absorbed by the portal vein (AmtPV-1) is higher in the IR VA tablet compared to the SR VA tablet (Figure 1A) as predicted in the program. This may be prudent to correlate the difference in the dissolution rate between IR and SR tablets. The IR tablet exhibits rapid drug dissolution in gastric content for the profound availability of the drug for absorption at the intestinal mucosa of the GIT lumen. On the other hand, the SR tablet follows a different dissolution process due to the rate-limiting membrane of the polymer coating on the tablet. Slow and sustained release of the drug caused slow and extended absorption as predicted in Figure 1B. The total amount of the drug absorbed from both tablets is approximately the same as predicted in the program (green bold color) suggesting no significant difference in the modified form of the tablet over a period of 24 h. This may be due to the slightly acidic nature of VA (pKa = 5.14) suitable for absorption from the intestinal area as the prime site of drug absorption. The therapeutic window of the drug is 30–100 mg/L after oral administration in humans [17]. The drug is rapidly absorbed from the oral dosage form and the drug access to the brain is limited due to the high protein binding capacity (90%). The low volume of distribution (0.125 L/Kg) is very similar to that found in humans suggesting no significant bounding of the drug to the brain. Therefore, this needs a high blood plasma level by administering high oral doses. The limited free drug in the plasma is available for brain access. In a previous study, it was observed that VA transport to the brain occurs via the monocarboxylic acid transport system. The plasma level of VA < 60 μg/mL results in a low level of the drug in the brain. For clinical effectiveness in humans, it is only possible with a relatively high plasma concentration above 55 μg/mL [5].

Table 1. Summary of input data for GastroPlus simulation and prediction of VA sodium.

Parameter	Values
Molecular formula	$C_8H_{15}NaO_2$
Molecular weight (g/mol)	166.19
Melting point (°C)	300
Aqueous solubility (mg/mL) at 25 °C	<1
Density (g/mL)	0.9
Pka	5.14
Log p	3.08
Apparent permeability coefficient (cm/h) across hCMEC/D3 and CC-2565 of in vitro blood brain barrier	0.625
Dose (mg)	200
Body weight (kg)	70
Dosing volume (mL)	1
Mean precipitation time (s)	30
AUC (μg. h/mL)	10–160
C_{max} (mg/L)	~120
T_{max} (h) (mean)	5
Elimination half-life (h)	8–15
Clearance (L/h)	0.206–1.154
Plasma protein binding (%)	90–95
V_d (L)	8.4–23.3
pH for reference solubility	7.0
Simulation time (h)	24

2.1.1. Prediction of Plasma Drug Concentration Time Profile

The program predicted the plasma drug concentration–time profile of the IR tablet and the SR tablet of VA. The result of the predicted PK profile is displayed in Figure 2 wherein C_{max} values of VA IR and VA SR tablets were predicted as 159.3 μg/mL and 82.5 μg/mL, respectively. The predicted values are quite interesting and convincing as explained before for therapeutic effectiveness. Both values are enough to produce a substantial level of the drug in the blood plasma for brain access (>55 μg/mL) [5]. The acidic form of the drug is suitable for solubility in water and an acidic medium (pKa 5.4). Therefore, the IR tablet showed rapid drug dissolution for immediate drug absorption. Therefore, the IR tablet (2.1 h) showed relatively low T_{max} compared to the SR tablet (5.2 h) in prediction. These predicted T_{max} values are in good agreement with the published report for the oral solution and the SR tablet [18]. This indicated that the model is a good fit (as observed by the high Akaike value) for simulation and prediction. The result can be correlated to the difference in oral bioavailability of VA in the drug solution and the SR formulation. In the literature, the drug solution and SR formulation resulted in 100% and 80–90% bioavailability for VA [19]. Thus, the predicted pattern of the VA SR tablet suggested slow and sustained delivery of VA for a long-term effect within the therapeutic window (200 mg). However, the drug is limited to brain access due to various possible reasons. These may be due to high protein binding capacity, high hepatic drug metabolism, the low solubility of the drug, and extra hepatic drug metabolism. The sustained-release tablet slightly decreased the drug absorption to the portal vein (Figure 1B).

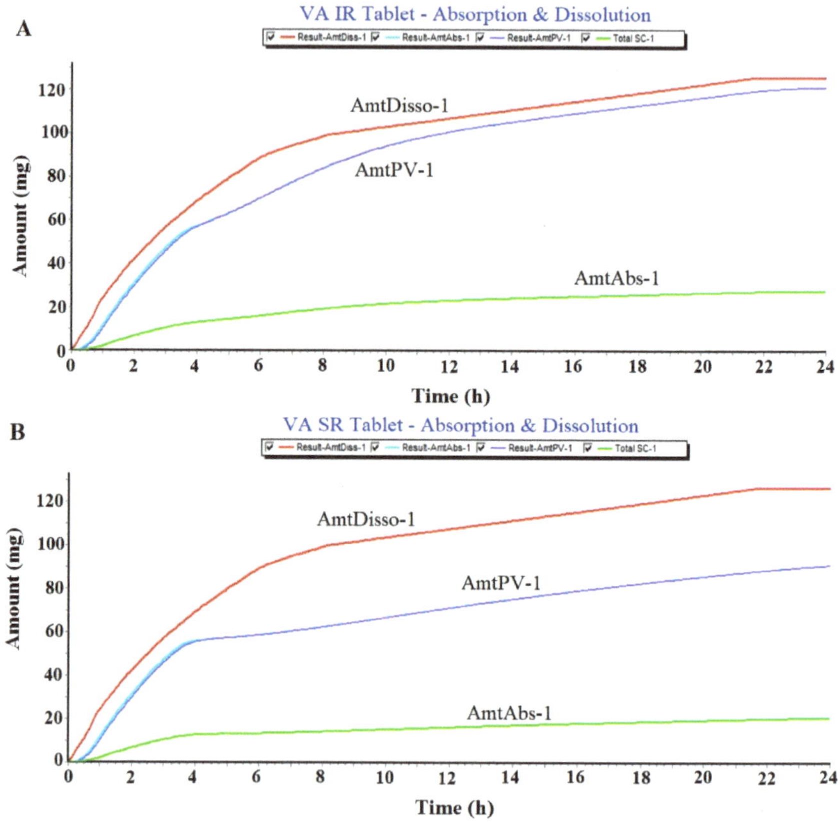

Figure 1. Simulation and prediction software (GastroPlus)-based analysis of IR tablet and SR tablet for oral administration. (**A**) Prediction of in vivo dissolution and in vivo absorption of VA IR Tablet (200 mg) for oral delivery (once a day in fast-state condition) and (**B**) prediction of in vivo dissolution and in vivo absorption of VA SR tablet (200 mg) for oral delivery (once a day in fast-state condition).

2.1.2. Regional Compartmental Absorption of Both Tablets

The program predicted nine compartmental absorption sites in GIT. Both IR and SR tablets were processed in the system to estimate the percent of regional absorption of the drug. The result is displayed in Figure 3A,B wherein the IR tablet and the SR tablet were predicted to have an overall total absorption of 95.3% and 86.6%, respectively. VA is a slightly acidic drug with a pKa value of 5.6. Therefore, the drug was predicted to be absorbed primarily from the proximal portion of GIT. Thus, the duodenum and jejunum are the main sites of oral absorption. The drug is considered poorly absorbed from the distal GIT region as shown in Figure 3A,B. The predicted values are in agreement with the published report of oral bioavailability for the drug solution (100%) and SR tablet (80–90%) [19].

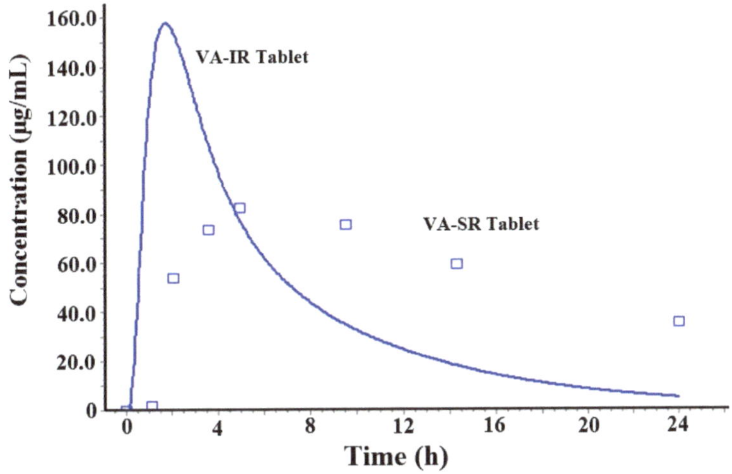

Figure 2. Plasma drug concentration–time profile predicted for VA IR tablet and VA SR tablet (200 mg). This was a predicted profile of a human in fast-state condition using GastroPlus.

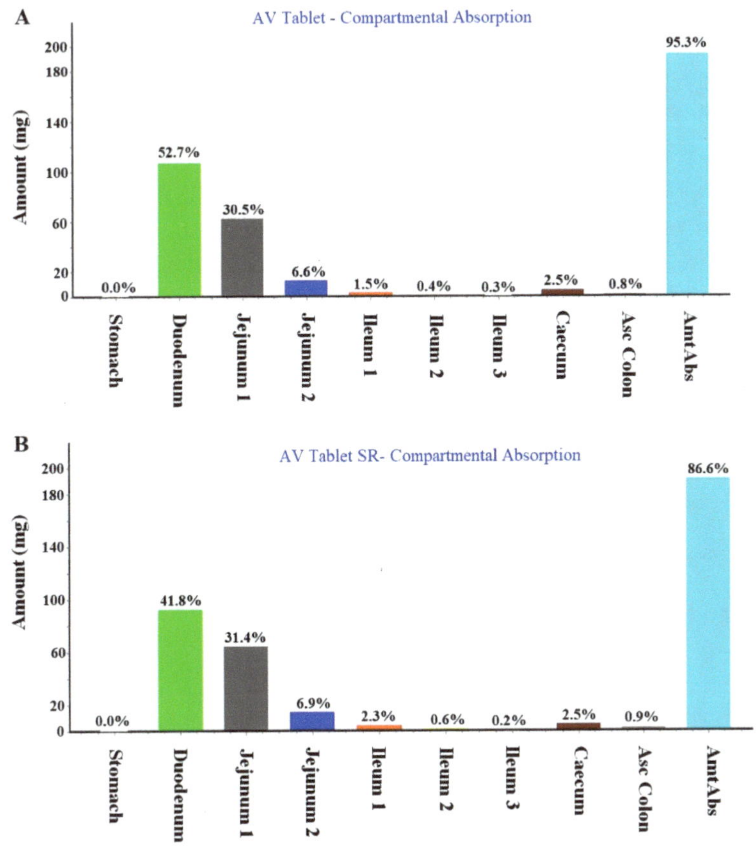

Figure 3. Regional compartmental absorption of the drug from human GIT (gastrointestinal tract) in fasted condition: (**A**) Prediction for the IR tablet and (**B**) predicted values for SR tablet.

2.1.3. PSA (Parameter Sensitivity Analysis) Assessment

PSA assessment was performed to identify the relevant factors responsible for affecting the PK parameters of the VA tablet on oral administration (Figure 4A–C). The analysis was carried out using the GastroPlus program considering the fasted-state condition of the subject. This avoided any interaction with food. In the study, we attempted to predict the impact of a nanocarrier system for oral drug delivery and its impact on PK parameters such as C_{max}, T_{max}, and AUC (area under the curve). It is clear from the prediction study and literature-based findings that conventional dosage forms of VA (tablet, SR tablet, and solution) do not have much difference in terms of bioavailability in humans [10,16,20–24]. Therefore, the predicted oral bioavailability values are almost similar to the reported values (as described before). However, the program predicted that the nano effect had no impact on PK parameters after oral delivery. This may be correlated with the lipophilic nature of the drug being absorbed and poor dissolution (BCS class II) [25,26]. Conclusively, GastroPlus simulation and the prediction program assisted in the understanding of the nanonized product of VA for oral delivery could be of no benefit for brain delivery. Therefore, it is better to formulate nanocarrier-based drug delivery for brain delivery using the nasal route of administration. The basal route contains the olfactory chamber directly linked to the brain for drug access. Thus, the purpose of GastroPlus-based prediction was to understand the feasibility of the oral nanocarrier for brain delivery using clinical data (obtained from the literature). The program provided various predicted in vivo values for a human trial.

2.1.4. Hansen Solubility Parameters for VA and Excipients

HSP values helped to select excipients possibly exhibiting maximum drug solubility via a cohesive interaction (cohesive forces) [27]. The program is well-exploited for solute miscibility/solubility in a particular solvent. The HSP values of the drug and each excipient are summarized in Table 2. It is easy to understand that the lipid, the surfactant, and the co-surfactant possessing HSP close to the values of the drug could be the most appropriate and suitable for drug solubility. The values of δ_d, δ_p, and δ_h of the drug are 16.1, 4.3, and 9.0 $MPa^{1/2}$, respectively. The δ_h value of tween 80 is quite close to the HSP values of VA compared to span 80 (δ_h of span 80 is 12.4 compared to 9 of tween 80). Therefore, a solute interacts with a solvent through these cohesive forces working together. Thus, the difference of any parameter between the solute and solvent close to zero is considered miscible or soluble. Thus, the program predicted relevant excipients based on these HSP values of each excipient close to the HSP values of the model drug. The program estimated these values as shown in Table 2. The HSP values of oils (safflower, Flaxseed oil, and grape seed oil), lecithin, and PC were obtained from the literature and calculated manually based on the percent composition of linoleic acid or phosphatidylcholine (PC) present [28]. Among these oils, safflower seed oil might be the most suitable for tailoring a cationic nanoemulsion due to the predicted miscibility of the drug in terms of HSP. The oil has been reported t have a high content of linoleic acid (78%), and linoleic acid is considered a promoter for the diffusion of the lipophilic drug across the blood–brain barrier [29].

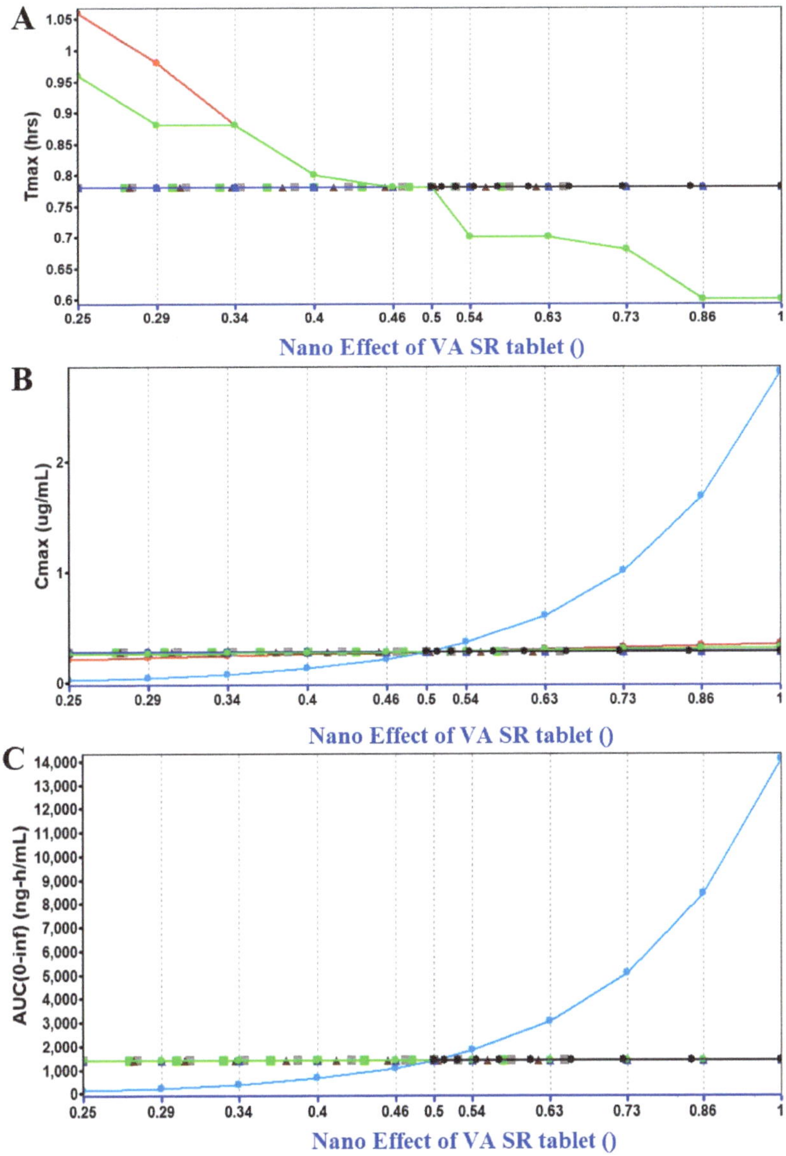

Figure 4. Parameter sensitivity analysis assessment using GastroPlus Version 9.8.3 simulation and prediction program. (**A**) Impact of nanonized VA product (oral delivery) on T_{max}, (**B**) impact of nanonized product on C_{max}, and (**C**) impact of nanonized product on AUC values. Blue bold line indicates the impact of Nano effect. Red bold line indicates "P_{eff}" permeability coefficient across mucosal membrane, and effect of duodenum ASF (absorption scale factor). The green curve indicates the impact of reference solubility. Sky blue curve indicates the dose effect. The brown triangle symbol represents particle density whereas grey curve indicated shape behaviour.

Table 2. Summary of HSP values for the drug and selected excipients.

Drug and Excipient	δ_d (MPa$^{1/2}$)	δ_p (MPa$^{1/2}$)	δ_h (MPa$^{1/2}$)
AV	16.1	4.3	9.0
Safflower seed oil (87%) *	14.5	2.7	5.3
Grape seed oil (70%) *	11.69	2.17	4.27
Flaxseed oil (60%) *	10.02	1.86	3.66
Tween 80	16.6	5.3	7.5
Span 80	16.7	6.1	12.4
Lecithin (PC as 20%) *	3.2	0.54	0.64
Linoleic acid *	16.7	3.1	6.1
Transcutol HP	16.0	2.8	6.2
PG $^{\Phi}$	16.8	10.4	21.3
PC *	16	2.7	3.2

* Estimated using reference [28]; $^{\Phi}$ [30].

2.1.5. Solubility of Valproate (VA) in Various Excipients

The result of the experimental solubility of VA is portrayed in Figure 5. The solubility of the drug was found to be the maximum in safflower seed oil (8.9 ± 0.11 mg/mL), tween 80 (5.3 ± 0.09 mg/mL), and transcutol (6.3 ± 0.08 mg/mL). This maximized solubility can be rationalized based on the HSP values predicted in Table 2. The difference value of $\Delta\delta_d$ is 1.6 (16.1–14.5) for the solute (VA) and the solvent (safflower oil), which is quite low for high miscibility/solubility. Similarly, the difference values of $\Delta\delta_p$ and $\Delta\delta_h$ are 1.6 (4.3–2.7) and 3.7 (9–5.3), respectively, for the drug in safflower. These differences are quite convincing for maximized drug solubility due to interactive forces (polarization, hydrogen bonding ability, and dispersion nature). Among the co-surfactants, transcutol was selected due to the highest solubility and suitability for the drug. Flaxseed (50–70%), safflower (70–87%), and grape seed oil (70%) are prime sources of linoleic acid. Linoleic acid-rich oils are gaining popularity in pharmaceutical and cosmeceutical industries due to possessing various skin benefits such as (a) anti-inflammatory, (b) acne-reductive, (c) skin-softening, and (d) moisture-retentive qualities, as well as possessing the ability to (e) facilitate drug diffusion across the blood–brain barrier (50–87% linoleic acid) and (f) biocompatibility [10,31]. Thus, safflower, tween 80, and transcutol were selected as the oil, surfactant, and co-surfactant. However, a blend of tween 80 and lecithin (1:1) was used for a stable and small-sized nanoemulsion. The combination was supposed to stabilize the nanoemulsion with a small size particle as compared to tween 80 as a standalone.

2.1.6. VA Loaded Cationic Nanoemulsions Prepared

To construct a cationic nanoemulsion, a constant amount of stearylamine (5 mg) was used in each formulation. A series of nanoemulsions (CVE as cationic and AVE as anionic nanoemulsion) were prepared as shown in Table 3. Formulation CVE5 exhibited unique characteristic features among them. The globular size, PDI, zeta potential, %T, and product strength (% *w*/*w*) were found to be 79 nm (the lowest value), 0.11 (the lowest value), +27.1 mV (optimal), 95%, and 0.5%, respectively. The lowest value of PDI is due to the lowest content of oil (9.8%) and the sufficient amount of S_{mix} (21.84%) responsible for efficient emulsification and resulted in the homogeneous nature of the globular distribution. However, %DC was found to be low (~0.5%) for CVE5, which may be related to the low content of oil (9.8%). Formulation CVE6 had an optimal content of oil (14.46%) and S_{mix} (17.15%) to render the optimal size (113 nm), high zeta potential (+34.7 mV) for enhanced stability, and high %DC (67%) as compared to others. Comparing CVE1, CVE3, and CVE4, it is clear that by increasing the relative concentration of S_{mix} compared to oil, the size values regularly decreased from 185 nm to 148 nm. This may be due to efficient emulsification by the surfactant mixture. Comparing CVE2 (189 nm) and CVE4 (148 nm), the size of CVE4 was substantially decreased due to the high content of the surfactant mixture, even decreasing the relative content of the co-surfactant transcutol (from 1:3 to 1:2) within the S_{mix}. AVE6 was anionic (zeta potential = −22 mV) in nature due to the lack of stearylamine in the formulation and served as a control group. The negative zeta potential is due to the lipid (triglycerides of fatty acids). The study aimed to address the impact of the charge on the nanocarrier for permeation behavior across nasal mucosa followed by blood–brain barrier. Notably, all of the nanoemulsions showed %T (%transmittance) higher than 96% suggesting the isotropic and transparent nature of the cationic and anionic nanoemulsions.

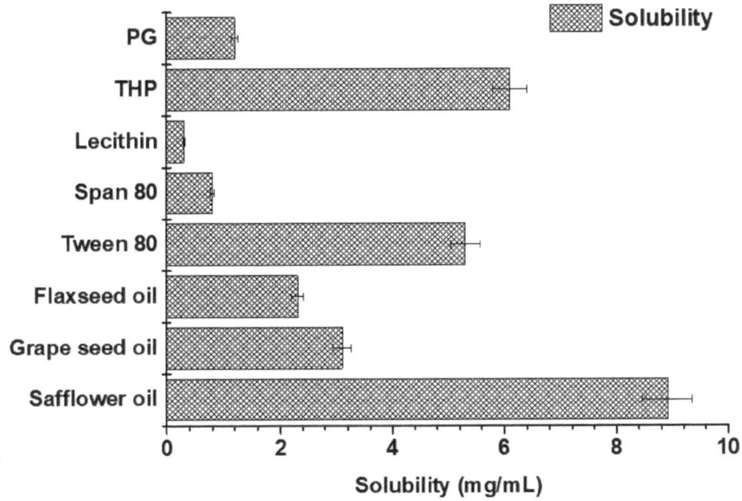

Figure 5. Experimental solubility of VA in various excipients at 40 °C. Data were expressed as mean ± standard deviation (n = 3).

Table 3. Summary of selected cationic/anionic VA-loaded nanoemulsions and their evaluated parameters.

Code	SO (%)	S_{mix} (%)	Water (%)	S_{mix} Ratio	ST (%)	Size (nm)	PDI	ZP (mV)	%T	Product Strength (% *w*/*w*)
CVE1	16.46	30.21	48.59	1:2	0.04	162	0.27	+24.7	98.5	0.4
CVE2	20.75	23.5	50.22	1:3	0.05	189	0.32	+26.8	96.8	0.5
CVE3	14.72	21.67	57.12	1:2	0.06	185	0.31	+31.6	97.2	0.6
CVE4	19.88	43.51	32.16	1:2	0.04	148	0.18	+23.9	96.9	0.4
CVE5	9.8	21.84	63.04	2:1	0.05	79	0.11	+27.1	95.3	0.5
CVE6	14.46	17.15	60.99	3:1	0.07	113	0.26	+34.7	97.8	0.7
AVE6	14.46	17.15	60.92	3:1	0.0	126	0.29	−22.8	95.6	0.7
	Nanoemulsion gel (0.5% *w*/*w*) composition (VA strength)					Evaluated parameters				
0.5% VE gel	NE (g)	Gel-blank (g)	Triethanolamine (g)	CVE6:gel ratio		Size (nm)	PDI	ZP (mV)	Viscosity (cP)	pH
CVE6 gel (0.35%)	1	0.95	0.05 g	1:1		129	0.24	+21.9	1837.3	6.8
AVE6 gel (0.35%)	1	0.95	0.05 g	1:1		142	0.31	−26.5	1907.1	7.1

Note: SO = Safflower, S_{mix} = Tween 80-lecithin: transcutol, ST = stearylamine (cationic charge inducer), PDI = Polydispersity index; ZP = zeta potential, NE = Nanoemulsion, CVE = Cationic NE, VE = Valproic acid loaded nanoemulsion, AVE6 = Anionic valproic acid loaded nanoemulsion.

2.1.7. Freeze–Thaw Cycle and Ultracentrifugation of Nanoemulsions

The developed formulations CVE1-CVE6 and AVE6 were subjected to ensure stability and capability to withstand thermal and physical stress during storage and transportation. The centrifugation step confirmed the physical stability to face attrition- and friction-triggered phase separation usually observed during transportation [32]. On the other hand, extreme temperatures (freeze and accelerated temperatures) assured stability against thermal-mediated instability in the nanoemulsion. The result is presented in Table 4. All of the formulations (cationic and anionic) were physically and thermally stable at the explored temperatures for the studied time period. A sequential series of thermal exposure from low to high via room temperature indicated that each product resumed its original state of the transparent isotropic nature of the nanoemulsion with good flowability, consistency, and elegancy. There were no signs of any instability over the explored period of time. It was imperative to corroborate the thermal and physical stability so that the developed nanoemulsion can be stored and transported accordingly.

Table 4. Freeze–thaw and centrifugation cycles and observation.

Formulations	Freezing (−21 °C)	Room Temperature	Thaw (40 °C)	Centrifugation	Inference *
CVE1	✓	✓	✓	✓	Stable
CVE2	✓	✓	✓	✓	Stable
CVE3	✓	✓	✓	✓	Stable
CVE4	✓	✓	✓	✓	Stable
CVE5	✓	✓	✓	✓	Stable
CVE6	✓	✓	✓	✓	Stable
AVE6	✓	✓	✓	✓	Stable

Note: * Recovery of original form/state of nanoemulsion at room temperature after exposure to the extreme temperature was considered stable in inference. Formulations exhibiting any signs of instability in terms of drug precipitation, phase separation, color development, and creaming were dropped out from further studies. The symbol "✓" means passed the test.

2.2. *Evaluation of Cationic and Anionic Nanoemulsions Gels*

CVE6 and AVE6 were used to incorporate 1% carbopol gel (1:1 ratio) into the respective nanoemulsion gel (0.5%) containing 0.35% *w/w* of VA in the gels. Thus, the final product strength was 0.35% *w/w* in each gel. Both gels were evaluated for size, PDI, ZP, viscosity, and final pH as shown in Table 3. It is apparent that the pH (from 7.4 to 6.8) and zeta potential (from +34.7 to +21.9 mV) values of the CVE6 gel were significantly reduced from the respective CVE6 nanoemulsion. This is obvious due to the acidic carbopol polymeric gel with free carboxylic acid in its structural backbone. However, globular size values were nearly similar to the CVE6 nanoemulsion, suggesting no globular aggregation in the gel carrier. Viscosity values of the CVE6 gel and the AVE6 gel were 1837 and 1907 cP, respectively. These findings are in good agreement with the reported 0.5% carbopol 934 gel for topical application [33]. The viscosity indicates good consistency and shear thinning behavior after topical application due to the oil in a water-based system. In the final selected formulations, a fixed amount of SA was used in CVE6 and CVE6-gel to achieve concerted positivity on the globular surface of the nanoemulsion, which may facilitate the mucoadhesive property (as a result of electrostatic interaction) after nasal administration to improve the residence time and absorption [34]. The final pH of CVE6, AVE6, CVE6-gel, and AVE-6 gel products was found to be in the range of 6.8–7.4, which provided agreeable consistency and compatibility with the nasal mucosa.

2.2.1. Morphological Evaluation of the Optimized Cationic Nanoemulsion and Respective Gel

CVE6, AVE6, CVE6-gel, and AVE6-gel were considered the most stable and optimized nanoemulsion and respective gels. Generally, the size, shape, and size distribution are expected to change after the incorporation of the nanoemulsion into a hydrogel carrier. Therefore, it was requisite to visualize CVE6, AVE6, CVE6-gel, and AVE6-gel. Thus, the morphology of nanoemulsions was compared after incorporation into the gel. The result is shown in Supplementary Figure S1 including the shape, size, and globular size distribution. The shape of the globular particle is approximately similar in the nanoemulsion and respective gel. However, the cationic nanoemulsion is found to be well dispersed in CVE6 compared to AVE6, which may be due to the imposed positive charge. AVE6 is slightly dispersed without forming any aggregation. A similar observation was obtained in the respective gel. Thus, hydrogel could not change the shape, size, or globular distribution of the nanoemulsion. Moreover, there was no observed drug precipitation even after the amalgamation of CVE6 or AVE6 into the carbopol hydrogel matrix. This suggested there was a substantially firm layer of S_{mix} coated on oil globules containing solubilized VA. It is noteworthy that the size obtained from DLS always differs from the size estimated using TEM. This happens due to instrumental error and differences in the working principle. Both techniques are quite different and followed different sample processes during analysis. Therefore, this error is defined as a "Fold error" and estimated using the following formula:

$$\text{Fold error (FE)} = 1/n\ [\log^{\text{size of DLS/size of TEM}}] \quad (1)$$

In general, the error is considered acceptable when it drops below 2 ($\leq$2) [35]. The values of FE for CVE6, AVE6, CVE6-gel, and AVE6 gel were found to be 1.4, 1.7, 1.3, and 1.9, respectively. For the gel, the sample was first diluted in water to a gel consistency similar to the respective nanoemulsion before analysis using the DLS technique. The same sample was scanned under TEM. In DLS analysis and TEM-based scanning, the temperature was kept constant to avoid any further errors in the results.

2.2.2. Drug Content Estimation

The percent drug contents of CVE1-CVE6, AVE6, AVE6-gel, and CVE6 gel were estimated using the HPLC method. The sample was dissolved in an acetonitrile-methanol mixture (30:70) to extract the drug. The sample was filtered and analyzed. The percent of drug content in each formulation was not less than 99.3%. There was a slight loss of drug content during the preparation and handling process. The percent strength of each nanoemulsion and gel is presented in Table 3.

2.2.3. In Vitro Drug Release Profile

The model drug is acidic in nature (pKa = 5.2) and poorly soluble in water (1.3 mg/mL). The drug is reported to be soluble in an alkaline medium such as sodium hydroxide and alcohol. The optimized nanoemulsions (CVE6 and AVE6) and their respective gels showed different release behavior at pH 6.8 and 7.4 (phosphate buffer solution). The result is presented in Figure 6A,B. The nasal fluid and mucosa pH is approximately 6.8 and systemic delivery across the blood–brain barrier is exposed to pH of 7.4. Therefore, it was mandatory to investigate the impact of mucosal pH and blood pH when formulations are expected to be transported across mucosal and BBB for brain delivery. The result showed two important findings. These were (a) the impact of gel and (b) the impact of the release medium pH. It is quite clear that the drug was rapidly released from cationic and anionic nanoemulsions through the dialysis membrane as compared to the respective gel. This may be correlated with the viscous nature of the gel and two drug-release-limiting factors. These drug release rate-limiting factors are the gel matrix and dialysis membrane slowing down nanoglobules diffusion from the matrix to the medium. In the case of the nanoemulsion, there is only the dialysis membrane as a drug-release rate-limiting factor. The low viscosity further facilitated drug diffusion from the nanoemulsion to the release medium [36]. The release medium chamber was maintained at a temperature of 32 ± 1 °C throughout the study. The drug suspension (7 mg/mL) was rapidly released (>90%) within 30 min due to its salt solubility (1.3 mg/mL) at pH 6.8 (Figure 6A). A similar pattern was observed at pH 7.4 (>78.4% within 30) (Figure 6B).

2.2.4. Ex Vivo Drug Permeation and Drug Deposition Using Goat Nasal Mucosal Tissue

Various reports have been published for drug delivery to the brain using the nasal route. The nasal mucosa composition, the viscosity of the nasal formulation, mucoadhesiveness, residence time, and nasal pH are major critical factors responsible for controlled drug release and permeation across the nasal epithelium [37,38]. The study was conducted for up to 6 h to avoid any loss of natural anatomical structural integrity of mucosal tissue and tissue viability [39]. The study was conducted using a simulated nasal fluid with pH 6.8 (to mimic nasal pH) to avoid nasal irritation and discomfort after application [40]. Gel products are relatively viscous and more mucoadhesive compared to CVE6 and AVE6. The cumulative amount of drug permeation is revealed in Figure 7A and the drug deposition into the nasal mucosal tissue is presented in Figure 7B. The values of permeation flux for CVE6, AVE6, CVE6-gel, and AVE6-gel were estimated as 67.64, 48. 01, 57.18, 31.74, and 3.15 μg/cm^2/h, respectively, across the nasal mucosa of goats. The steady-state permeation flux values of the cationic nanoemulsion and its gel exhibited 21.47- and 18.15-fold higher flux rates as compared to the control suspension, which may be correlated with cationic and mucoadhesive gel carriers providing an electrostatic interaction with a negatively charged mucosal surface, extended residence time, and linoleic acid reported to facilitate drug permeation across the blood–brain barrier [41,42]. Moreover, the gel is mucoadhesive, biocompatible, and slightly acidic comparable to nasal fluid pH and drug pKa value (5.2–5.6). The flux value of CVE6 is very comparable to the published report of flux (~73 μg/cm^2/h) for a VA-loaded niosomal in situ gel across a goat mucosal membrane [43]. Slightly high flux values may be attributed to a niosomal loading efficiency greater than the nanoemulsion. Fortunately, these parameters are suitable for maximized nasal permeation of the drug in the explored carrier for brain delivery. The drug is supposed to remain unionized at nasal pH due to the comparable pKa value for enhanced permeation and drug deposition. In addition, considering poorly vascularized (anterior third of each nasal cavity) and highly vascularized anatomical areas (the respiratory epithelium and two-thirds of the posterior portion of the cavity) of the nose, inhaled particles or nanoglobules were thought to be lodged by three prime mechanisms, namely (a) gravitational sedimentation, (b) inertial impaction, and (c) Brownian diffusion (if spayed) [43]. To understand the mechanistic perspective of drug delivery from the nose to the brain, it is imperative to consider the interplay of various critical factors such as formulation characteristics, the device, and patient-related conditions. These factors are directly involved in the drug-laden nanodroplets for maximized permeation and drug deposition within nasal cavities and, subsequently, the drug access to the brain. Notably, the exact localization of the drug for deposition is recognized as key to the success or failure of the nasal product [44]. The sites of drug localization within the nose dictate the purpose of local, systemic, and brain drug delivery. For drug delivery to the brain, the nasal cavities (innervated with olfactory and trigeminal nerves) are the most ideal site for drug localization and constitute a potential target for nose-to-brain delivery using a cationic nanoemulsion and gel formulation. Moreover, these cavities rapidly absorb the lodged drug through the thin membrane to achieve faster onset of action at a low dose, high patient compliance, reduced dose and metabolite (4-eve-VPA)-based side effects (hepatic toxicity due to the reticuloendothelial system) without hepatic metabolism and maximized drug access to the brain [45–47]. Greater uptake by RES indicates greater drug metabolism and incidence of side effects. Considering formulation-related factors such as the globular size, shape, zeta potential, viscosity, and mucoadhesiveness, the drug solubility, polarity, hydrophilicity, and composition (surfactant and oil) are complementary factors. Linoleic acid-rich oils are gaining popularity in pharmaceutical and cosmeceutical industries due to possessing various skin benefits such as (a) anti-inflammatory, (b) acne-reductive, (c) skin-softening, and (d) moisture-retentive abilities, as well as (e) facilitating drug diffusion across the blood–brain barrier (50–70% linoleic acid) and (f) biocompatibility [10,31]. Tween 80 possessed high hydrophilicity due to the high HLB value (14.5) and it is anticipated to achieve maximized emulsification in the hydrophilic mucosal layer to keep nanoglobules in an emulsified form within the mucosal matrix for prolonged systemic circulation time (likely due to the long fatty acid chain in lipid. such as linoleic acid) in the brain or reduced RES uptake. The surfactant is reported to have several benefits for nasal nanoemulsion for VA delivery to the brain. These are (a) protection of the drug from enzymatic degradation, (b) improved brain bioavailability, and (c) prolonged circulation time in the brain due to the long fatty acid and polyunsaturated fatty acid (PUFA) of the present oil [10].

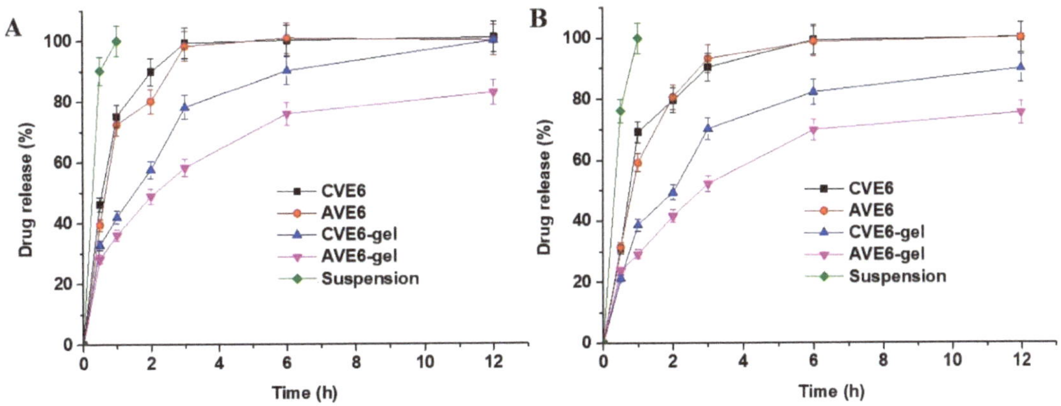

Figure 6. In vitro drug release using a dialysis membrane: (**A**) The drug release profile at pH 6.8 and (**B**) the drug release profile at pH 7.4 and 37 ± 1 °C (data are expressed as mean ± standard deviation, $n = 3$).

The result of the drug deposition is presented in Figure 7B wherein CVE6, AVE6, CVE6-gel, AVE6-gel, and the suspension showed percent drug depositions of 67.64, 48.0, 57.18, 31.74, and 3.15%, respectively. It is quite clear that greater drug deposition means greater permeation flux as observed in CVE-6 as compared to the respective gel and other nanoemulsions. The gel matrix slightly delayed permeation and drug deposition, which is good for prolonged drug release and an extended effect to control epileptic fits and seizures. However, considering the types of patients and working or traveling schedules, both formulations are important. For immediate relief, it is better to spray a cationic nanoemulsion as it is aqueous and free-flowing due to its low viscosity. In the case of a planned traveling schedule, a gel product is better and more suitable as a prophylactic dose for prolonged relief from seizure attacks. Globular size, surface charge, and pH are other factors controlling drug deposition and, subsequently, drug flux. The nanoemulsion size depends on the oil content (the oil content is inversely proportional to the globular size of the nanoemulsion) and the content and type of surfactant. Tan et al. revealed reduced globular sizes of the nanoemulsion from 142 nm to 80 nm due to the reduced content of oil from 6% to 1.5%, respectively [10]. In the literature, it was reported that VA transport and nanoemulsion permeation across the blood–brain barrier is mediated via the organic anion transporter and the LDL-mediated endocytosis due to the presence of tween 80, respectively [48–50]. This may explain the significant difference in permeation profiles between the drug suspension and formulations.

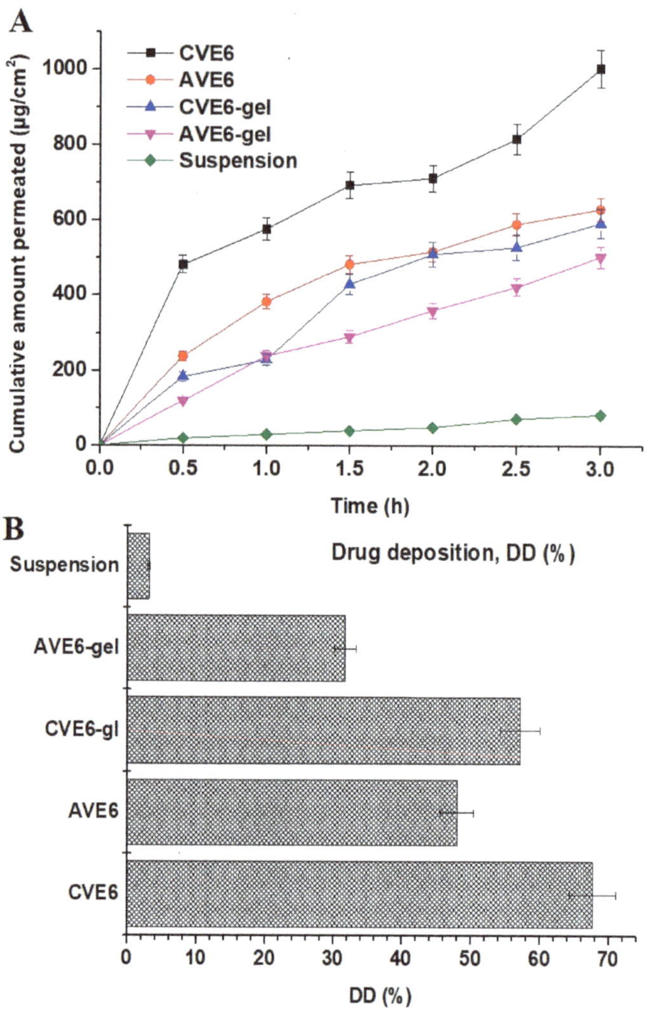

Figure 7. (**A**) Ex vivo cumulative drug permeation across nasal mucosa of goat over a period of 6 h in simulated nasal fluid, and (**B**) drug deposition of the drug in nasal mucosa after 6 h of ex vivo permeation at 37 ± 1 °C (data are expressed as mean ± standard deviation, $n = 3$).

2.2.5. Confocal Laser Scanning Microscopy (CLSM)

To evaluate the degree of penetration and permeation across the superior nasal concha (nasal membrane), we scanned the nasal mucosa treated with the formulations under CLSM. For comparison, the R123 solution was used as the control. The result is provided in Figure 8A–F. It is obvious from the result that the dye solution and suspension were not penetrable across the hydrophilic (approximately 90–95% water and glycoprotein, providing a gel-like structure) nasal mucosal membrane as evidenced by the poor fluorescence intensity [51]. The drug suspension containing the dye showed approximately similar intensity due to the drug insolubility and poor permeation behavior. The fluorescence intensity values of the dye solution, suspension, AVE6-R, AVE6-R-gel, CVE6-R, and CVE6-R-gel were obtained as 11.6, 17.3, 65.6, 75.62, 84.7, and 96.11%, respectively. The lowest fluorescence intensity associated with the dye solution and suspension could be attributed to poor dye and drug permeation across the hydrophilic nasal mucosa as a result of low solubility. However, high-intensity values were observed for both the nanoemulsions (AVE6-R and CVE6-R) and gels (AVE6-R-gel and CVE6-R-gel) as shown in Figure 8. A high degree of intense fluorescence by the gel and cationic nanoemulsion can be correlated with mucoadhesiveness and prolonged residence time on the nasal mucosa of goats. Carbopol gel is known for its good mucoadhesive nature at compatible pH for nasal delivery (4.5–6.8) without producing any nasal irritation [52]. Nasal pH (4.5–6.8) is very suitable for the gel consistency maintained after nasal application. Moreover, the drug is slightly acidic to ensure it is in a stable and non-ionized form if it comes into contact with the nasal fluid and mucosal membrane. The drug- and formulation-related properties provide suitability for the drug permeation, penetration, and compatibility for intranasal delivery of the drug to control convulsion in patients. Moreover, the imposed positive charge on the cationic nanoemulsion facilitated the nanoemulsion penetration as compared to the anionic counterpart as evidenced by the remarkably high fluorescence intensity. This can be correlated to the electrostatic interaction-mediated improved permeation and, subsequently, the drug deposition within the submucosal region of nasal tissues. In addition, intranasal delivery of the nanocarrier-based drug offers several advantages over oral administration of the drug.

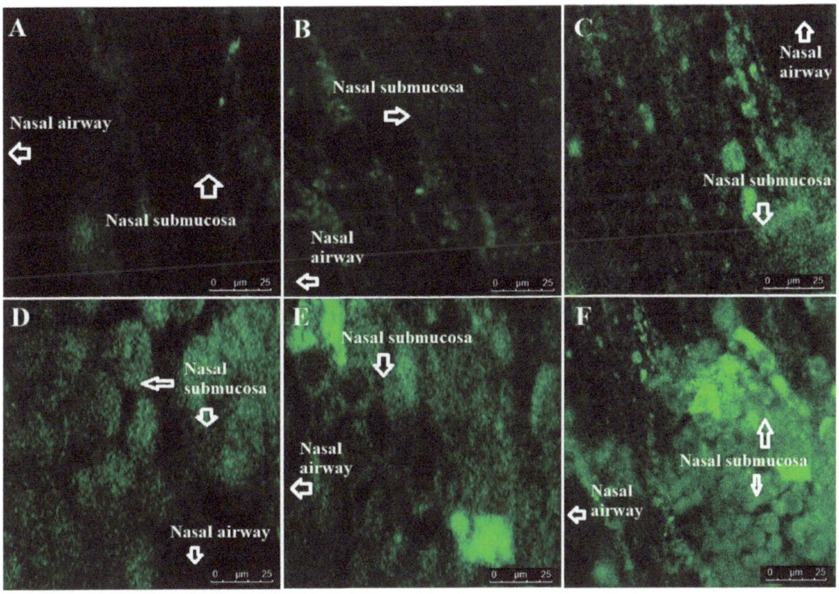

Figure 8. Penetration and permeation of the optimized nanoemulsions and respective gels across nasal epithelium to submucosal and mucosal regions using CLSM (confocal laser scanning microscopy). (**A**) Control using R123 solution, (**B**) R123-probed drug suspension, (**C**) AVE6-R nanoemulsion, (**D**) AVE6-R-gel, (**E**) CVE6-R nanoemulsion, and (**F**) CVE6-R-gel. Mean intensity measured using image J software 1.54f (**E**).

Conclusively, the dye solution and the drug suspension itself are not capable of being penetrated. Both nanoemulsions were relatively less viscous as compared to the gel formulation. This caused slightly lower residence time in the mucosal region. The gel carrier provided hydration and high residence time for nanoemulsion penetration. Finally, the cationic globular electrostatic interaction with the negatively charged nasal membrane rendered the investigated nanoemulsion suitable for maximized permeation and penetration [53]. Thus, it was hypothesized that the optimized viscosity, imposed cationic charge, reduced globular size, and mucoadhesive gel could be working in tandem for drug delivery to the brain through nasal administration.

3. Conclusions

The conventional dosage form of VA is associated with multiple challenges. These challenges are related to the physicochemical properties, pharmacokinetic behavior, and pharmacodynamics properties of the drug. Low bioavailability to the brain, high hepatic metabolism, and severe side effects upon oral and parenteral delivery gained widespread attention from formulation scientists for alternative and high therapeutic benefits. The GastroPlus program assisted us to understand the in vivo behavior of the drug in the human body at the explored dose, dosing frequency, and dosage form. Moreover, the program predicted various factors responsible for affecting in vivo pharmacokinetics and drug dissolution. HSPiP software predicted various excipients based on HSP parameters to reduce the experimental screening duration and development stage. Cationic nanoemulsions may be a promising option for maximized drug access to the nasal cavity due to their small size (113 nm), high mucoadhesiveness (high positive zeta potential and mucoadhesive carbopol gel), and linoleic acid (as high content in the oil)-mediated drug permeation across the blood–brain barrier. Ex vivo permeation flux, the enhancement ratio, drug deposition, and the penetration property of CVE6 and CVE6 gel confirmed electrostatic and mucoadhesiveness worked in tandem for extended residence time in the nasal mucosa and, subsequently, augmented the drug's access to the brain. Conclusively, this strategy is a promising and suitable alternative to conventional cream or oral tablets to control seizures with high therapeutic effectiveness and patient compliance.

4. Materials and Methods

4.1. Materials

Valproic acid sodium salt (VA, 98.0% pure) and polysorbate-80 were procured from Sigma Aldrich (Merck), Mumbai, Maharashtra, India). Soya lecithin powder (97%) was purchased from Otto Chemie Pvt. Ltd., Mumbai, India. HPLC (high-performance liquid chromatography)-grade solvents (methanol, ethanol, acetonitrile, and buffering reagents) were purchased from Merck, Mumbai, India. Edible safflower, flaxseed, and grape seed oils were purchased from a local medical shop. Buffer reagents (potassium dihydrogen phosphate, sodium chloride, and sodium hydroxide) were procured from S.D. Fine, Mumbai, India. In-house-distilled water was used as an aqueous solvent. For HPLC mobile phase preparation, Milli-Q water was used (Millipore, Burlington, MA, USA).

4.2. Methods

4.2.1. Prediction and Simulation Study Using GastroPlus for Oral Tablet

The program was used to predict pharmacokinetic parameters (PK) of orally delivered VA tablets for adult patients with a dose of 250 mg. In the literature and on the DEPAKENE tablet label, varied bioavailability, absorption rate, and PK parameters have been described depending on the patient's body weight. To avoid preclinical and clinical studies due to expensive and tedious investigations, the program assisted in predicting various PK parameters in a targeted patient for the desired dose, dosage form, dosage volume, and frequency of dosing frequency. For this, the program used three basic tabs such as (a) the compound tab, (b) the formulation tabs, and (c) the pharmacokinetic tabs. We used various literature data, experimental values, and by-default program-suggested values to run the simulation and prediction (as shown in Table 1). Moreover, parameter sensitivity assessment (PSA) was used to determine the impact of various factors (physicochemical properties of the drug and physiological conditions such as intestinal lumen and related factors) affecting the PK parameters of the drug. Physicochemical properties of the drug include the reference solubility, particle size, volume, density, logP, pKa, and molecular weight. Physiological factors include gastrointestinal pH, stomach volume, residence time, and radius. Formulation factors are the nanosize, shape, and solubility. The regional compartmental model predicts regional absorption of the drug through nine different GIT (gastrointestinal tract) sections (stomach, duodenum, ileum-1,2, jejunum-1,2, ascending colon, colon, and caecum). Total absorption indicates the sum of absorption from the GIT of patients. Prediction and simulation were carried out considering fast subjects to avoid a food interaction in the prediction model. The simulation time was 24 h for each run of prediction and simulation [25,26].

Hansen Solubility Parameters for VA and Excipients

Hansen solubility parameters have been used for various solvents, co-solvents, drugs, and human skin (normal and abnormal). The parameters were estimated using the HSPiP program. The fundamentals of the software are based on the physicochemical interactions (in terms of cohesive energy) of a solute for a particular solvent. These parameters are dispersion energy (δ_d), polarity (δ_p), and hydrogen bonding energy (δ_h) [54,55]. Therefore, a solute interacts with a solvent through these cohesive forces working together. Thus, the difference in any parameter between a solute and solvent close to zero is considered miscible or soluble. Thus, the program predicted relevant excipients based on these HSP values of each excipient close to the HSP values of the model drug. The program estimated the values shown in Table 1. The HSP values of oils, lecithin, and PC were obtained from the literature, and these were calculated manually based on the percent composition of linoleic acid or phosphatidylcholine (PC) present [28].

Solubility of Valproate Sodium in Various Excipients

The solubility of VA was determined in various lipids, surfactants, and co-surfactants to identify the most suitable and biocompatible excipients for nasal nanoemulsion. Stearylamine was added to the organic phase to impose the cationic charge on a globular surface for adhesive purposes [13]. Flaxseed (50–70%), safflower (70–78%), and grape seed oil (70%) are prime sources of linoleic acid. Tween 80, Span-80, transcutol, propylene glycol, and lecithin were used as surfactants and co-surfactants. Briefly, a fixed amount of each excipient was transferred to a clean glass vial. A weighed amount of the drug was added to each vial containing the individual excipients. The glass vials were closed and sealed for the solubility study. The vial was placed inside a water shaker bath (Remi Shaker, Mumbai, India) set at a fixed temperature (40 °C) and shaking rate (75 rpm). The study was continued for 72 h to achieve equilibrium. Then, the mixture was centrifuged to obtain the supernatant liquid. The amount of the drug dissolved was assayed using a UV Vis spectrophotometer (U 1800, Japan) at 210 nm [10]. The study was repeated to obtain a mean and standard deviation (n = 3).

Pseudo Ternary Phase Diagram, Cationic Nanoemulsions, and Nanoemulsion Gel

To prepare a cationic nanoemulsion, a constant amount of stearylamine (0.1%) was used in the formulation. Based on HSP values and the experimental solubility of AV, excipients were selected. The excipient possessing HSP values close to the HSP values of AV and excipients with the highest solubility of AV were selected for cationic nanoemulsion. Thus, safflower seed oil, tween 80 + lecithin (1:1), and PG were selected as the oil, surfactant, and co-surfactant, respectively. To impose a cationic charge, a constant amount (0.1%) of SA (stearylamine as the cationic lipid) was incorporated into the organic phase of each formulation [34]. Various pseudoternary phase diagrams were constructed to identify the correct ratio of the surfactant to the co-surfactant (S_{mix}). A slow and spontaneous titration method was adopted to prepare the nanoemulsion by varying the lipid-to-S_{mix} ratio [25]. A transparent and isotropic cationic nanoemulsion was selected for further characterization. To prepare a nanoemulsion gel, the cationic nanoemulsion was incorporated into a carbopol gel (1%). The final strength of the gel was 0.5% w/w. Each nanoemulsion and respective gel contained a constant amount of VA. For this, a weighed amount of carbopol 934 was dispersed into warm distilled water to obtain the final strength of 1% w/w. The dispersed gel was vigorously stirred using a mixer at high speed (10,000 rpm). The obtained gel was treated with a few drops (3–5 drops) of triethanolamine (base) as a cross-linking agent. The acidic solution of the carbopol dispersion was triggered for cross-linking under triethanolamine and become a transparent viscous gel. Equal weights of gel and lyophilized formulation were mixed together using a homogenizer to obtain a gel of 0.5% gel strength. The final concentration of AV in the gel product was approximately 5% *w*/*w*. The final pH of each formulation was adjusted to 6.8 to obtain good consistency and compatibility with nasal mucosa.

Thermodynamic Stability of Cationic Nanoemulsion: Freeze–Thaw Cycle and Ultracentrifugation

Each developed nanoemulsion was subjected to extreme physical (ultracentrifugation) and thermal stress (extreme low and extreme high temperatures). For this, each cationic nanoemulsion was stored in a clear glass vial, labelled, and sealed. Each formulation was separately stored in the stability chamber at the set temperature. A cycle of exposure to as low as freeze (−21 °C) and as high as thaw (40 °C) temperatures was repeated thrice followed by room temperature conditions. Each sample was withdrawn from both temperatures and kept at room temperature (25 °C) to resume its original stable form (isotropic liquid). In the second phase, each stable formulation was subjected to an ultracentrifugation step (22,000 rpm for 5 min). Any sign of physical instability (drug precipitation, color, creaming, and phase separation) was considered an unstable product and dropped out from further study. This freeze–thaw cycle was mandatory to identify the most stable product.

4.2.2. Evaluation of Cationic Nanoemulsions and Gels

Nanoemulsions were characterized by globular size, size distribution, and zeta potential. These parameters were determined using a Zetasizer (Malvern Instrument Limited, Malvern, Worcestershire, UK). Formulations were diluted with distilled water before scanning for size analysis. In the case of zeta potential, the formulations were analyzed without dilution to obtain tangible zeta potential values. This value was expected to be positive for the cationic nanoemulsion, whereas the nanoemulsion without stearylamine was anticipated to be negative. The analysis was carried out at room temperature. The viscosity of each formulation was determined using a viscometer (Bohlin visco88, Malvern Instrument Ltd., Worcestershire, UK). The sample was processed at room temperature (25 °C). The study was replicated for the mean and standard deviation (n = 3). The values of pH were estimated using a calibrated digital pH meter.

Morphological Evaluation of the Optimized Cationic Nanoemulsion and Respective Gel

The optimized cationic nanoemulsion and respective gel were observed under cryogenic transmission electron microscopy (cryo-TEM) [56,57]. The tool was used to visualize the globular size, size distribution, and shape. For this, the sample was placed on a glass coverslip previously coated with poly-L-lysin (a fixative) [56]. Then, the sample was processed for cryogenic TEM by placing the sample on a copper screen with lacey carbon film and a blotting time of 5 s. Scanning was conducted using Thermofisher Krios G3 (Thermo Fisher Scientific India, Private Limited, Mumbai, India) equipment (low energy consumption method) coupled with a Bioquantum K3 detector [57,58]. The images were processed at an operating voltage power of 80 Kv (a Gatan cryoholder system) (Gatan, Inc. Corporate Headquarters 5794 W. Las Positas Blvd. Pleasanton, CA 94588, United States of America). Finally, the dried sample was scanned at various magnifications and resolutions. A fixed location was located and scanned for the sample. The process was conducted at room temperature. The wet sample was avoided in the scan due to poor scanning, and the resolution of images was observed by an interfered electronic beam.

Drug Content Estimation

The drug content was estimated from the optimized anionic nanoemulsion, cationic nanoemulsion, and respective gel formulations. In brief, a weighed amount of the formulation was dissolved in a methanol-chloroform mixture (1:10). The mixture was stirred for 10 min to extract the drug. The mixture was centrifuged for 15 min at 12,000 rpm to separate the low-density nanoemulsion from the insoluble drug and water. The supernatant and settled pellet were separately estimated to find the total drug and the entrapped drug. The drug was assayed using the validated HPLC method at 210 nm. The experiment was repeated to obtain the mean and standard deviation. In the case of gel formulation, a weighed amount of gel was dispersed into a water–ethanol mixture (1:2) to obtain the extracted drug. Then, the mixture was stirred for 15 min followed by centrifugation. The supernatant was used to estimate the drug content.

4.2.3. In Vitro Drug Release Profile

The in vitro drug release profile for each nanoemulsion and the respective gel was determined using a dialysis membrane with a molecular weight cut-off of 12–14K Dalton (HiMedia, Mumbai, India). For this, a fixed dimension of the dialysis membrane was cut from ribbon and soaked in saline for 12 h before use. The activated dialysis bag was filled with the test sample and both ends were clipped with a plastic clipper. This maintained a constant effective surface area for drug release. The release medium (500 mL) was phosphate buffer at pH 6.8 and pH 7.4. A glass beaker containing the release medium was used for the drug release. The test sample bag was suspended in the release medium already placed on a heating magnetic stirrer. A Teflon-coated magnetic bead was used to maintain the temperature and uniform drug distribution within the bulk volume released from the bag. The sample was collated at different time points (0.5, 1, 2, 3, and 6 h). The withdrawn volume was replaced with the fresh-release medium. The withdrawn sample was filtered (0.22 μ as pore size) and used for the drug content released after each time point. The drug was analyzed using an HPLC method. The release medium chamber was maintained at a temperature of 32 ± 1 °C throughout the study. The effective surface area for passive diffusion of the drug was 1.34 cm^2 functional at 32 ± 1 °C [36,59].

4.2.4. Ex Vivo Drug Permeation and Drug Deposition Using a Goat Nasal Mucosa

Drug permeation and deposition studies were performed using an excised goat nasal mucosa obtained from a local slaughterhouse. The excised tissue was used 20 min after sacrifice to avoid tissue damage and death. The intact nose was obtained, and the skin was removed. Then, the nose was stored in a cold phosphate buffer solution (pH 7.4) [37]. The nasal mucosa was removed using surgical scissors and forceps without making any surgical cut in the desired area of the mucosal membrane. The obtained mucosal tissue was immersed in a freshly prepared Ringer's solution with proper aeration [38]. The excised tissue has a dimension of 0.2 mm × 10 mm with an effective surface area of diffusion of 1.78 cm^2. The tissue was mounted between the receptor and the donor chambers. The receptor chamber was filled with SNF (simulated nasal fluid) at pH 6.8 [40]. The release medium was maintained at 37 ± 1 °C by circulating hot water through a jacketed system around the chamber. A rice bead was placed inside the receptor chamber rotating at 300 rpm on a magnetic stirrer. A constant amount of the sample was placed on the mucosal adhesive side for drug permeation. Four groups were categorized as (a) CVE6-1, AVE6, CVE6-gel, and AVE6-gel. For comparison, the drug solution was used as a control in gel formulation. In each case, an equivalent amount of the drug was loaded on the effective permeation area. The sampling (1 mL using a syringe) was conducted at different time points such as 0.5, 1, 1.5, 2, 2.5, and 3 h. The withdrawn sample volume was replaced with an equal volume of the fresh medium. The sample was filtered using a membrane filter (0.2 μm) and the content of the drug was estimated using the HPLC method. The study was replicated to obtain the mean and standard values. The result was expressed as the percent of the drug permeated for the brain delivery or percent diffusion for the brain access or availability of the drug to the brain (ex vivo). The permeation parameters (steady state flux, targeted flux, permeability coefficient, and enhancement ratio) were estimated using the following equation:

$$J_{ss} = (dM/dt) \times (1/A) = PC \quad (2)$$

where A, P, and C represent the effective surface area for diffusion, the permeability diffusion coefficient, and the initial loaded content of the drug, respectively. Jss indicates the steady state flux of the solution as per Fick diffusion in Equation (2). The value of dM indicates the amount of the drug diffused across the mucosal membrane within a given time point (dt). The study was conducted for up to 360 min to avoid the loss of the natural integrity of tissue and tissue viability [39].

A drug deposition study was conducted after the completion of the ex vivo permeation study. The mounted tissue was removed from the diffusion cell. Each tissue was separately sliced into small pieces. The sliced pieces were transferred to a vial containing methanol and chloroform (1:2). The mixture was stirred for 12 h under closed conditions using a magnetic bead. The drug was extracted from the tissue and subjected to centrifugation. The fatty debris and tissues settled at the bottom as pellets and the supernatant (clear solution) was removed for the drug analysis. The supernatant was filtered using a membrane filter and analyzed using the HPLC method [10,60].

4.2.5. Confocal Laser Scanning Microscopy (CLSM)

To visualize the degree of drug penetration, the same formulations and control were reformulated using rhodamine 123 as a probe in the formulation. The composition and experimental conditions were kept constant as in the ex vivo permeation and drug deposition study section. The dye was present as 0.01% *w*/*v* in each formulation. The Franz diffusion cell, tissue mounting, release SNF, volume, pH, and loaded dose were constant for 3 h. After 3 h of permeation study, the tissue was removed for each group (five groups), and the adhered material was washed with running water. The tissue was sliced as per CLSM requirements. The treated and untreated skin was sliced into small pieces using a microtome. The tissue specimen was placed on the glass coverslip and air-dried for 12 h. Each tissue was visualized under CLSM and evaluated for globular penetration across the mucosal membrane (Fluorescence Correlation Microscope-Olympus FluoView FV1000, Olympus, Melville, NY, USA) with an argon laser beam with excitation at 488 nm and emission at 590 nm [15,61].

Supplementary Materials: The following supporting information can be downloaded at: https://www.mdpi.com/article/10.3390/gels9080603/s1, Figure S1: Representative images of cryo-TEM micrographs: (A) CVE6 nanoemulsion, (B) AVE6 nanoemulsion, (C) CVE6 gel, and (D) AVE6 gel. Magnification 49000X.

Author Contributions: A.H.: Conceptualization, methodology, drafting, and writing, M.A.A.: Drafting, review, and analysis, M.A.M.: Review and validation, M.R.: Analysis, data curation, and resources, T.K.: Editing and visualization. All authors have read and agreed to the published version of the manuscript.

Funding: The authors extend their appreciation to the Deputyship for Research and Innovation, "Ministry of Education" in Saudi Arabia for funding this research (IFKSUOR3-129-1).

Institutional Review Board Statement: Not applicable for the present study.

Informed Consent Statement: Not applicable.

Data Availability Statement: Not applicable.

Acknowledgments: The authors extend their appreciation to the Deputyship for Research and Innovation, "Ministry of Education" in Saudi Arabia for funding this research (IFKSUOR3-129-1).

Conflicts of Interest: Authors report no conflict of interest.

References

1. World Health Organization Epilepsy. Available online: https://www.who.int/news-room/fact-sheets/detail/epilepsy (accessed on 20 December 2021).
2. Available online: https://www.ninds.nih.gov/current-research/focus-disorders/focus-epilepsy-research/curing-epilepsies-promise-research (accessed on 20 December 2021).
3. Al Rajeh, S.; Awada, A.; Bademosi, O.; Gunniyi, A. The prevalence of epilepsy and other seizure disorders in an Arab population: A community-based study. *Seizure* **2001**, *10*, 410–414. [CrossRef] [PubMed]
4. Ishizue, N.; Niwano, S.; Saito, M.; Fukaya, H.; Nakamura, H.; Igarashi, T.; Fujiishi, T.; Yoshizawa, T.; Oikawa, J.; Satoh, A.; et al. Polytherapy with Sodium Channel-Blocking Antiepileptic Drugs Is Associated with Arrhythmogenic ST-T Abnormality in Patients with Epilepsy. *Seizure* **2016**, *40*, 81–87. [CrossRef]
5. Hammond, E.J.; Perchalski, R.J.; Villarreal, H.J.; Wilder, B.J. In vivo uptake of valproic acid into brain. *Brain Res.* **1982**, *240*, 195–198. [CrossRef] [PubMed]
6. US FDA (US Food and Drug Administration). Available online: https://www.accessdata.fda.gov/drugsatfda_docs/label/2016/018081s065_018082s048lbl.pdf (accessed on 1 July 2023).
7. Eskandari, S.; Varshosaz, J.; Minaiyan, M.; Tabbakhian, M. Brain delivery of valproic acid via intranasal administration of nanostructured lipid carriers: In vivo pharmacodynamic studies using rat electroshock model. *Int. J. Nanomed.* **2011**, *6*, 363–371.
8. Wang, H.; Huang, Q.; Chang, H.; Xiao, J.; Cheng, Y. Stimuli-responsive dendrimers in drug delivery. *Biomater. Sci.* **2016**, *4*, 375–390. [CrossRef]
9. Lopez, T. Biocompatible Titania Microtubes Formed by Nanoparticles and its Application in the Drug Delivery of Valproic Acid. *Opt. Mater.* **2006**, *29*, 70–74. [CrossRef]
10. Kirby, T.B.P.; Stanslas, J.; Basri, H.B. Characterization, in-vitro and in-vivo evaluation of valproic acid-loaded nanoemulsion for improved brain bioavailability. *J. Pharm. Pharmacol.* **2017**, *69*, 1447–1457.
11. Mori, N.; Ohta, S. Comparison of anticonvulsant effects of valproic acid entrapped in positively and negatively charged liposomes in amygdaloid-kindled rats. *Brain Res.* **1992**, *593*, 329–331. [CrossRef]
12. Tan, S.F.; Masoumi, H.R.F.; Karjiban, R.A.; Stanslas, J.; Kirby, B.P.; Basri, M.; Basri, H.B. Ultrasonic emulsification of parenteral valproic acid-loaded nanoemulsion with response surface methodology and evaluation of its stability. *Ultrason. Sonochemistry* **2016**, *29*, 299–308. [CrossRef] [PubMed]
13. Yadav, S.; Gandham, S.K.; Panicucci, R.; Amiji, M.M. Intranasal brain delivery of cationic nanoemulsion-encapsulated TNFα siRNA in prevention of experimental neuroinflammation. *Nanomed. Nanotechnol. Biol. Med.* **2016**, *12*, 987–1002. [CrossRef]
14. Azambuja, J.H.; Schuh, R.S.; Michels, L.R.; Gelsleichter, N.E.; Beckenkamp, L.R.; Iser, I.C.; Lenz, G.S.; de Oliveira, F.H.; Venturin, G.; Greggio, S.; et al. Nasal Administration of Cationic Nanoemulsions as CD73-siRNA Delivery System for Glioblastoma Treatment: A New Therapeutical Approach. *Mol. Neurobiol.* **2020**, *57*, 635–649. [CrossRef] [PubMed]
15. Hussain, A.; Samad, A.; Singh, S.K.; Ahsan, M.N.; Haque, M.W.; Faruk, A.; Ahmed, F.J. Nanoemulsion gel-based topical delivery of an antifungal drug: In vitro activity and in vivo evaluation. *Drug Deliv.* **2016**, *23*, 642–647. [CrossRef] [PubMed]
16. Teixeira-da-Silva, P.; Pérez-Blanco, J.S.; Santos-Buelga, D.; Otero, M.J.; García, M.J. Population pharmacokinetics of valproic acid in pediatric and adult caucasian patients. *Pharmaceutics* **2022**, *14*, 811. [CrossRef] [PubMed]
17. Zaccara, G.; Messori, A.; Moroni, F. Clinical Pharmacokinetics of Valproic Acid—1988. *Clin. Pharmacokinet.* **1988**, *15*, 367–389. [CrossRef]
18. Johannessen, C.U.; Johannessen, S.I. Valproate: Past, present, and future. *CNS Drug Rev.* **2003**, *9*, 199–216. [CrossRef]
19. Winter, M.E. *Basic Clinical Pharmacokinetics*, 5th ed.; Lippincott Williams & Wilkins Health: Philadelphia, PA, USA, 2010.
20. Loscher, W. Serum protein binding and pharmacokinetics of valproate in man, dog, rat and mouse. *J. Pharmacol. Exp. Ther.* **1978**, *204*, 255–261.
21. Dickinson, R.G.; Harland, R.C.; Ilias, A.M.; Rodgers, R.M.; Kaufman, S.N.; Lynn, R.K.; Gerber, N. Disposition of valproic acid in the rat: Dose dependent metabolism, distribution, enterohepatic recirculation and choleretic effect. *J. Pharmacol. Exp. Ther.* **1979**, *211*, 583–595. [PubMed]
22. Hansen, S.; Lehr, C.-M.; Schaefer, U.F. Improved input parameters for diffusion models of skin absorption. *Adv. Drug Deliv. Rev.* **2013**, *65*, 251–264. [CrossRef] [PubMed]
23. Methaneethorn, J. A systematic review of population pharmacokinetics of valproic acid. *Br. J. Clin. Pharmacol.* **2018**, *84*, 816–834. [CrossRef] [PubMed]
24. Gugler, R.; von Unruh, G.E. Clinical Pharmacokinetics of Valproic Acid1. *Clin. Pharmacokinet.* **1980**, *5*, 67–83. [CrossRef] [PubMed]
25. Hussain, A.; Alshehri, S.; Ramzan, M.; Afzal, O.; Altamimi, A.S.A.; Alossaimi, M.A. Biocompatible solvent selection based on thermodynamic and computational solubility models, in-silico GastroPlus prediction, and cellular studies of ketoconazole for subcutaneous delivery. *J. Drug Deliv. Sci. Technol.* **2021**, *65*, 102699. [CrossRef]
26. Alsarra, I.A.; Al-Omar, M.; Belal, F. Valproic Acid and Sodium Valproate: Comprehensive Profile. Profiles of Drug Substances. *Excip. Relat. Methodol.* **2005**, *32*, 209–240. [CrossRef]
27. Vay, K.; Scheler, S.; Frie, W. Application of Hansen solubility parameters for understanding and prediction of drug distribution in microspheres. *Int. J. Pharm.* **2011**, *416*, 202–209. [CrossRef] [PubMed]
28. De La Peña-Gil, A.; Toro-Vazquez, J.F.; Rogers, M.A. Simplifying Hansen Solubility Parameters for Complex Edible Fats and Oils. *Food Biophys.* **2016**, *11*, 283–291. [CrossRef]

29. Hamilton, J.A.; Brunaldi, K. A model for fatty acid transport into the brain. *J. Mol. Neurosci.* **2007**, *33*, 12–17. [CrossRef] [PubMed]
30. Hussain, A.; Altamimi, M.A.; Afzal, O.; Altamimi, A.S.A.; Ali, A.; Ali, A.; Martinez, F.; Siddique, M.U.M.; Acree, W.E., Jr.; Jouyban, A. Preferential solvation study of the synthesized aldose reductase inhibitor (SE415) in the {PEG400 (1) + Water (2)} cosolvent mixture and GastroPlus-based prediction. *ACS Omega* **2022**, *7*, 1197–1210. [CrossRef]
31. Han, X.; Cheng, L.; Zhang, R.; Bi, J. Extraction of safflower seed oil by supercritical CO_2. *J. Food Eng.* **2009**, *92*, 370–376. [CrossRef]
32. Afzal, O.; Alshammari, H.A.; Altamimi, M.A.; Hussain, A.; Almohaywi, B.; Altamimia, A.S. Hansen solubility parameters and green nanocarrier based removal of trimethoprim from contaminated aqueous solution. *J. Mol. Liq.* **2022**, *361*, 119657. [CrossRef]
33. Abdullah, G.Z.; Abdulkarim, M.F.; Mallikarjun, C.; Mahdi, E.S.; Basri, M.; Sattar, M.A.; Noor, A.M. Carbopol 934, 940 and Ultrez 10 as viscosity modifiers of palm olein esters based nano-scaled emulsion containing ibuprofen. *Pak. J. Pharm. Sci.* **2013**, *26*, 75–83. [PubMed]
34. Jones, M.N.; Song, Y.-H.; Kaszuba, M.; Reboiras, M.D. The Interaction of Phospholipid Liposomes with Bacteria and Their Use in the Delivery of Bactericides. *J. Drug Target.* **1997**, *5*, 25–34. [CrossRef] [PubMed]
35. Altamimi, M.A.; Hussain, A.; Alshehri, S.; Imam, S.S.; Alnemer, U.A. Development and evaluations of transdermally delivered luteolin loaded cationic nanoemulsion: In vitro and ex vivo evaluations. *Pharmaceutics* **2021**, *13*, 1218. [CrossRef] [PubMed]
36. Christensen, J.M.; Chuong, M.C.; Le, H.; Pham, L.; Bendas, E. Hydrocortisone diffusion through synthetic membrane, mouse skin, and Epiderm™ cultured skin. *Arch. Drug Inf.* **2011**, *4*, 10–21. [CrossRef]
37. Yuwanda, A.; Surini, S.; Harahap, Y.; Jufri, M. Study of valproic acid liposomes for delivery into the brain through an intranasal route. *Heliyon* **2022**, *8*, e09030. [CrossRef] [PubMed]
38. Basu, S.; Maity, S. Preparation and Characterisation of Mucoadhesive Nasal Gel of Venlafaxine Hydrochloride for Treatment of Anxiety Disorders. *Indian J. Pharm. Sci.* **2012**, *74*, 428–433. [CrossRef] [PubMed]
39. Khuroo, T.; Khuroo, A.; Hussain, A.; Mirza, M.A.; Panda, A.K.; Iqbal, Z. Qbd based and Box-Behnken design assisted oral delivery of stable lactone (active) form of topotecan as polymeric nanoformulation: Cytotoxicity, pharmacokinetic, in vitro, and ex vivo gut permeation studies. *J. Drug Deliv. Sci. Technol.* **2022**, *77*, 103850. [CrossRef]
40. Trenkel, M.; Scherlie, R. Nasal Powder Formulations: In-Vitro Characterisation of the Impact of Powders on Nasal Residence Time and Sensory Effects. *Pharmaceutics* **2021**, *13*, 385. [CrossRef] [PubMed]
41. Edmond, J. Essential Polyunsaturated Fatty Acids and the Barrier to the Brain: The Components of a Model for Transport. *J. Mol. Neurosci.* **2001**, *16*, 181–194. [CrossRef] [PubMed]
42. Kaur, P.; Garg, T.; Rath, G.; Goyal, A.K. In situ nasal gel drug delivery: A novel approach for brain targeting through the mucosal membrane. *Artif. Cells Nanomed. Biotechnol.* **2015**, *44*, 1167–1176. [CrossRef] [PubMed]
43. Shilpa, P.; Vibhavari, C.; Chatur, M. Development of Valproic Acid Niosomal in situ Nasal Gel Formulation for Epilepsy. *Indian J. Pharm. Educ. Res.* **2013**, *47*, 31–41.
44. Ehrick, J.D.; Shah, S.A.; Shaw, C.; Kulkarni, V.S.; Coowanitwong, I.; De, S.; Suman, J.D. Considerations for the Development of Nasal Dosage Forms. *AAPS Adv. Pharm. Sci. Ser.* **2013**, *6*, 99–144.
45. Vidgren, M.T.; Kublik, H. Nasal delivery systems and their effect on deposition and absorption. *Adv. Drug Deliv. Rev.* **1998**, *29*, 157–177. [PubMed]
46. Dhuria, S.V.; Hanson, L.R.; Frey, W.H. Intranasal delivery to the central nervous system: Mechanisms and experimental considerations. *J. Pharm. Sci.* **2010**, *99*, 1654–1673. [CrossRef]
47. Kumar, S.; Wong, H.; Yeung, S.A.; Riggs, K.W.; Abbott, F.S.; Rurak, D.W. Disposition of valproic acid in maternal, fetal, and newborn sheep II: Metabolism and renal elimination. *Drug Metab. Dispos.* **2000**, *28*, 845–856. [PubMed]
48. Tang, W.; Borel, A.G.; Fujimia, T.; Abbott, F.S. Fluorinated analogs as mechanistic probes in valproic acid hepatotoxicity: Hepatic microvesicular steatosis and glutathione status. *Chem. Res. Toxicol.* **1995**, *8*, 671–682. [CrossRef] [PubMed]
49. Rossi, M.A. Targeting anti-epileptic drug therapy without collateral damage: Nanocarrier-based drug delivery. *Epilepsy Curr.* **2012**, *12*, 199–200. [CrossRef] [PubMed]
50. Alexis, F.; Pridgen, E.; Molnar, L.K.; Farokhzad, O.C. Factors affecting the clearance and biodistribution of polymeric nanoparticles. *Mol. Pharm.* **2008**, *5*, 505–515. [CrossRef]
51. Andersen, I.; Proctor, D.F. Measurement of nasal mucociliary clearance. *Eur. J. Respir. Dis. Suppl.* **1983**, *127*, 37–40. [PubMed]
52. Patel, R.B.; Patel, M.R.; Bhatt, K.K.; Patel, B.G. Formulation consideration and characterization of microemulsion drug delivery system for transnasal administration of carbamazepine. *Bull. Fac. Pharm. Cairo Univ.* **2013**, *51*, 243–253. [CrossRef]
53. Yang, Y.; Jing, Y.; Yang, J.; Yang, Q. Effects of intranasal administration with Bacillus subtilis on immune cells in the nasal mucosa and tonsils of piglets. *Exp. Ther. Med.* **2018**, *15*, 5189–5198. [PubMed]
54. Ezati, N.; Roberts, M.S.; Zhang, Q.; Moghimi, H.R. Measurement of Hansen Solubility Parameters of Human Stratum Corneum. *Iran J. Pharm. Res.* **2020**, *19*, 572–578.
55. Hansen, C.M.; Andersen, B. The affinities of organic solvents in biological systems. *Am. Ind. Hyg. Assoc.* **1988**, *49*, 301–308. [CrossRef]
56. Hussain, A.; Singh, S.K. Evidences for anti-mycobacterium activities of lipids and surfactants. *World J. Microbiol. Biotechnol.* **2015**, *32*, 7. [CrossRef] [PubMed]
57. Wolf, L.; Hoffmann, H.; Talmon, Y.; Teshigawara, T.; Watanabe, K. Cryo-TEM imaging of a novel microemulsion system of silicone oil with an anionic/nonionic surfactant mixture. *Soft Matter* **2010**, *6*, 5367. [CrossRef]

58. Baghdasaryan, A.; Wang, F.; Ren, F.; Ma, Z.; Li, J.; Zhou, X.; Grigoryan, L.; Xu, C.; Dai, H. Phosphorylcholine-conjugated gold-molecular clusters improve signal for Lymph Node NIR-II fluorescence imaging in preclinical cancer models. *Nat. Commun.* **2022**, *13*, 5613. [CrossRef]
59. Hansen, C.M. The significance of the surface condition in solutions to the diffusion equation: Explaining "anomalous" sigmoidal, Case II, and Super Case II absorption behavior. *Eur. Polym. J.* **2010**, *46*, 651–662. [CrossRef]
60. Nava, G.; Piñon, E.; Mendoza, L.; Mendoza, N.; Quintanar, D.; Ganem, A. Formulation and in vitro, ex vivo and in vivo evaluation of elastic liposomes for transdermal delivery of ketorolac tromethamine. *Pharm. Times* **2011**, *3*, 954–970.
61. Chen, H.; Chang, X.; Du, D.; Liu, W.; Liu, J.; Weng, T.; Yang, Y.; Xu, H.; Yang, X. Podophyllotoxin-loaded solid lipid nanoparticles for epidermal targeting. *J. Control. Release* **2006**, *110*, 296–306. [CrossRef]

gels

MDPI

Article

Photodynamic Therapy with Aminolevulinic Acid Enhances the Cellular Activity of Cells Cultured on Porcine Acellular Dermal Matrix Membranes Used in Periodontology

Morena Petrini [1,*,†], Emira D'Amico [1,†], Tania Vanessa Pierfelice [1], Gitana Maria Aceto [1], Maryia Karaban [2], Pietro Felice [2], Adriano Piattelli [3], Antonio Barone [4,5] and Giovanna Iezzi [1]

1 Department of Medical, Oral and Biotechnological Sciences, University G. d'Annunzio of Chieti-Pescara, 66100 Chieti, Italy; emira.damico@unich.it (E.D.); tania.pierfelice@unich.it (T.V.P.); gitana.aceto@unich.it (G.M.A.); gio.iezzi@unich.it (G.I.)
2 Department of Biomedical and Neuromotor Sciences, University of Bologna, 40126 Bologna, Italy; maryia.karaban@studio.unibo.it (M.K.); pietro.felice@unibo.it (P.F.)
3 School of Dentistry, Saint Camillus International University of Health and Medical Sciences, 00131 Rome, Italy; apiattelli51@gmail.com
4 Department of Surgical, Medical, Molecular Pathologies and of the Critical Needs, School of Dentistry, University of Pisa, 56126 Pisa, Italy; antonio.barone@unipi.it
5 Complex Unit of Stomatology and Oral Surgery, University Hospital of Pisa, 56126 Pisa, Italy
* Correspondence: morena.petrini@unich.it
† These authors contributed equally to this work.

Abstract: This study aims to test a photodynamic protocol based on a gel containing aminolevulinic acid followed by red-LED (ALAD-PDT) irradiation on human gingival fibroblasts (hGFs) and osteoblasts (hOBs) cultured on a porcine acellular dermal matrix membrane (PADMM). In the previous literature, ALAD-PDT showed solid antibacterial activity and proliferative induction on HGFs cultured on plates and HOBs cultured on a cortical lamina. PADMMs are used in dentistry and periodontology to treat gingival recessions and to increase the tissue thickness in the case of a thin biotype without the risks or postoperative discomfort associated with connective tissue grafts. However, one of the possible complications in this type of surgery is represented by bacterial invasion and membrane exposition during the healing period. We hypothesized that the addition of ALAD-PDT to PADMMs could enhance more rapid healing and decrease the risks connected with bacterial invasion. In periodontal surgery, PADMMs are inserted after a full-thickness flap elevation between the bone and the flap. Consequently, all procedures were performed in parallel on hOBs and hGFs obtained by dental patients. The group control (CTRL) was represented by the unexposed cells cultured on the membranes, group LED (PDT) were the cells subjected to 7 min of red LED irradiation, and ALAD-PDT were the cells subjected to 45 min of ALAD incubation and then to 7 min of red LED irradiation. After treatments, all groups were analyzed for MTT assay and subjected to histological examination at 3 and 7 days and to the SEM observations at 3, 7, and 14 days. Different bone mineralization assays were performed to quantify the effects of ALAD-PDT on hOBs: ALP activity, ALP gene expression, osteocalcin, and alizarin red. The effects of ALAD-PDT on hGFs were evaluated by quantifying collagen 1, fibronectin, and MMP-8. Results showed that ALAD-PDT promoted cellular induction, forming a dense cellular network on hOBs and hGFs, and the assays performed showed statistically significantly higher values for ALAD-PDT with respect to LED alone and CTRLs. In conclusion, ALAD-PDT could represent a promising aid for enhancing the healing of gingival tissues after PADMM applications.

Keywords: photodynamic therapy; aminolevulinic acid; red light; dermal matrix membrane; gingival fibroblasts; oral osteoblasts

Citation: Petrini, M.; D'Amico, E.; Pierfelice, T.V.; Aceto, G.M.; Karaban, M.; Felice, P.; Piattelli, A.; Barone, A.; Iezzi, G. Photodynamic Therapy with Aminolevulinic Acid Enhances the Cellular Activity of Cells Cultured on Porcine Acellular Dermal Matrix Membranes Used in Periodontology. *Gels* **2023**, *9*, 584. https://doi.org/10.3390/gels9070584

Academic Editors: Adina Magdalena Musuc, Magdalena Mititelu and Mariana Chelu

Received: 5 June 2023
Revised: 17 July 2023
Accepted: 18 July 2023
Published: 20 July 2023

1. Introduction

Aminolevulinic acid (5-ALA) is a precursor of protoporphyrin IX, a photosensitive substance that, after activation with red light at specific wavelengths, enhances the production of free radicals and singlet oxygen molecules [1]. It is currently applied in dermatology for medical and cosmetic purposes, including treating precancerous lesions [2]. The 5-ALA-mediated photodynamic therapy (PDT) is a non-invasive emerging method in dentistry as a diagnostic and therapeutic tool [3]. It has been demonstrated that ROS production is the critical factor in the effectiveness of PDT, and it depends on light dose and photosensitizer concentrations [4,5]. However, a recent study has demonstrated better tolerability of 5-ALA at a concentration of less than 20%, representing the indicated concentration PDT guidelines for managing oral leucoplakia [6,7]. In addition, the hydrophilic nature of 5-ALA limits its ability to cross the cellular membranes and penetrate the skin [8]. However, a novel gel for oral cavity applications has been recently formulated with a mixture of poloxamers to convey its active ingredient better. This thermosensitive gel containing 5% of 5-delta aminolevulinic acid (ALAD), associated with red-LED irradiation (ALAD-PDT), has shown antimicrobial activity against Gram-negative and positive bacteria against *C. albicans* and oral biofilm [9–11]. Clinical case reports have recently been published that show the application of ALAD-PDT during root canal disinfection in endodontic treatments and in periodontology [12,13]. In particular, for the treatment of periodontal pockets and peri-implantitis sites, the addition of this protocol respects the traditional gold standard scaling root planning (SRP), and was associated with a reduction in the total bacterial load. Despite the antibacterial and anticancer activity of 5-ALA-PDT, a pro-regenerative nature of 5-ALA-mediated PDT is recently emerging in the literature. Several studies reported the capacity of photodynamic therapy to accelerate skin wound healing by promoting epithelial stem cell functions [14–17]. However, little is known about the potential healing effects of 5-ALA-PDT in periodontal-related tissues. Our previous study shows a pro-proliferative activity of the gel ALAD in osteoblasts (hOBs) and fibroblasts (hGFs) on cell culture on plates and a cortical membrane for bone regeneration. Results showed increased hOBs proliferation, ALP activity, and bone mineralization proportional to ALAD concentration [18,19]. Several efforts have been made to find a biomaterial to be used in tissue and bone regeneration [20,21]. During periodontal and bone regeneration procedures, the membrane and the biomaterial interact with the host's cells, and their chemical and surface features play a fundamental role in accelerating the healing of the tissues and long-term success. It has been shown that the topographical characteristics of Porcine acellular dermal matrix membranes (PADMs) can influence cellular proliferation and behavior [22]. PADMs are composed of a three-dimensional acellular network of collagen types I and III and elastin of heterologous origin. PADMs permit the treatment of gingival recessions or to increase the tissue thickness in the case of a thin biotype without the risks or postoperative discomfort associated with connective tissue grafts (CTGs). Therefore, based on our previous in vitro studies, we hypothesized that applying ALAD-PDT protocol to PADMMs could enhance the healing process. At the same time, the risks connected with bacterial invasion could be decreased. In periodontal surgery, PADMMs are inserted after a full-thickness flap elevation between the bone and the flap. Consequently, this study evaluated the response of hOBs and hGFs, extracted from dental patients, cultured on the membranes, and subjected to ALAD-PDT. The effects of ALAD-PDT were investigated in terms of adhesion, growth, mineralization activity, and gene expression.

2. Results and Discussion

2.1. Cell Attachment on the PADMM after ALAD-PDT

Figure 1 shows the membrane without cells. Its surface appeared smoothed. Cell adhesion was evaluated by SEM at 3, 7, and 14 days (Figures 2 and 3). At 3 days, HGFs colonized the surface of the matrix, especially after the ALAD-PDT protocol, as observed in the different magnifications (Figure 2). At 1000×, the presence of elongated and spindle-shaped with cytoplasmic extensions and lamellipodia were recognizable. Confluence was reached after 7 days; at 14 days, the cells covered the matrix entirely. At 14 days, the morphology of cells was not notable because of the high density of cells.

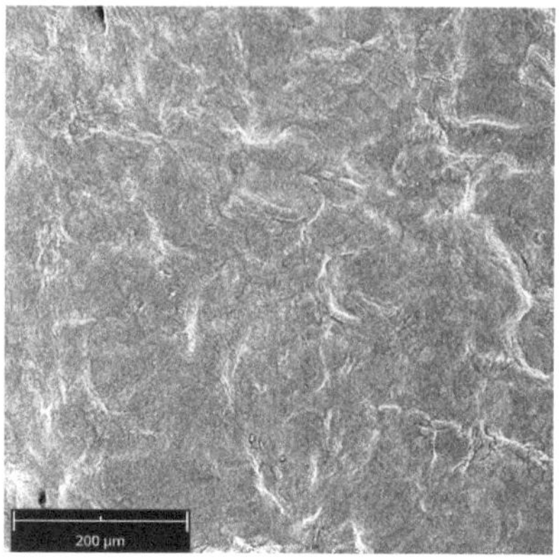

Figure 1. Scanning electron microscopy (SEM) images of the membrane (PADMM) without cells. Magnification 390×.

Regarding hOBs, ALAD-PDT promoted the attachment of cells to the membrane at 3 days, as observed at each magnification (Figure 3). After 7 days of culture, hOBs colonized the matrix surface. At 3000×, cellular extensions among osteoblasts were observed at 3 and 7 days. In contrast, at 14 days, it was impossible to recognize the cells' morphology because of their full confluence reached. The increased adhesion of cells to the PADMM membrane after ALAD-PDT may be due to the formulation of ALAD gel based on the mixture of poloxamers. Indeed, the topical application of 5-ALA is limited by its hydrophilic nature, preventing it from entering easily into the cells. The poloxamers-rich compound permits the thermosensitive gel ALAD to improve this limiting characteristic and makes it ideal for the application of oral mucosa. In addition, the formulation is temperature-dependent, allowing the transient state of ALAD from liquid to gel at a temperature above 28 °C; thus, ALAD acts like a glue that increases the attachment of cells, and it also counteracts the continuous secretion of saliva that could obstacle the adhesion of cells.

2.2. Gingival Fibroblasts and Oral Osteoblast Proliferation

The proliferation rate of gingival fibroblasts and oral osteoblasts cultured on the PADMMs was evaluated by MTT at 3, 7, and 14 days (Figures 2 and 3). As observed in Figure 2D, the growth of hGFs is time dependent. LED irradiation in the PDT group exerted a pro-proliferative activity compared to the control group after 7 and 14 days of culture. This is in line with a recent study that demonstrated that 655 and 808 nm diode lasers speed up the proliferation of dermal fibroblasts [23]. Recently, Rossi et al. documented an opposite effect, both inhibitory and stimulatory, in different doses, by treating dermal fibroblasts with a blue LED light at 420 nm [24]. In the present study, although LED was able to enhance cell proliferation, the maximum stimulatory effect was given by ALAD-PDT at each time. However, the results were statistically significant only at 7 and 14 days compared to CTRL and PDT groups. Jang et al. reported that PDT-induced intracellular ROS in dermal fibroblasts leads to increased proliferation via ERK pathway activation [25]. In hOBs, ALAD-PDT also increased cell growth, but not in a time-dependent manner (Figure 3D), and it was statistically significant only after 14 days. In the literature, a contradictory effect of 5-ALA-PDT on osteoblasts was reported. Kushibiki et al. demonstrated that 5-ALA, combined with low-dose light, can promote osteoblast differentiation via the activation of AP-1 [26]. On the contrary, Egli et al. showed an inhibitory effect of 5-ALA-PDT on fibroblasts and osteoblasts viability [27].

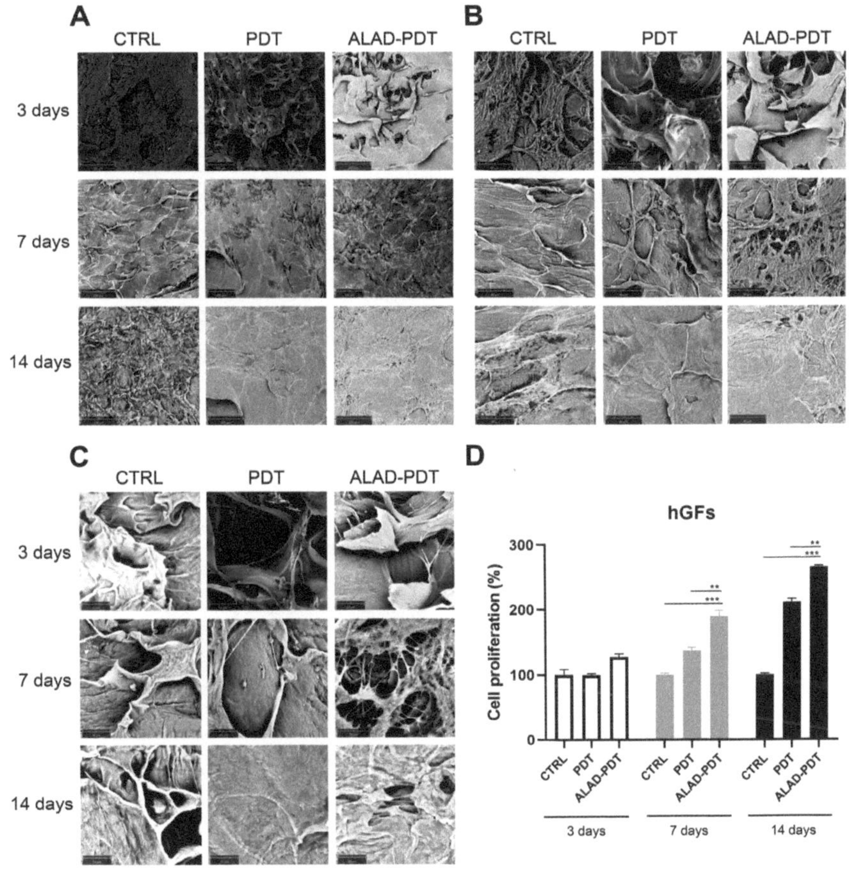

Figure 2. Scanning electron microscopy (SEM) images of hGFs cultured on the PADMM and exposed to ALAD-PDT at 3, 7, and 14 days. (**A**) Magnification 390×; (**B**) Magnification 1000×; (**C**) Magnification 3000×; (**D**) Cell proliferation of hGFs cultured on the PADMM at 3, 7, and 14 days. (** $p < 0.001$; *** $p < 0.0001$).

2.3. Cell Interaction with the PADMM after ALAD-PDT

The interaction of specialized cells to a membrane is an essential characteristic of its physiological functions, and in this study, it was evaluated by histology at 400× at 3 and 7 days (Figure 4 and 5). At 3 days, after ALAD-PDT protocol, hGFs grew on the edges of the membrane by establishing connections among cells (Figure 4C). After 7 days, PDT also promoted the interaction of hGFs (Figure 4E), and after ALAD-PDT cells colonized the inside of the membrane (Figure 4F). At 3 days, hOBs appeared roundish in shape, and after 7 days their shape became elongated (Figure 5). A recent publication reported an inhibitory effect of toluidine blue-mediated PDT, used at a concentration of 50% on the migratory activity of gingival fibroblasts [28]. In contrast, the thermosensitive gel used in the present study and which contains 5% of 5-ALA as the pro-drug able to induce the accumulation of PpIX inside the cells, seemed to enhance the interaction of the cells and favor their migratory activity.

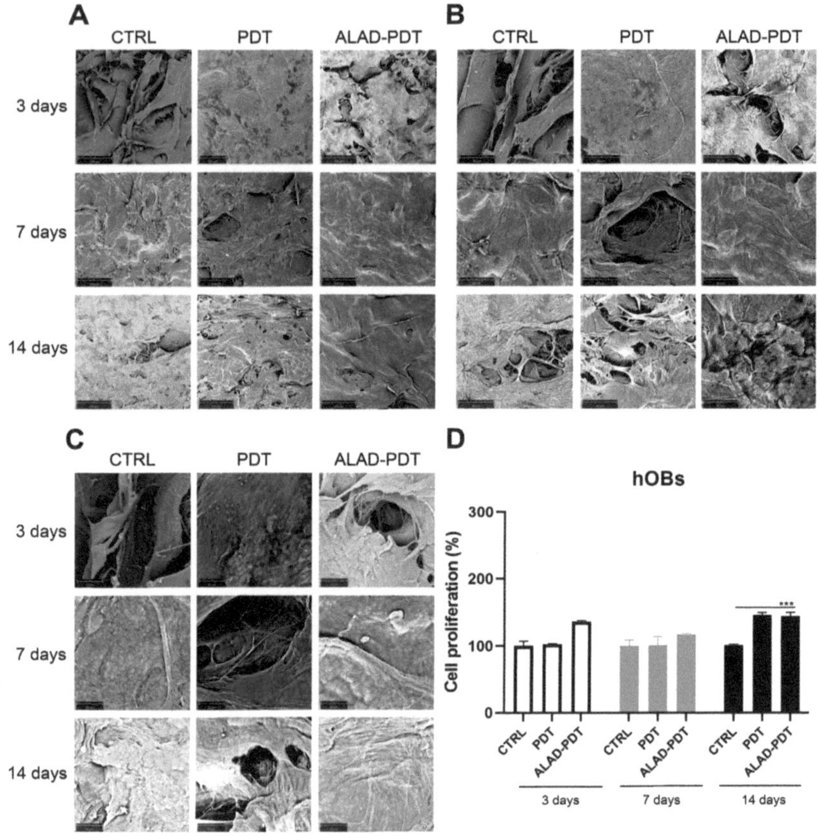

Figure 3. Scanning electron microscopy (SEM) images of hOBs cultured on the PADMM and exposed to ALAD-PDT at 3, 7, and 14 days. (**A**) (Magnification = 390×; (**B**) Magnification, 1000×; (**C**) Magnification and 3000×; (**D**). Cell proliferation of hOBs cultured on the PADMM at 3, 7, and 14 days. (*** $p < 0.0001$).

2.4. Mineralization

To determine the presence of calcific deposition in osteoblast cultures, calcium deposits were evaluated qualitatively by Alizarin Red staining (ARS) and quantitatively by CPC after 14 days of culture (Figure 7). Brighter red mineralized nodules were observed after ALAD-PDT protocol compared to CTRL (Figure 7A), indicating more mineralization activity in osteoblasts treated with ALAD-PDT. The PDT group also showed a similar intensity of red to the ALAD-PDT group. Quantization with CPC confirmed the qualitative results. Therefore, the percentage of calcium deposition was statistically higher in PDT and ALAD-PDT groups than in the CTRL group ($p < 0.0001$). However, there was no difference in calcium deposits between PDT and ALAD-PDT (Figure 7B). Yang et al. conversely observed more mineralized nodule formation in laser-treated cells [29]. In addition, in our previous study, an increase in the mineralization and calcium deposits was observed when osteoblasts were cultured on a cortical, rigid, and collagenated bone lamina and subjected to ALAD-PDT [30].

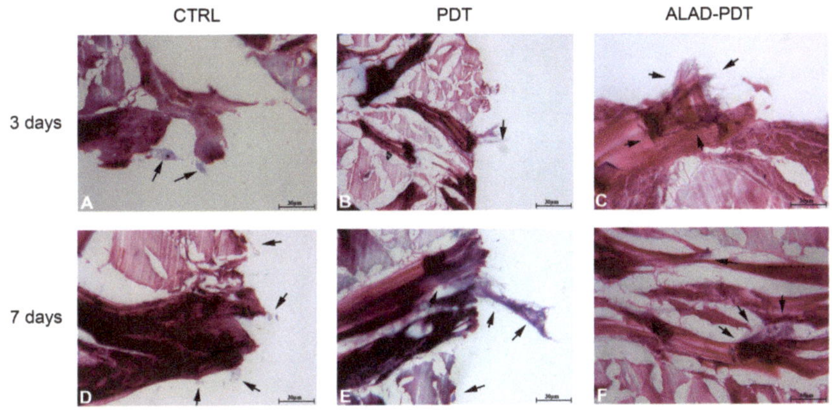

Figure 4. hGFs interaction with PADMM. hGFs grew on the edges of the membrane at 3 days (**A–C**). At 7 days, they colonize the inside of the membrane (**D–F**). Magnification: 400×. The arrows pointed to the cells.

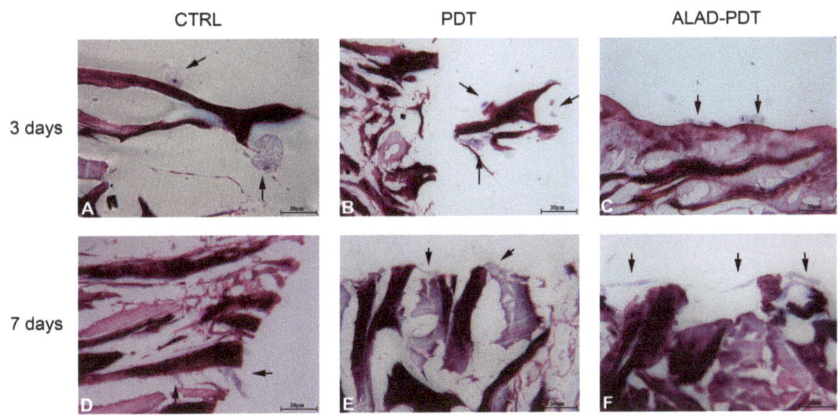

Figure 5. Interaction between hOBs and PADMM. At 3 days, osteoblast showed a round shape (**A–C**); at 7 days, they appeared more elongated (**D–F**). Magnification: 400×. The arrows pointed to the cells.

2.5. ALP Activity

The levels of Alkaline Phosphatase (ALP), as the main osteoblastic marker, were evaluated at 7 days in hOBs (Figure 6). ALP relative activity increased after Led irradiation in the PDT group, although the ALAD-PDT induced a significative increment of ALP compared to both CTRL group ($p < 0.0001$) and to PDT group ($p < 0.0001$). Yang et al. reported a similar result by treating gingival fibroblasts with 5 μM methylene blue PDT. However, they obtained a higher ALP activity in the laser treatment group [29]. Whereas in our study, only the synergy between ALAD gel and light-LED significantly promoted the activity of ALP in osteoblasts.

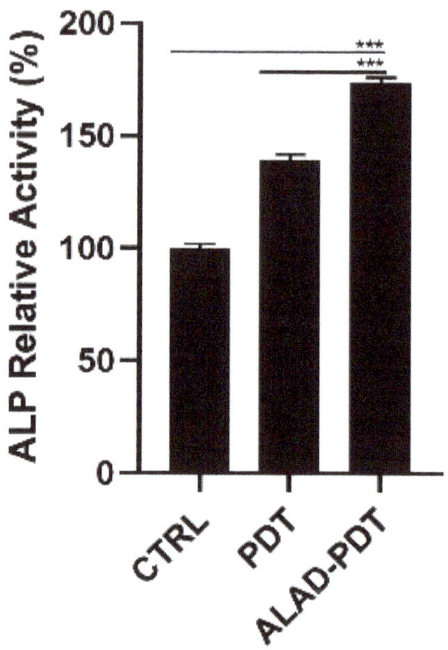

Figure 6. ALP activity of hOBs cultured for 7 days and exposed to ALAD-PDT protocol. Increased levels were observed after the treatment (*** $p < 0.0001$).

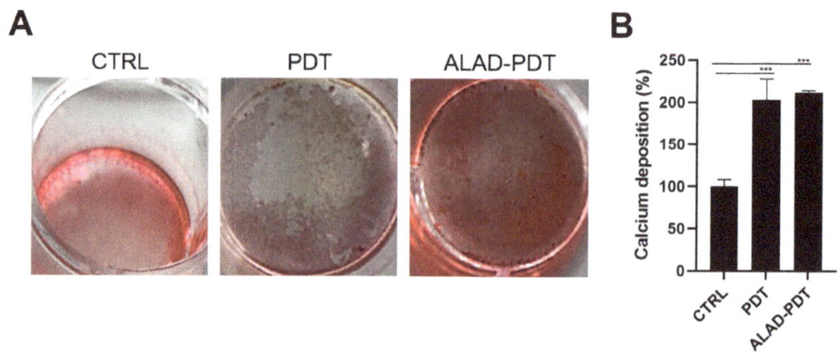

Figure 7. Mineralization was evaluated at 14 days in hOBs seeded on the membrane and exposed to ALAD-PDT. The qualitative analysis was performed by ARS (**A**), while the quantization was carried out by CPC (**B**). ALAD-PDT promoted the highest calcium deposits compared to CTRL. (*** $p < 0.0001$).

2.6. Gene Expression of Gingival Fibroblasts and Oral Osteoblast Cultured on the PADMM and Exposed to Photodynamic Therapy

The gene expression of gingival fibroblasts and oral osteoblasts cultured on the matrix and exposed to ALAD-PDT protocol was evaluated at 3, 7, and 14 days. Fibronectin 1 (FN1), Collagen 1 (COL-1), and metalloproteinase 8 (MMP8) expression were evaluated for hGFs (Figure 8), while ALP and osteocalcin (OCN) for hOBs (Figure 9). Fibronectin and Collagen I are well known to have a tight relationship both related to connective tissue regulation. The comparative analysis showed that the expression of FN1 and COL-1 was statistically higher in PDT and ALAD-PDT groups than in the CTRL group at every time point (Figure 8A,B). Although, the highest values for both FN1 and COL-1 have been observed at 14 days. Any statistical difference was observed between PDT and ALAD-PDT. MMP8 is a member of the metalloproteinase family, specifically involved in both the degradation of matrix and the wound-healing-processes [31]. MMP8 was not modulated at 3 and 7 days with respect to CTRL group (Figure 8C). ALAD-PDT and PDT slightly increased MMP8 expression only at 14 days ($p < 0.05$). Karrer et al. found that 5-ALA-mediated PDT induced in fibroblast the expression of metalloproteinases MMP1, MMP2, and MMP3 but not Collagen I mRNA expression [32]. In contrast, Yang et al. reported an enhancement of fibroblast-related genes FN1 and COL1 after applying 5 μM methylene blue-PDT on cells. Although, in the same study, an inhibitory effect on gene expression of fibroblasts treated with higher concentrations than 5 μM was reported [29]. Our results concerning FN1, COL1, and MMP8 expression could indicate a stimulatory effect of ALAD-PDT for the renewal of the extracellular matrix (ECM).

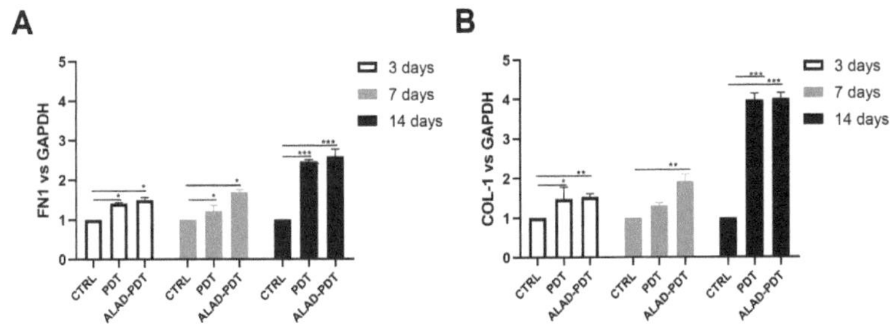

Figure 8. Real-time PCR of fibroblasts (HGFs) cultured on the matrix and treated with ALAD-PDT for Fibronectin 1 (FN1) (**A**), Collagen 1 (COL-1) (**B**), Metalloprotease 8 (MMP8) (**C**) at 3, 7, and 14 days post-seeding. ALAD-PDT induced increased expression of FN1, COL-1, and MMP8 at 14 days (* $p < 0.05$; ** $p < 0.001$; *** $p < 0.0001$).

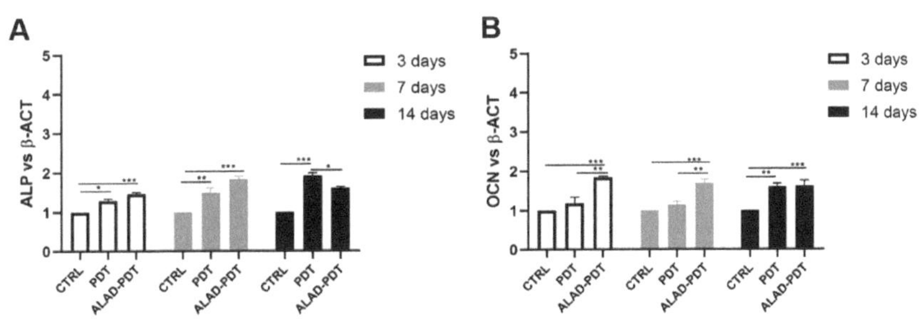

Figure 9. Real-time PCR of osteoblasts (hOBs) seeded on the PADMM and treated with ALAD-PDT for genes encoding Alkaline Phosphatase (ALP) (**A**) and Osteocalcin (OCN) (**B**) at 3, 7, and 14 days post-seeding. ALP and OCN were more expressed in the ALAD-PDT group than in CTRL. (* $p < 0.05$; ** $p < 0.001$; *** $p < 0.0001$).

The expression of bone tissue-specific genes ALP and OCN, which are tightly regulated at different stages of osteoblasts, was evaluated at 3, 7, and 14 days (Figure 9). Although ALP gene expression was statistically augmented after ALAD-PDT protocol compared to CTRL at every time points, the expression was not modulated during time with the highest value at 7 days (Figure 9A). Interestingly, PDT group showed the maximum expression at 14 days. Osteocalcin expression is usually correlated to the ALP expression, and in this study, OCN showed a similar trend to ALP. OCN-relative mRNA was up-regulated in the ALAD-PDT group compared to CTRL and PDT, with the highest value at an early stage of osteoblasts. Although, at 14 days, OCN expression showed similar levels in PDT and ALAD-PDT groups (Figure 9B).

Together, these findings indicated that the ALAD-PDT protocol may be applied during guided tissue regeneration (GTR) and guided bone regeneration (GBR) to improve the performance of PADMM in the periodontal tissue's augmentation. The thermosensitive

ALAD gel, formulated to improve the topical application of 5-ALA on the oral mucosa, where the saliva represents an additional limit to the retention of 5-ALA, enhanced the adhesion and the interaction of periodontal-related cells cultured on the membrane.

The results of this in vitro study could represent the bases for an in vivo study in which ALAD-PDT could be applied as an adjuvant during GBR and GTR.

3. Conclusions

In conclusion, ALAD-PDT applied on the gingival fibroblasts and oral osteoblasts cultured on a porcine dermal matrix membrane promoted the proliferation, mineralization, and expression of functional genes such as FN1 and COL1 in hGFs, and ALP and OCN in hOBs.

4. Materials and Methods

4.1. Study Design

The effects of the photodynamic protocol (ALAD-PDT) on the human gingival fibroblasts (hGFs) and human oral osteoblast (hOBs) cultured on the PADMM (Cellis Dental, La Rochelle, France) were performed using MTT assay, SEM and gene expression at 3, 7 and 14 days, histology at 3 and 7 days, Alkaline Phosphatase levels (ALP) at 7 days and Alizarin Red Staining (ARS) at 14 days. The following experimental conditions were used:

i. PDT: cells cultured on the membrane and exposed to 630 nm LED for 7 min;
ii. ALAD-PDT: cells cultured on the membrane and treated with a gel containing 5% of 5-aminolevulinic acid (ALAD) for 45 min and irradiated with red light (LED) for 7 min;
iii. And CTRL: cells cultured on the membrane PADM.

The PADMM, used in the in vitro tests, was cut into squares of 5 mm $\times$ 5 mm under sterilized conditions, and they were hydrated with NaCl 0.9% three times before the culture of cells.

Photodynamic therapy is based on the use of a gel containing 5-aminolevulinic acid at the concentration of 5% (ALAD) (AlphaStrumenti, Melzo (MI), Italy) and a 630 nm LED (PDT) (AlphaStrumenti, Melzo (MI), Italy). ALAD gel is a thermosetting product, protected by a patent (PCT/IB2018/060368, 12.19.2018), that remains liquid at temperatures below 28 °C, becoming gel at higher temperatures. Further, 630 nm LED has an intensity of 380 mW/cm^2 with a light dose of 23 J/cm^2.

4.2. Cell Culture

Primary human gingival fibroblasts (hGFs) were purchased from ATCC (Manassas, VA, USA), and human oral osteoblasts (hOBs) were extracted from bone fragments of patients treated at Dental Clinic of University Gabriele d'Annunzio (Ethical Committee reference numbers: BONEISTO N. 22 of 10.07.2021) according to the protocol described by Pierfelice TV and co-workers [33]. In particular, during the procedures of dental implant insertion, during the implant site preparation, a trephine bur was used to sample a bone fragment. Briefly, bone fragments were subjected to three enzymatic digestions at 37 °C for 20, 40, and 60 min using collagenase type 1A (Sigma-Aldrich, St. Louis, MO, USA) and trypsin-EDTA 0.25% (Corning, New York, NY, USA). After each digestion, this solution was centrifuged at 1200 rpm for 10 min, and the pellet obtained was transferred into a T25 culture flask with low-glucose (1 g/L) DMEM supplemented with 10% FBS (SIAL, Rome, Italy), 1% antibiotics (100 μg/mL^{-1} streptomycin and 100 IU/mL^{-1} penicillin), and 1% L-glutamine (Corning) at 5% CO_2 and 37 °C. The medium was changed every 4–5 days. hGFs and hOBs were cultured in DMEM low glucose (Corning) supplemented with 10% fetal bovine serum (FBS) (SIAL), 1% penicillin, and streptomycin (Corning) at 37 °C and 5% CO_2. Both cell lines, hOBs and hGFs, were used from the 3rd and 5th passages. Figure 10 shows hOBs at the optical microscope.

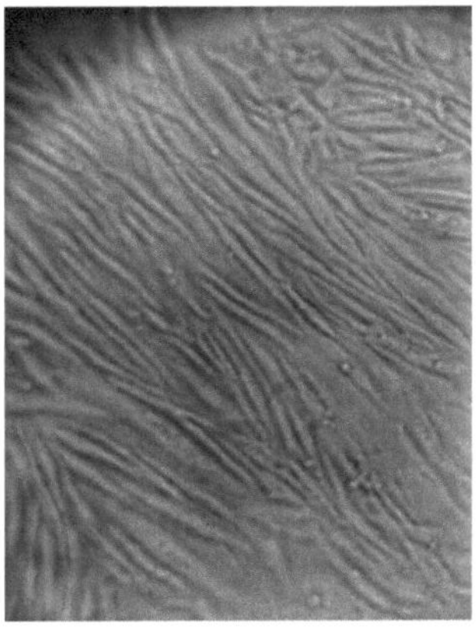

Figure 10. Human oral osteoblasts (hOBs) at optical microscopy at 5th passage. Magnification: 10×.

4.3. Cell Proliferation Assay

MTT evaluated the growth of cells seeded on the matrix. A total of 10^4 cells/membrane were seeded on the matrix, exposed to ALAD-PDT, and cultured for 3, 7, and 14 days. The choice of cell density was based on previous studies and considering the area of the specimen [34,35]. Then, the MTT assay (Sigma Aldrich, St. Louis, MO, USA) was used according to the manufacturer's instructions. At the end of each incubation period, a solution of 0.5 mg/mL MTT (Sigma Aldrich, St. Louis, MO, USA) was added to each well, and then the cells were incubated for 4 h at 37 °C and 5% CO_2. A solubilization solution was added to each well to dissolve the insoluble formazan. Then, the plate was read at 570 and 630 nm by a microplate reader (Synergy H1 Hybrid BioTek Instruments, Winooski, VT, USA) to determine the absorbance (A). Then, a subtraction A = A570 − A630 was performed. The results were expressed as percentages and calculated with respect to control (CTRL).

4.4. Cell Attachment

The adhesion of cells was observed using scanning electron microscope (SEM) images. The cells at 10^4 cells/membrane density were seeded on the matrix, treated with ALAD-PDT, and cultured for 3, 7, and 14 days. Loosely adherent cells were removed from the experiment wells by washing twice with 0.1 M PBS (pH 7.4). Thereafter, cells were fixed with 2.5% glutaraldehyde for 1 h and dehydrated using sequential concentrations of ethanol (40, 50, 75, 95, 100%). Before the observation, they were sputtered with gold and observed at 390×, 1000×, and 3000× using SEM (Philips XL20; Philips Inc., Eindhoven, the Netherlands) at 15 kV.

4.5. Histological Analysis

The interaction of cells with the membrane was evaluated by histological analysis. A total of 10^4 cells/membranes were seeded on the top of the matrix. Cells were exposed to ALAD-PDT protocol, and the culture was carried out for 7 days. Each specimen was fixed with 10% buffered formalin and dehydrated in an ascending alcohol series. They were then polymerized in a glycol methacrylate resin (Technovit 7200 VLC; Kulzer, Wehrheim, Germany). The sections, about 30 μm in width, were stained with fuchsin and toluidine blue. The images were taken by an optical microscope (Leica, Wetzlar, Germany) at 400×.

4.6. ALP Assay

ALP assay was performed to evaluate ALP levels in hOBs using a colorimetric kit AB83369 (Abcam Inc., Cambridge, UK) based on the cleavage of *p*-nitrophenyl phosphate (pNPP). hOBs at a density of 5 × 10^4 were seeded on top of the matrix, treated with

ALAD-PDT, and cultured for 7 days. Thereafter, the assay was performed according to the manufacturer's instructions. After 7 days, the cells were washed three times with PBS and resuspended in assay buffer. The cell suspension was then homogenized by a Tissue Rupture device (QIAGEN, Hilden, Germany) and centrifuged at 10,000× *g* for 15 min. The relative ALP activity of the supernatant was measured using pNPP, as the substrate, for 1 h. After incubation, the reaction was stopped and the relative ALP activity was quantified as an absorbance value at 405 nm.

4.7. Mineralization

The deposition of calcium nodules was evaluated by ARS (Sigma Aldrich, St. Louis, MO, USA) and quantized with 10% Cetylpyridinium Chloride (CPC) solution (Sigma-Aldrich). Then, 5×10^4 hOBs/membrane were seeded on the membrane, exposed to ALAD-PDT protocol, and cultured for 14 days. The cells were fixed with 2.5% glutaraldehyde and then stained with ARS solution for 1h. After 1 h, deionized water was used to remove the excess dye, and the presence of mineral nodules stained by red color was observed. Then, to quantize calcium deposits, 1 mL of 10% CPC solution was added, and after 1 h the absorbance was read at 540 nm in a microplate reader (Synergy H1 Hybrid BioTek Instruments).

4.8. Gene Expression

The gene expression of different markers of osteoblastic cells (ALP and OCN) and markers of fibroblast activity (COL1, FN1, and MMP8) were evaluated by RT-PCR. Total RNA was isolated using the Trifast reagent (EuroClone, Pero (MI), Italy), and RNA was quantified on a Nanophotometer NP80 spectrophotometer (Implen NanoPhotometer, Westlake Village, CA, USA) for analysis of RNA integrity, purity, and concentration. Then, the GoTaq®2 Step RT-qPCR Kit (Promega, Madison, WI, USA) was used to obtain complementary DNA (cDNA), and SYBR Green (GoTaq® 2 Step RT-qPCR Kit, Promega) was used to perform RT-qPCR according to manufacturer's instructions. Gene expression was determined using Quant Studio 7 Pro Real-Time PCR System (ThermoFisher, Waltham, MA, USA). The results were normalized to Glyceraldehyde-3-Phosphate Dehydrogenase (GAPDH for hGFs and to β-actin (β-ACT) for hOBs using the $2^{-\Delta\Delta ct}$ method. Primer sequences are reported in Table 1.

Table 1. Primer sequences used in RT-qPCR.

Gene	Forward Primer (5′-3′)	Reverse Primer (5′-3′)
OCN	TCAGCCAACTCGTCACAGTC	GGCGCTACCTGTATCAATGG
ALP	AATGAGTGAGTGACCATCCTGG	GCACCCCAAGACCTGCTTTAT
COL1	AGTCAGAGTGAGGACAGTGAATTG	CACATCACACCAGGAAGTGC
FN1	GGAAAGTGTCCCTATCTCTGATACC	AATGTTGGTGAATCGCAGGT
MMP8	ATGTTCTCCCTGAAGACGCT	AGACTGATACTGGTTGCTTGGT
B-ACT	CCAGAGGCGTACAGGGATAG	GAGAAGATGACCCAGGACTCTC
GAPDH	ACGGGAAGCTTGTCATCAAT	GGAGGGATCTCGCATTTCTT

4.9. Statistical Analysis

The data are reported as means ± standard deviation (SD). Statistical analyses were performed using the GraphPad Prism8 (GraphPad Software San Diego, CA, USA), and ANOVA and post hoc Tukey tests were adopted. A p-value < 0.05 was considered significant. All experiments were performed in biological triplicates and repeated three times.

Author Contributions: Conceptualization: M.P., G.I., A.B. and P.F. Methodology: E.D. and T.V.P. Validation: M.P., G.I., M.K. and P.F. Formal analysis: G.M.A., G.I., A.B., P.F. and A.B. Investigation: E.D. and T.V.P. Resources: M.P. and G.I. Data curation: E.D., T.V.P., M.K. and P.F. Writing—original draft preparation: E.D., M.P. and T.V.P. Writing—review and editing: E.D., M.P., G.I., G.M.A. and A.P.

Supervision: G.I., M.P. and A.P. Project administration: G.I. and P.F. Funding acquisition: G.I. and M.P. All authors have read and agreed to the published version of the manuscript.

Funding: This work was supported by G.I. and M.P., and the FAR GRANT University of Chieti–Pescara Fund.

Institutional Review Board Statement: The study was conducted in accordance with the Declaration of Helsinki and approved by the Institutional Review Board (or Ethics Committee) of the Ethical Committee of The University G. d'Annunzio Chieti-Pescara, protocol BONEISTO N. 22 of 10 July 2021.

Informed Consent Statement: Informed consent was obtained from all subjects involved in the study.

Data Availability Statement: MDPI Research Data Policies.

Acknowledgments: Tania Vanessa Pierfelice has a doctorate fellowship (code n. DOT1353500) in the framework of PON RI 201/2020, Action I.1-"Innovative PhDs with industrial characterization", funded by the Ministry of University and Research (MUR), Italy, FSE-FESR. Aladent gel was supported by ALPHA Strumenti s.r.l. (MelzoMI), Italy. The membrane was supported by Cellis Dental (La Rochelle, France).

Conflicts of Interest: The authors declare no conflict of interest.

References

1. Rick, K.; Sroka, R.; Stepp, H.; Kriegmair, M.; Huber, R.M.; Jacob, K.; Baumgartner, R. Pharmacokinetics of 5-Aminolevulinic Acid-Induced Protoporphyrin IX in Skin and Blood. *J. Photochem. Photobiol. B* **1997**, *40*, 313–319. [CrossRef] [PubMed]
2. Ablon, G. Phototherapy with Light Emitting Diodes: Treating a Broad Range of Medical and Aesthetic Conditions in Dermatology. *J. Clin. Aesthet. Dermatol.* **2018**, *11*, 21. [PubMed]
3. Stájer, A.; Kajári, S.; Gajdács, M.; Musah-Eroje, A.; Baráth, Z. Utility of Photodynamic Therapy in Dentistry: Current Concepts. *Dent. J.* **2020**, *8*, 43. [CrossRef] [PubMed]
4. Harris, F.; Pierpoint, L. Photodynamic Therapy Based on 5-Aminolevulinic Acid and Its Use as an Antimicrobial Agent. *Med. Res. Rev.* **2012**, *32*, 1292–1327. [CrossRef] [PubMed]
5. Pesce, M.; Tatangelo, R.; La Fratta, I.; Rizzuto, A.; Campagna, G.; Turli, C.; Ferrone, A.; Franceschelli, S.; Speranza, L.; Patruno, A.; et al. Aging-Related Oxidative Stress: Positive Effect of Memory Training. *Neuroscience* **2018**, *370*, 246–255. [CrossRef]
6. Serini, S.M.; Cannizzaro, M.V.; Dattola, A.; Garofalo, V.; Del Duca, E.; Ventura, A.; Milani, M.; Campione, E.; Bianchi, L. The Efficacy and Tolerability of 5-Aminolevulinic Acid 5% Thermosetting Gel Photodynamic Therapy (PDT) in the Treatment of Mild-to-Moderate Acne Vulgaris. A Two-Center, Prospective Assessor-Blinded, Proof-of-Concept Study. *J. Cosmet. Dermatol.* **2019**, *18*, 156–162. [CrossRef]
7. Chen, Q.; Dan, H.; Tang, F.; Wang, J.; Li, X.; Cheng, J.; Zhao, H.; Zeng, X. Photodynamic Therapy Guidelines for the Management of Oral Leucoplakia. *Int. J. Oral. Sci.* **2019**, *11*, 14. [CrossRef]
8. Zheng, Y.; Fan, W.; Jiang, L.; Lu, Y. Sonophoresis Enhances the Skin Penetration of 5-Aminolevulinic Acid: A Promising Pretreatment for Photodynamic Therapy. *Exp. Dermatol.* **2022**, *31*, 1939–1943. [CrossRef]
9. Greco, G.; Di Piazza, S.; Chan, J.; Zotti, M.; Hanna, R.; Gheno, E.; Zekiy, A.O.; Pasquale, C.; De Angelis, N.; Amaroli, A. Newly Formulated 5% 5-Aminolevulinic Acid Photodynamic Therapy on Candida Albicans. *Photodiagn. Photodyn. Ther.* **2020**, *29*, 101575. [CrossRef]
10. Lauritano, D.; Moreo, G.; Palmieri, A.; Della Vella, F.; Petruzzi, M.; Botticelli, D.; Carinci, F. Photodynamic Therapy Using 5-Aminolevulinic Acid (Ala) for the Treatment of Chronic Periodontitis: A Prospective Case Series. *Appl. Sci.* **2022**, *12*, 3102. [CrossRef]
11. Radunović, M.; Petrini, M.; Vlajic, T.; Iezzi, G.; Di Lodovico, S.; Piattelli, A.; D'Ercole, S. Effects of a Novel Gel Containing 5-Aminolevulinic Acid and Red LED against Bacteria Involved in Peri-Implantitis and Other Oral Infections. *J. Photochem. Photobiol. B* **2020**, *205*, 111826. [CrossRef] [PubMed]
12. Carlesi, T.; Petrini, M.; Plotino, G.; Piattelli, A.; D'Ercole, S. Photodynamic Therapy with 5-Aminolevulinic Acid and Red LED, 660 in Vitro and in Vivo Studies. *J. Endod.* **2021**, *7*, 1–32.
13. Rossi, R.; Rispoli, L.; Lopez, M.A.; Netti, A.; Petrini, M.; Piattelli, A. Photodynamic Therapy by Mean of 5-Aminolevulinic Acid for the Management of Periodontitis and Peri-Implantitis: A Retrospective Analysis of 20 Patients. *Antibiotics* **2022**, *11*, 1267. [CrossRef] [PubMed]
14. Yang, Z.; Hu, X.; Zhou, L.; He, Y.; Zhang, X.; Yang, J.; Ju, Z.; Liou, Y.C.; Shen, H.M.; Luo, G.; et al. Photodynamic Therapy Accelerates Skin Wound Healing through Promoting Re-Epithelialization. *Burn. Trauma* **2021**, *9*, tkab008. [CrossRef]
15. Huang, J.; Wu, S.; Wu, M.; Zeng, Q.; Wang, X.; Wang, H. Efficacy of the Therapy of 5-Aminolevulinic Acid Photodynamic Therapy Combined with Human Umbilical Cord Mesenchymal Stem Cells on Methicillin-Resistant Staphylococcus Aureus-Infected Wound in a Diabetic Mouse Model. *Photodiagn. Photodyn. Ther.* **2021**, *36*, 102480. [CrossRef]

16. Pallavi, P.; Girigoswami, A.; Girigoswami, K.; Hansda, S.; Ghosh, R. Photodynamic Therapy in Cancer. In *Handbook of Oxidative Stress in Cancer: Therapeutic Aspects*; Springer: Berlin/Heidelberg, Germany, 2022; pp. 1–24. [CrossRef]
17. Pallavi, P.; Sharmiladevi, P.; Haribabu, V.; Girigoswami, K.; Girigoswami, A. A Nano Approach to Formulate Photosensitizers for Photodynamic Therapy. *Curr. Nanosci.* **2021**, *18*, 675–689. [CrossRef]
18. Pierfelice, T.V.; D'Amico, E.; Iezzi, G.; Petrini, M.; Schiavone, V.; Santalucia, M.; Pandolfi, A.; D'Arcangelo, C.; Piattelli, A.; Di Pietro, N. Effect of a 5-Aminolevulinic Acid Gel and 660 Nm Red LED Light on Human Oral Osteoblasts: A Preliminary in Vitro Study. *Lasers Med. Sci.* **2022**, *37*, 3671–3679. [CrossRef]
19. Pierfelice, T.V.; D'Amico, E.; Petrini, M.; Pandolfi, A.; D'Arcangelo, C.; Di Pietro, N.; Piattelli, A.; Iezzi, G. The Effects of 5% 5-Aminolevulinic Acid Gel and Red Light (ALAD-PDT) on Human Fibroblasts and Osteoblasts. *Gels* **2022**, *8*, 491. [CrossRef]
20. Jamnezhad, S.; Asefnejad, A.; Motififard, M.; Yazdekhasti, H.; Kolooshani, A.; Saber-Samandari, S.; Khandan, A. Development and Investigation of Novel Alginate-Hyaluronic Acid Bone Fillers Using Freeze Drying Technique for Orthopedic Field. *Nanomed. Res. J.* **2020**, *5*, 306–315.
21. Foroutan, S.; Hashemian, M.; Khosravi, M.; Nejad, M.G.; Asefnejad, A.; Saber-Samandari, S.; Khandan, A. A Porous Sodium Alginate-$CaSiO_3$ Polymer Reinforced with Graphene Nanosheet: Fabrication and Optimality Analysis. *Fibers Polym.* **2021**, *22*, 540–549. [CrossRef]
22. Felice, P.; D'Amico, E.; Pierfelice, T.V.; Petrini, M.; Barausse, C.; Karaban, M.; Barone, A.; Iezzi, G. Osteoblasts and Fibroblasts Interaction with a Porcine Acellular Dermal Matrix Membrane. *Int. J. Mol. Sci.* **2023**, *24*, 3649. [CrossRef]
23. Topaloglu, N.; Özdemir, M.; Çevik, Z.B.Y. Comparative Analysis of the Light Parameters of Red and Near-Infrared Diode Lasers to Induce Photobiomodulation on Fibroblasts and Keratinocytes: An in Vitro Study. *Photodermatol. Photoimmunol. Photomed.* **2021**, *37*, 253–262. [CrossRef] [PubMed]
24. Rossi, F.; Magni, G.; Tatini, F.; Banchelli, M.; Cherchi, F.; Rossi, M.; Coppi, E.; Pugliese, A.M.; Rossi Degl'Innocenti, D.; Alfieri, D.; et al. Photobiomodulation of Human Fibroblasts and Keratinocytes with Blue Light: Implications in Wound Healing. *Biomedicines* **2021**, *9*, 41. [CrossRef] [PubMed]
25. Jang, Y.H.; Koo, G.B.; Kim, J.Y.; Kim, Y.S.; Kim, Y.C. Prolonged Activation of ERK Contributes to the Photorejuvenation Effect in Photodynamic Therapy in Human Dermal Fibroblasts. *J. Investig. Dermatol.* **2013**, *133*, 2265–2275. [CrossRef]
26. Kushibiki, T.; Tu, Y.; Abu-Yousif, A.O.; Hasan, T. Photodynamic Activation as a Molecular Switch to Promote Osteoblast Cell Differentiation via AP-1 Activation. *Sci. Rep.* **2015**, *5*, 13114. [CrossRef]
27. Egli, R.J.; Schober, M.; Hempfing, A.; Ganz, R.; Hofstetter, W.; Leunig, M. Sensitivity of Osteoblasts, Fibroblasts, Bone Marrow Cells, and Dendritic Cells to 5-Aminolevulinic Acid Based Photodynamic Therapy. *J. Photochem. Photobiol. B* **2007**, *89*, 70–77. [CrossRef]
28. WD, G. Therapeutic Window of Photodynamic Treatment (PDT) in Conservative Periodontal Therapy-Analysis of Cell Migration within A Three Dimensional Collagen Matrix-. *Online J. Dent. Oral Health* **2021**, *4*. [CrossRef]
29. Yang, R.; Guo, S.; Xiao, S.; Ding, Y. Enhanced Wound Healing and Osteogenic Potential of Photodynamic Therapy on Human Gingival Fibroblasts. *Photodiagn. Photodyn. Ther.* **2020**, *32*. [CrossRef] [PubMed]
30. Pierfelice, T.V.; D'Amico, E.; D'Ercole, S.; Lepore, S.; Piattelli, A.; Barone, A.; Iezzi, G.; Petrini, M. Functionalization of A Cortical Membrane with a Photodynamic Protocol. *J. Funct. Biomater.* **2023**, *14*, 133. [CrossRef]
31. Zhu, L.; Yao, Y.; Liu, J.; Wang, J.; Xie, H. Expression of B-catenin and MMP-8 in Gingival Crevicular Fluid and Gingival Tissue Indicates the Disease severity of Patients with Chronic Periodontitis. *Exp. Ther. Med.* **2019**, *18*, 2131–2139. [CrossRef]
32. Karrer, S.; Bosserhoff, A.K.; Weiderer, P.; Landthaler, M.; Szeimies, R.M. Influence of 5-Aminolevulinic Acid and Red Light on Collagen Metabolism of Human Dermal Fibroblasts. *J. Investig. Dermatol.* **2003**, *120*, 325–331. [CrossRef] [PubMed]
33. Pierfelice, T.V.; D'amico, E.; Iezzi, G.; Piattelli, A.; Di Pietro, N.; D'arcangelo, C.; Comuzzi, L.; Petrini, M. Nanoporous Titanium Enriched with Calcium and Phosphorus Promotes Human Oral Osteoblast Bioactivity. *Int. J. Environ. Res. Public. Health* **2022**, *19*, 6212. [CrossRef] [PubMed]
34. Rodrigues, A.Z.; de Oliveira, P.T.; Novaes, A.B.; Maia, L.P.; de Souza, S.L.S.; Palioto, D.B. Evaluation of in Vitro Human Gingival Fibroblast Seeding on Acellular Dermal Matrix. *Braz. Dent. J.* **2010**, *21*, 179–189. [CrossRef] [PubMed]
35. Jubeli, E.; Khzam, A.; Yagoubi, N. Cells Integration onto Scaffolds Prepared from Polyester Based Polymers—Importance of Polymer Thermal Properties in Addition to Hydrophilicity. *Int. J. Polym. Mater. Polym. Biomater.* **2019**, *68*, 1068–1077. [CrossRef]

gels

MDPI

Article

Gel Formulations with an Echinocandin for Cutaneous Candidiasis: The Influence of Azone and Transcutol on Biopharmaceutical Features

Noelia Pérez-González [1], Lupe Carolina Espinoza [2], María Rincón [3,4], Lilian Sosa [5], Mireia Mallandrich [4,6,*], Joaquim Suñer-Carbó [4,6], Nuria Bozal-de Febrer [7], Ana Cristina Calpena [4,6] and Beatriz Clares-Naveros [1,4,8]

1 Department of Pharmacy and Pharmaceutical Technology, Faculty of Pharmacy, University Campus of Cartuja, University of Granada, 18071 Granada, Spain
2 Departamento de Química, Universidad Técnica Particular de Loja, Loja 1101608, Ecuador
3 Departament de Ciència de Materials i Química Física, Facultat de Química, Universitat de Barcelona (UB), C. Martí i Franquès 1-11, 08028 Barcelona, Spain
4 Institut de Nanociència i Nanotecnologia (IN2UB), Universitat de Barcelona (UB), 08028 Barcelona, Spain
5 Pharmaceutical Technology Research Group, Faculty of Chemical Sciences and Pharmacy, National Autonomous University of Honduras (UNAH), Tegucigalpa 11101, Honduras
6 Departament de Farmàcia, Tecnologia Farmacèutica, i Fisicoquímica, Facultat de Farmàcia i Ciències de l'Alimentació, Universitat de Barcelona (UB), 08028 Barcelona, Spain
7 Departament de Biologia, Sanitat i Medi Ambient, Facultat de Farmàcia i Ciències de l'Alimentació, Universitat de Barcelona (UB), 08028 Barcelona, Spain
8 Biosanitary Institute of Granada (ibs.GRANADA), 18012 Granada, Spain
* Correspondence: mireia.mallandrich@ub.edu

Citation: Pérez-González, N.; Espinoza, L.C.; Rincón, M.; Sosa, L.; Mallandrich, M.; Suñer-Carbó, J.; Bozal-de Febrer, N.; Calpena, A.C.; Clares-Naveros, B. Gel Formulations with an Echinocandin for Cutaneous Candidiasis: The Influence of Azone and Transcutol on Biopharmaceutical Features. *Gels* **2023**, *9*, 308. https://doi.org/10.3390/gels9040308

Academic Editors: Adina Magdalena Musuc, Magdalena Mititelu and Mariana Chelu

Received: 26 February 2023
Revised: 29 March 2023
Accepted: 31 March 2023
Published: 6 April 2023

Abstract: Caspofungin is a drug that is used for fungal infections that are difficult to treat, including invasive aspergillosis and candidemia, as well as other forms of invasive candidiasis. The aim of this study was to incorporate Azone in a caspofungin gel (CPF-AZ-gel) and compare it with a promoter-free caspofungin gel (CPF-gel). An in vitro release study using a polytetrafluoroethylene membrane and ex vivo permeation into human skin was adopted. The tolerability properties were confirmed by histological analysis, and an evaluation of the biomechanical properties of the skin was undertaken. Antimicrobial efficacy was determined against *Candida albicans*, *Candida glabrata*, *Candida parapsilosis*, and *Candida tropicalis*. CPF-AZ-gel and CPF-gel, which had a homogeneous appearance, pseudoplastic behavior, and high spreadability, were obtained. The biopharmaceutical studies confirmed that caspofungin was released following a one-phase exponential association model and the CPF-AZ gel showed a higher release. The CPF-AZ gel showed higher retention of caspofungin in the skin while limiting the diffusion of the drug to the receptor fluid. Both formulations were well-tolerated in the histological sections, as well as after their topical application in the skin. These formulations inhibited the growth of *C. glabrata*, *C. parapsilosis*, and *C. tropicalis*, while *C. albicans* showed resistance. In summary, dermal treatment with caspofungin could be used as a promising therapy for cutaneous candidiasis in patients that are refractory or intolerant to conventional antifungal agents.

Keywords: caspofungin; echinocandin; Azone; candidiasis; permeation enhancer; gel

1. Introduction

Candidiasis is a fungal infection caused by the yeast of the genera *Candida* spp., which is the most important cause of opportunistic mycoses worldwide and can proliferate in the skin and mucosal surface, even causing a systemic infection [1,2]. *Candida albicans* is the most prevalent pathogen that is responsible for about 70% of fungal infections [2].

Cutaneous candidiasis is caused by *Candida* proliferation in the skin, mainly in the intertriginous areas, producing inflammation, dryness, erosions, and pustules [3,4]. Despite the available pharmacological treatments, such as azoles and polyenes, complications from *Candida* infections are frequent, especially in patients that are refractory or intolerant to these antifungal drugs or those in hospital conditions, including immunosuppressed populations such as patients with malignancies or HIV infections for whom the mortality rates are high [4,5].

Therefore, the treatment of candidiasis becomes a challenge in this population, and consequently, other pharmacological options, such as echinocandins, have been addressed [6]. Echinocandin drugs are formed by a cyclic nucleus composed of several amino acid residues and N-linked acyl lipophilic fatty acid tails, which act as an anchor for the drug at the pathogen cell wall. These antifungal drugs inhibit the biosynthesis of β-(1,3)-D-glucan, which is present in the fungal cell walls, including in Aspergillus and *Candida* spp. [5,7,8]. Currently, there are some clinically used echinocandins, such as caspofungin, micafungin, and anidulafungin, all of which have a lipopeptide structure that is synthetically modified from the fermentation broths of different fungi [8]. Caspofungin is a semi-synthetic lipopeptide obtained from a fermentation product of *Glarea lozoyensis* [9]; its chemical structure is depicted in Figure 1. In 2001, caspofungin was approved by the US Food and Drug Administration (FDA) and the European Medicines Agency (EMEA). This drug is used as therapy for fungal infections that are difficult to treat, including invasive aspergillosis, as well as candidemia and other forms of invasive candidiasis [6,10]. However, caspofungin is only used for intravenous administration due to its high molecular weight (1093.31 g/mol) and poor oral bioavailability, and consequently, the need arises to investigate alternative formulations and routes of administration to improve the efficacy of the drug and facilitate its use [9]. Dermal treatment with caspofungin could be used as a promising therapy for cutaneous candidiasis in patients that are refractory or intolerant to conventional antifungal agents.

In recent years, formulations combining polymers have been designed to produce multifunctional gels [11]. The combination of polymers of a natural origin, such as chitosan, with polymers of a synthetic origin, such as poloxamer 407, produces gels with a biocompatible and biodegradable nature, with a high permeability capacity and, thus, better therapeutic effect [12]. Chitosan is a cationic polysaccharide consisting of glucosamine and N-acetylglucosamine units obtained by the alkaline deacetylation of the natural polysaccharide chitin and is known for its antimicrobial properties. It is biocompatible, biodegradable, and capable of forming gels in situ with negatively charged macromolecules [13–15]. On the other hand, poloxamer 407 is a hydrophilic nonionic surfactant consisting of a hydrophobic polypropylene oxide (PPO) core block flanked by hydrophilic poly (ethylene oxide) (PEO-PPO-PEO) blocks and is known to be thermoreversible [16]. To that end, skin penetration promoters can be used in order to facilitate the drug permeation through the stratum corneum and achieve an effective drug concentration [17]. Several permeation enhancers, including pyrrolidones, Azone, alcohols, sulfoxides, surfactants, essential oils, and their chemical constituents, have been proposed in cosmetic and pharmaceutical formulations to improve transdermal drug delivery and increase drug retention in the skin [18,19]. The ideal purpose of these permeation enhancers should be to temporarily and reversibly perturb the barrier function of the stratum corneum [19]. The reported mechanisms of action for these skin penetration promoters include the (i) reversible alteration of the intercellular lipid matrix or on the intracellular keratin domains, (ii) changes in the drug/tissue partition coefficient, and (iii) the disturbance of skin metabolism [20]. Azone (1-dodecylazacycloheptan-2-one) was the first synthetic compound specifically designed as a skin permeation promoter [21]. The chemical structure of Azone (Figure 1) shows a polar headgroup (within a seven-membered ring) attached to a C_{12} chain, which possibly allows this molecule to interact directly with the skin lipid domains to disturb the organized lipid packing [19,22]. Transcutol-P (diethylene glycol monoethyl ether) is a compound that is extensively used as a penetration enhancer in topical delivery systems, as well as for its solubilizing action [23–26]; Figure 1 depicts its chemical structure.

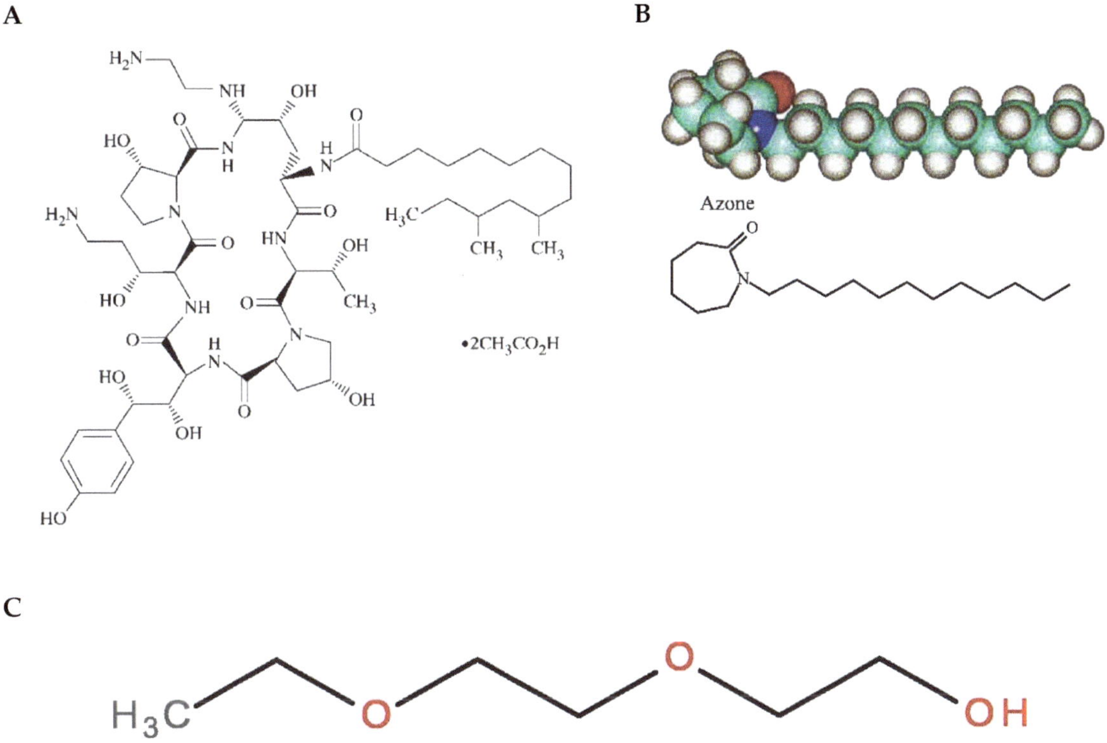

Figure 1. Chemical structure of drug and skin promoters. (**A**) Caspofungin [27], (**B**) Azone, [20] and (**C**) diethylene glycol monoethyl ether (Transcutol) [28].

Considering these remarkable findings, the aim of the study was to design and develop a topical delivery system for caspofungin. Two formulations were developed: a caspofungin gel (CPF-gel) and a caspofungin gel with the addition of two promoters: Azone and Transcutol-P (CPF-AZ-gel). The physicochemical, biopharmaceutical, and efficacy properties of both formulations in the treatment of cutaneous candidiasis were compared. Figure 2 illustrates a schematic representation of the development and evaluation of the hydrogels composed of pluronic F-127 and chitosan, with and without permeation enhancers loading caspofungin.

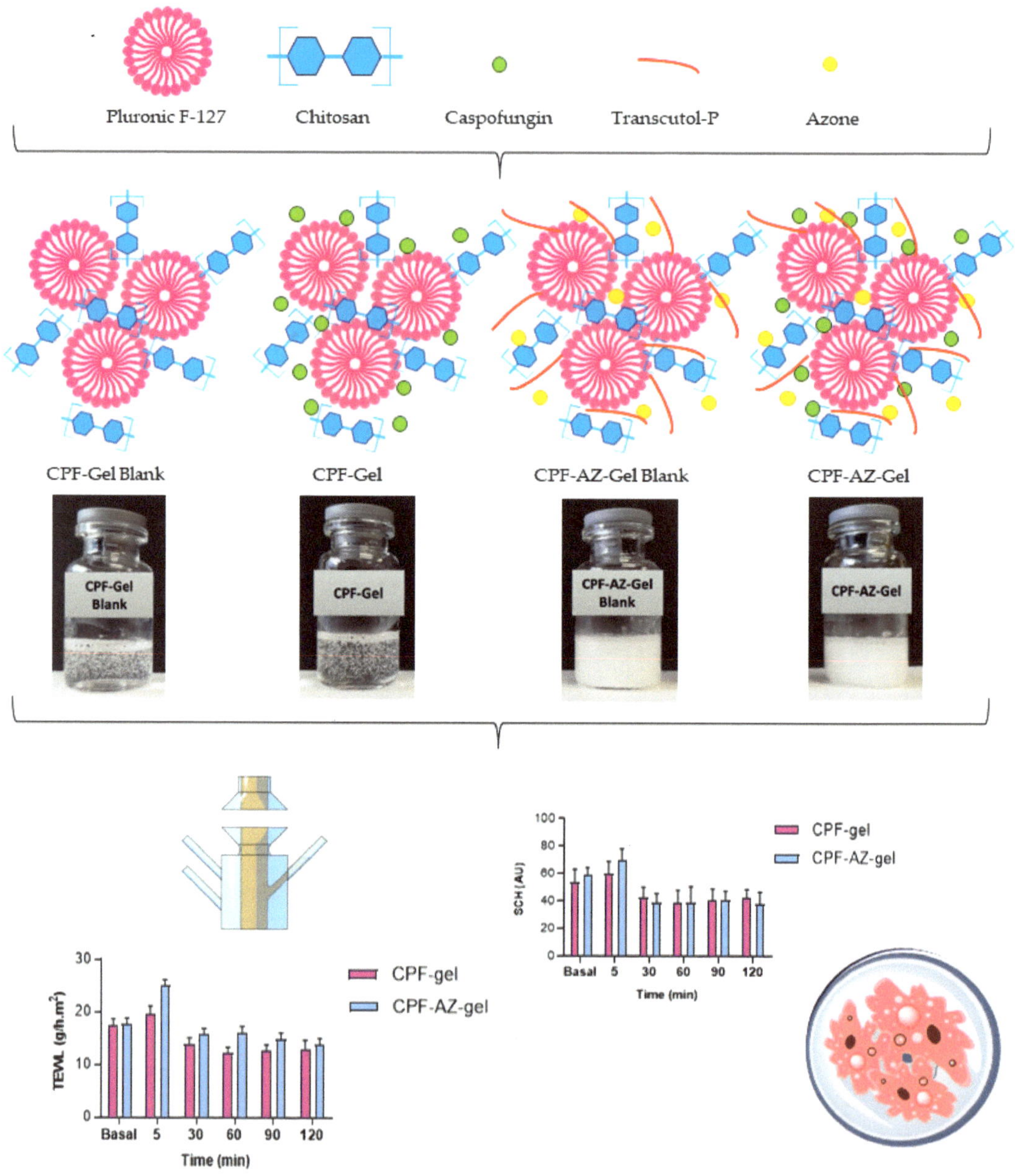

Figure 2. Schematic representation of the development of the loaded and unloaded hydrogels composed of pluronic F-127 and chitosan, with and without permeation enhancers, and the evaluation of the formulations.

2. Results and Discussion

2.1. Preparation of Gel Formulations

The composition of CPF-gel and CPF-AZ-gel obtained in this work are described in Table 1. CPF-gel was obtained by dissolving Pluronic F-127 (18%) and caspofungin (2%)

in cold water (58%), then chitosan (2%) that was previously dissolved in 0.1 M aqueous acid acetic solution (20%) was added. On the other hand, CPF-AZ-gel was obtained by incorporating a mixture of Azone (5%) and Transcutol-P (5%) into CPF-gel that was previously prepared by considering water adjustment. Both gels (CPF-gel and CPF-AZ-gel) had a uniform and homogeneous appearance, free of visible particles, lumps, and precipitates (Figure 3).

Table 1. Composition formula of caspofungin-loaded gels.

Ingredients	CPF-Gel (% *w/w*)	CPF-AZ-Gel (% *w/w*)
Caspofungin	2	2
Chitosan	2	2
Pluronic F-127	18	18
0.1 M Acid acetic solution	20	20
Water	58	48
Azone	-	5
Transcutol-P	-	5

Figure 3. Photograph of drug-loaded hydrogels and nondrug-loaded hydrogels. (**A**) CPF-Gel without drug; (**B**) CPF-Gel; (**C**) CPF-AZ-Gel without drug, and (**D**) CPF-AZ-Gel.

The gelling of CPF-gel and CPF-AZ-gel occurs because of the Pluronic F-127 and chitosan content. In an aqueous solution, the amphiphilic properties of Pluronic F-127 cause self-aggregation-forming micelles at concentrations above its critical micelle concentration, where PPO hydrophobic groups interact together via van der Waals forces to form the inner core, whereas the PEO hydrophilic groups interact with water molecules via hydrogen bonds to form the outer shell [29,30]. Consequently, gelation occurs at sufficient concentrations of Pluronic F-127, which, in this study, was 18% due to micelles packing [31]. On the other hand, the gelling with chitosan was produced by physical cross-linking, where ion interactions and hydrogen bonds are the driving forces for the formation of the entangled networks of this gel [32,33]. Figure 4 illustrates the mechanism of hydrogel formation.

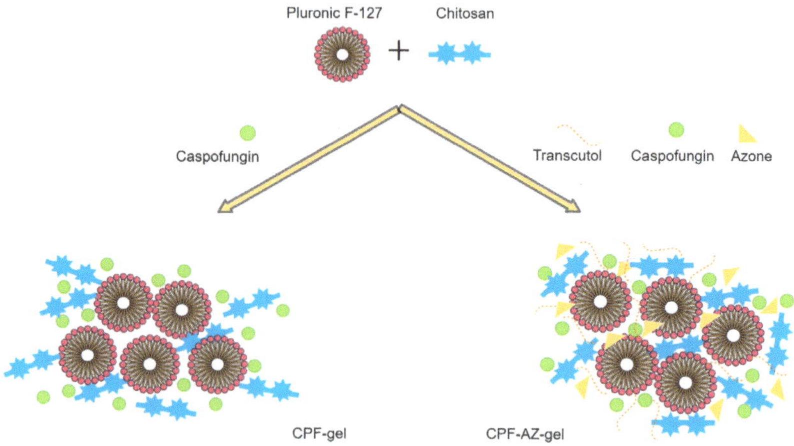

Figure 4. Schematic illustration of the mechanism of hydrogel formation.

2.2. Thermosensitive Properties

The incorporation of the drug into the hydrogels did not change the temperature sensitivity, as shown in Figure 5. No significant differences were observed when the gels were acclimatized to the different temperatures after the addition of the Azone and Transcutol-P promoters. As the temperature increased, the gels seemed to flow better. However, at no time was the gelation state reached.

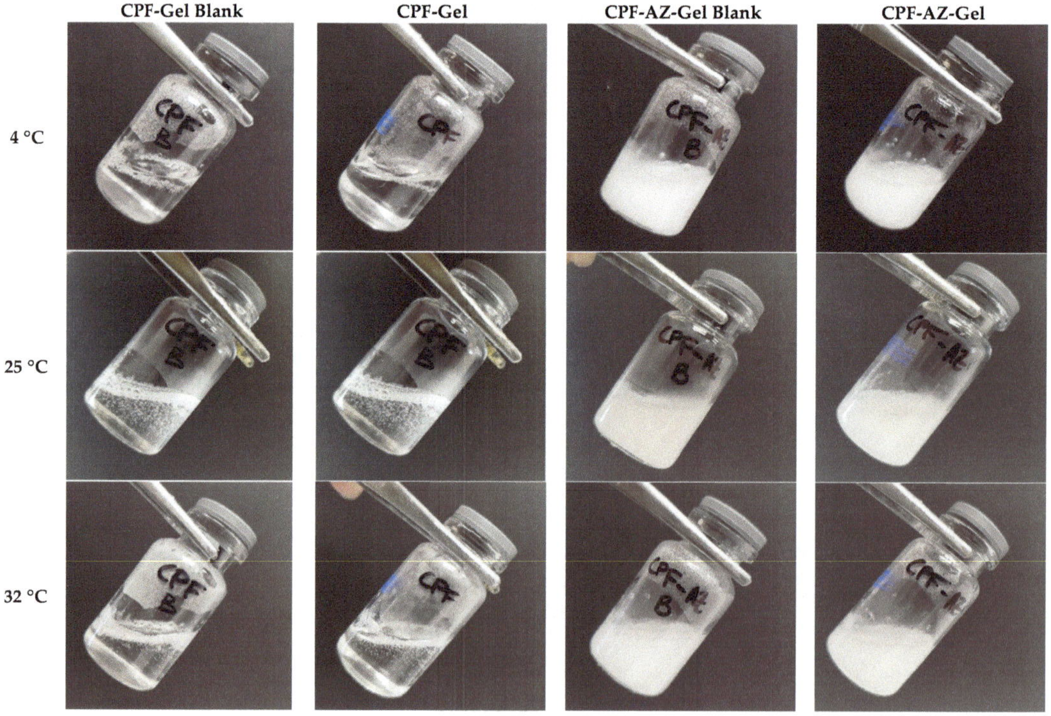

Figure 5. Physical appearance of the hydrogels at different temperatures: 4 °C, 25 °C, and 32 °C.

Combinations of polymers, such as CTS and P407, can modify the properties of the resulting system. The micellar corona increases in thickness because the number of hydrophilic chains increases. This is due to the interaction between CTS and the polyethylene oxide chains of P407. As a result of the interaction, the apparent viscosity increases, leading to a decrease in temperature, whereby the SOL-GEL transition could be reached [34,35].

2.3. Fourier Transform Infrared

Fourier transform infrared (FTIR) was performed to investigate any possible interaction between the drug and the gel matrix. Figure 6 shows the FTIR spectra for the loaded and unloaded gels of CPF-gel and CPF-AZ-gel. The peaks corresponding to different functional groups of caspofungin can be seen at about 1600 cm^{-1}: carbonyl and amide. Within the range 3200–3600 cm^{-1}, there are peaks related to O-H bond vibrations from the alcohol groups present in the structure, as well as from the amide groups (Figure 6c). When analyzing the hydrogels loaded with caspofungin, a remarkable decrease in the absorption is observed within the range 3000–3600 cm^{-1} (OH stretching) compared to the caspofungin spectrum. This fact indicates that caspofungin interacts with the hydrogel via the hydrogen bonds, especially between the hydroxyl groups in both caspofungin and the polymers.

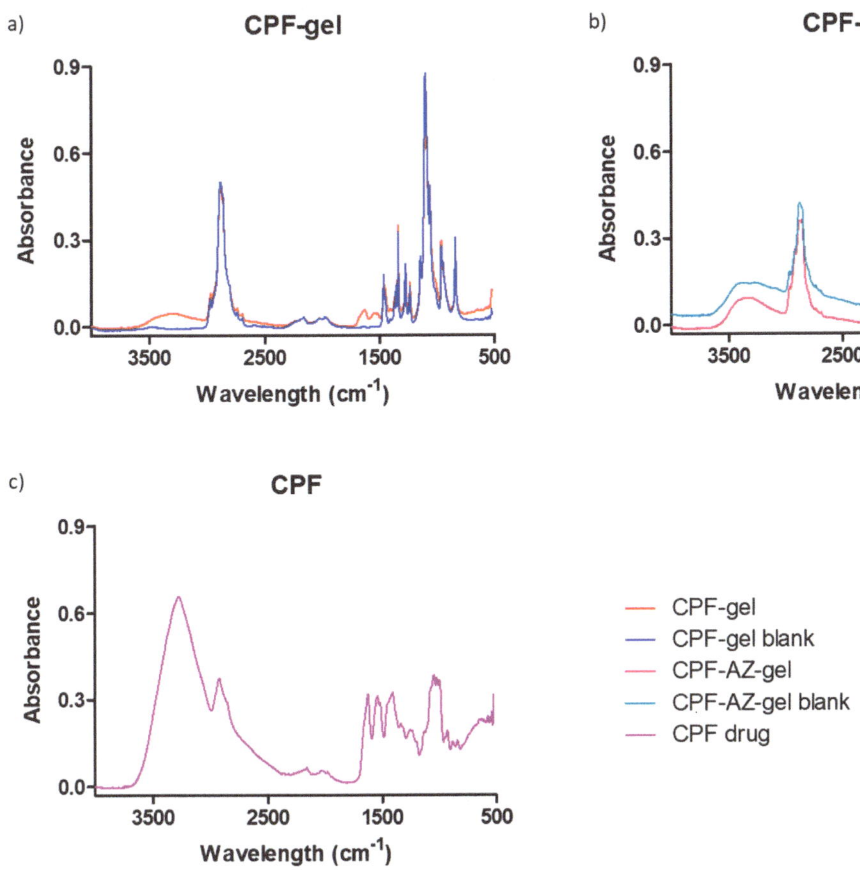

Figure 6. FTIR spectra for the gels: (**a**) CPF gel (red) and placebo (blue), (**b**) CPF-AZ-gel (pink) and placebo (cyan), and (**c**) caspofungin acetate salt Sun-Pharma (purple).

2.4. Morphological Study and Determination of the Porosity of the Hydrogels

The structure of the gels was evaluated via scanning electron microscopy (SEM) after desiccating the gels. Figure 7 shows the SEM images of the unloaded and loaded hydrogels. CPF-gel showed a denser and more compact structure than CPF-AZ-gel. The addition of Transcutol-P and Azone resulted in a more rounded shape structure, which was especially remarkable for the unloaded gels.

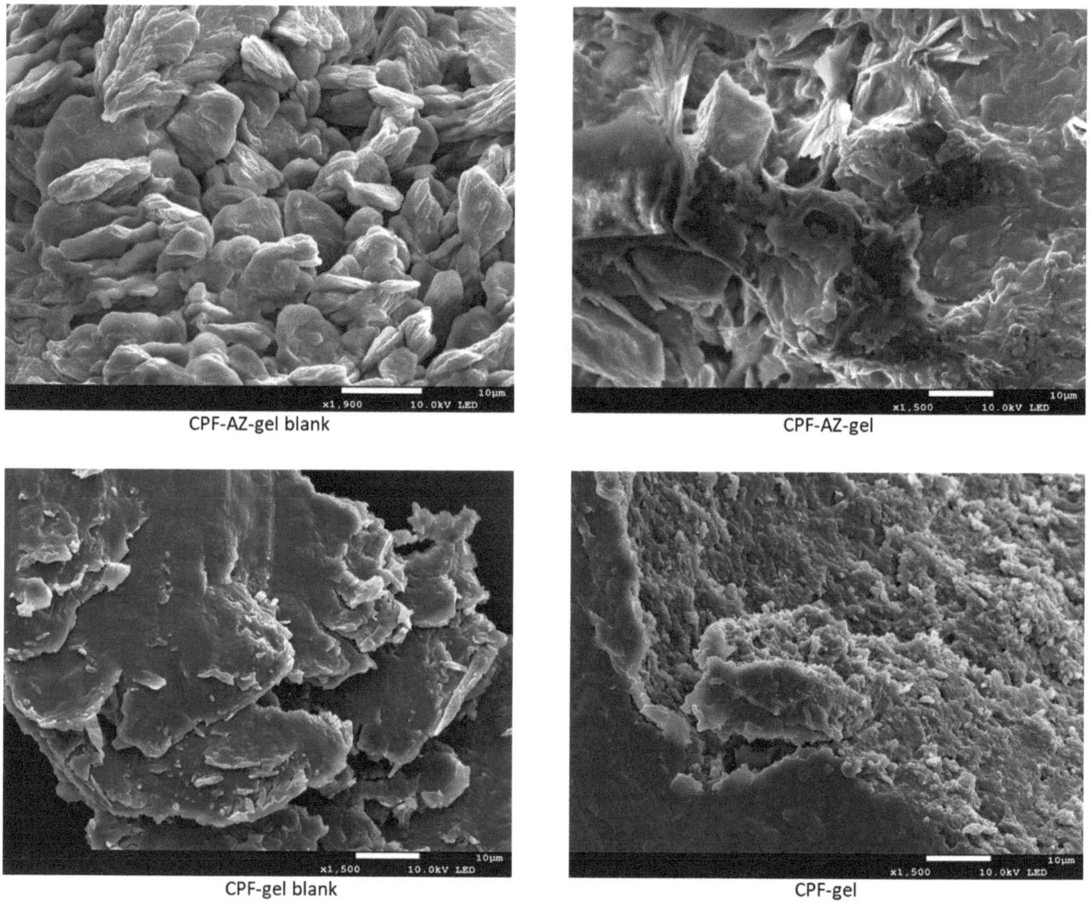

Figure 7. SEM images of the hydrogels: left panels show the blank hydrogels with or without permeation enhancers, and the right panels display the loaded hydrogels.

Hydrogels have the capacity to absorb large amounts of water or biological fluids; when the gels absorb more than 10 times their weight (in the dried state), they are classified as superabsorbent hydrogels. There are different techniques to evaluate gel porosity; in this work, the porosity was determined by the density of the solvent uptake. CPF-AZ-gel showed higher porosity than CPF-gel (Table 2). The addition of Transcutol-P and Azone in the hydrogel increased the porosity by about four-fold. The porosity of the gels has an impact on the drug release from the gel matrix [36]. This is in line with the drug release behavior observed in Section 2.7; CPF-AZ-gel released greater amounts of caspofungin than CPF-gel.

Table 2. Porosity of caspofungin-loaded gels.

Formulation	Porosity (%)
CPF-gel	5.7 ± 0.3
CPF-AZ-gel	22.4 ± 1.4

2.5. Rheological Behavior

Figure 8 shows the rheograms of CPF-gel and CPF-AZ-gel at 10, 25, and 32 °C. Rheological behavior plays a critical role in topical formulations, as it determines sensory and dosage characteristics, as well as modulates the biopharmaceutical parameters, such as the drug release rate from its vehicle [37]. CPF-AZ-gel exhibited a higher viscosity than CPF-gel due to its Azone and Transcutol-P content, providing greater consistency to the final product. The viscosity of both formulations decreased as the temperature increased, showing values of 1.48, 1.16, and 0.95 Pa·s at 10, 25, and 32 °C, respectively, for CPF-gel and 1.66, 1.52, and 1.09 Pa·s at 10, 25, and 32 °C, respectively, for CPF-AZ-gel.

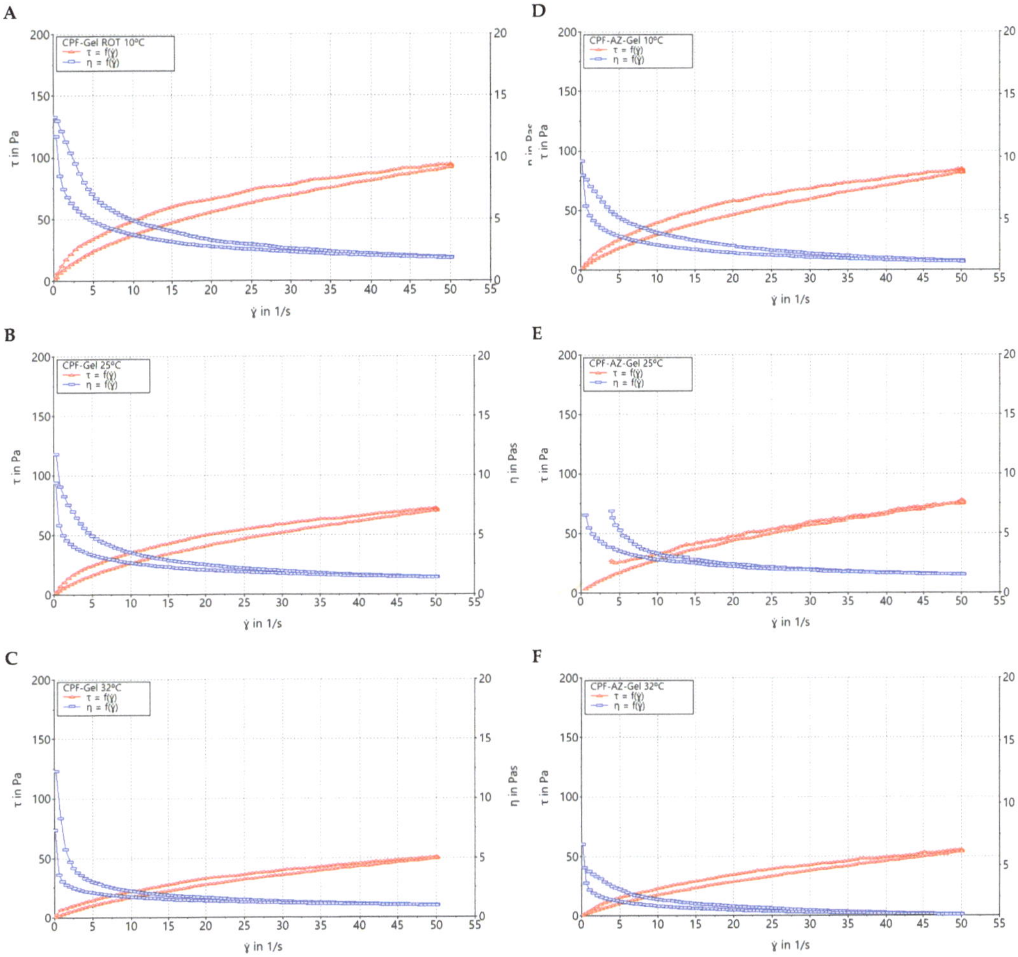

Figure 8. Rheograms of caspofungin formulations. (**A**) CPF-gel at 10 °C, (**B**) CPF-gel at 25 °C, (**C**) CPF-gel at 32 °C, (**D**) CPF-AZ-gel at 10 °C, (**E**) CPF-AZ-gel at 25 °C, and (**F**) CPF-AZ-gel at 32 °C.

Table 3 summarizes the results obtained from the rheological analysis of both formulations and the mathematical modeling of the experimental data. Both CPF-gel and CPF-AZ-gel at 10, 25, and 32 °C exhibit shear-thinning (pseudoplastic) behavior, according to the Cross model for the stretch ramp-up and stretch ramp-down. This behavior is characterized by a decrease in viscosity as friction or light massage is applied to the treated area facilitating its spreadability on the skin; however, when the friction is stopped, the viscosity of the product returns to its original state, favoring the product residence time [38].

Table 3. Viscosity and mathematical fitting of caspofungin formulations at 10, 25 and 32 °C.

	Temperature Conditions	10 °C	25 °C	32 °C
CPF-gel	Viscosity (Pa·s) at 100 s^{-1}	1.48 ± 0.01	1.16 ± 0.01	0.95 ± 0.03
	Mathematical model (stretch ramp up/down)	Cross	Cross	Cross
CPF-AZ-gel	Viscosity (Pa·s) at 100 s^{-1}	1.66 ± 0.02	1.52 ± 0.01	1.09 ± 0.01
	Mathematical model (stretch ramp up/down)	Cross	Cross	Cross

2.6. Spreadability Analysis

Sample spreading properties represent decisive parameters in the assessment of topical forms as they affect the process of incorporation into the container as well as the uniformity of the dose and application, thereby affecting therapeutic efficacy [39,40]. CPF-gel and CPF-AZ-gel followed a hyperbola one-site model (Figure 9), and similar values were observed regarding the extensibility of the two formulas (Table 4). The results obtained indicate that both formulas can be easily spread on the skin.

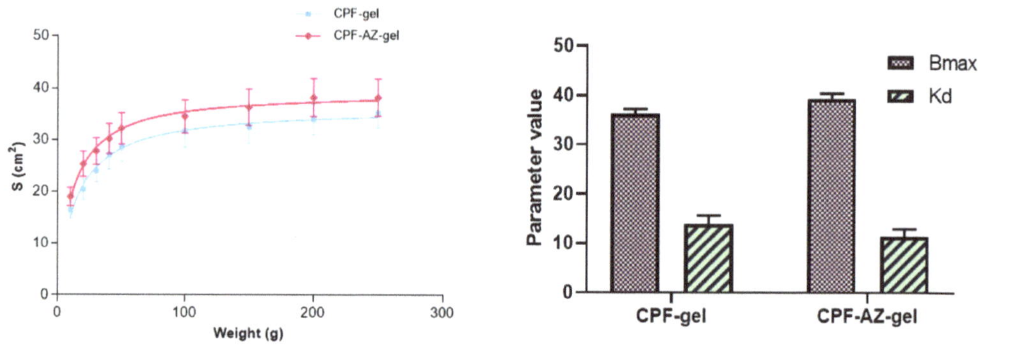

Figure 9. Evaluation of the spreadability of CPF-gel and CPF-AZ-gel, fitting a one-site hyperbola equation: (**a**) Spreading area (S, cm^2) as a function of the applied mass (g) at 22 ± 2 °C; 60 ± 5% RH, and (**b**) plot of the parameters of the formulas of CPF-gel, and CPF-AZ-gel fitting a one-site hyperbola equation. The data are expressed as the mean ± standard deviation (SD) of the three replicates (n = 3).

2.7. Evaluation of the Drug-Release Kinetics

The amount of caspofungin that was released from the formulations was assessed by in vitro release tests performed by Franz cells. A suitable receptor medium for the in vitro drug release studies must provide sink conditions so as to ensure that the concentration of the drug in the receiver compartment will not exceed 30% of the concentration to the saturation of the given drug in that medium [41]. Keeping sink conditions is crucial and prevents the inhibition of drug diffusion through the membrane/skin due to the saturation of the medium [42]. In this study, 0.9% physiological saline solution provided the sink conditions for the study.

Table 4. Fitting and goodness of fit for the parameters estimated according to one-site hyperbola equations for the formulas CPF-gel and CPF-AZ-gel.

One Site Binding (Hyperbola)	CPF-Gel	CPF-AZ-Gel
Best-fit values		
Bmax	36.23	39.41
Kd	13.98	11.50
Std. Error		
Bmax	0.9566	1.008
Kd	1.719	1.500
95% CI (profile likelihood)		
Bmax	34.29 to 38.29	37.37 to 41.57
Kd	10.67 to 17.92	8.612 to 14.92
Goodness of Fit		
Degrees of Freedom	25	25
R squared	0.8725	0.8480
Sum of Squares	142.0	176.0
Sy.x	2.383	2.653
AICc	51.86	57.66

Figure 10 shows the release profile of caspofungin using a polytetrafluoroethylene (PTFE) membrane. The selection of the membrane in the drug-release studies is also crucial since they must be inert and must not limit the drug release, nor can any binding between the drug and the membrane occur [41].

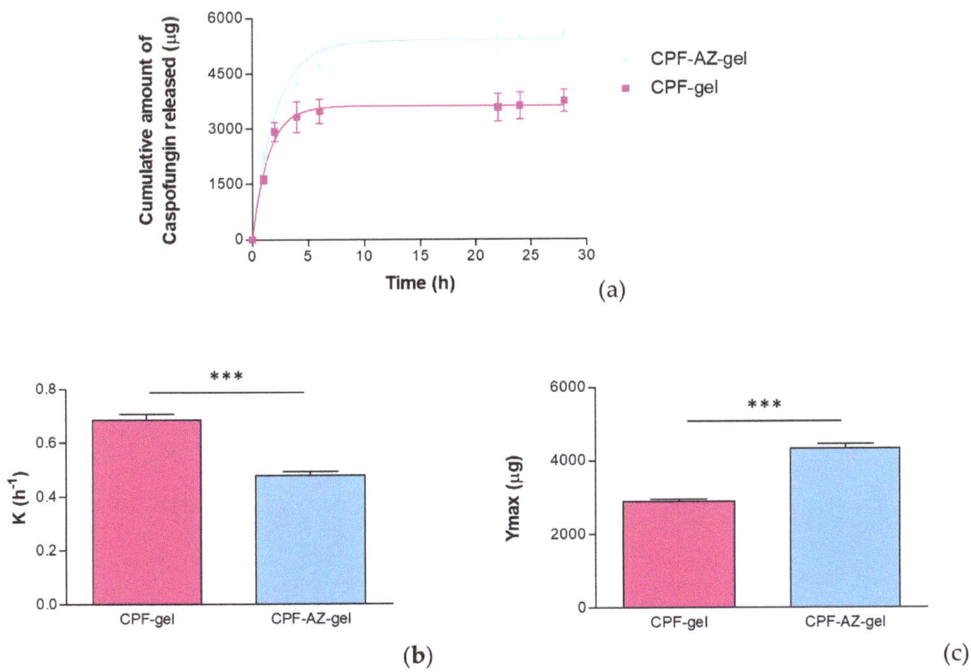

Figure 10. In vitro drug release of caspofungin from the CPF-gel and CPF-ZA gel formulations: (**a**) release profile of caspofungin over time; (**b**) release constant (K), estimated by modeling to the best-fit model, corresponding to a one-phase exponential association model; (**c**) maximum amount released (Y_{max}) estimated by modeling to the best-fit model, corresponding to a one-phase exponential association model. *** significant statistical differences between the two gels ($p < 0.0001$).

The release data were fitted to the kinetic models to describe the release profile of caspofungin. The best fit was obtained for a one-phase exponential association model (r^2 = 0.9986 for CPF-gel and r^2 = 0.9971 for CPF-AZ-gel), for which the mathematical equation is

$$Y = Y_{max} \cdot \left(1 - e^{-K \cdot X}\right) \quad (1)$$

where K is the release rate, Y_{max} is the maximum amount of drug released, and X is the time. In this study, the parameters K and Y_{max} were estimated and statistically compared by a *t*-test. According to the one-phase exponential association kinetics, the release of caspofungin is directly proportional to the concentration of the drug remaining in the donor chamber, resulting in a fast drug release in the early stages when the concentration of the drug is high in the donor chamber, and followed by a slowing down of the release rate as the concentration in the donor chamber decreases.

Although CPF-gel showed a greater release rate, it resulted in lower amounts of drug released than CPF-AZ-gel, which suggests that Azone and Transcutol-P improve the availability of the drug, and higher amounts of caspofungin would be delivered on the skin from the CPF-AZ-gel formulation.

2.8. Evaluation of the Permeation Capacity of Caspofungin through Ex Vivo Human Skin

The capacity of caspofungin to diffuse through the skin was evaluated by an ex vivo permeation study. In the ex vivo permeation studies, keeping sink conditions correct is as essential as in the in vitro drug release study. Furthermore, the receptor fluid, apart from providing sink conditions, must be biocompatible with the skin; for this reason, a 0.9% physiological saline solution was selected. Another important critical factor in permeation studies is the integrity of the ex vivo skin because using impaired skin or skin with altered barrier functions affects the permeability results. Hence, it is important to assess the skin integrity of the discs included in the study. Transepidermal water loss (TEWL) is one technique to evaluate skin integrity since there is a high correlation between TEWL values and skin integrity. Dey and co-workers investigated the TEWL values of intact skin, and then the researchers damaged the skin at different severities by removing stratum corneum with tape strips. They observed that the TEWL values for intact skin were below 13 $g/m^2/h$; moderately compromised skin was prepared with 20 consecutive tape strips and showed TEWL values up to 15.6 $g/m^2/h$; a total of 25–30 tape strips corresponded to highly compromised skin, for which the TEWL values increased up to 35 $g/m^2/h$. The maximum damage was reached by separating the epidermis from the dermis with heat treatment; this extremely damaged skin resulted in TEWL values up to 44.8 $g/m^2/h$ [43]. The stratum corneum is the outermost layer of the skin and is the main factor responsible for the barrier function. Skins with altered barrier functions are usually more permeable than intact skins. For this reason, the integrity of the skin discs used in this study was considered, and only those with TEWL values below 13 $g/m^2/h$ were included in the study.

Figure 11 shows the amounts of caspofungin permeated over 28 h and those retained in the skin at the end of the experiment.

Similar amounts of caspofungin permeated through the human skin for most of the sampling time points; contrarily, the formulation CPF-AZ-gel led to a major drug retention within the skin at 28 h. This is probably due to the presence of Azone and Transcutol-P in the formulation. Azone is known to be a penetration enhancer that is useful for increasing the permeation of both hydrophilic and lipophilic drugs. Azone increases the fluidity of the skin lipids by disrupting the lipid bilayers of the skin [44]. Transcutol-P is widely used in topical and oral formulations to improve the solubility of poorly soluble drugs; it disrupts the stratum corneum of the skin [45,46].

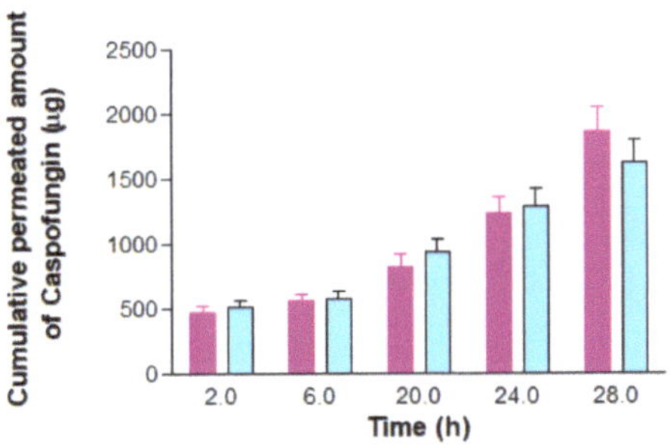

(a)

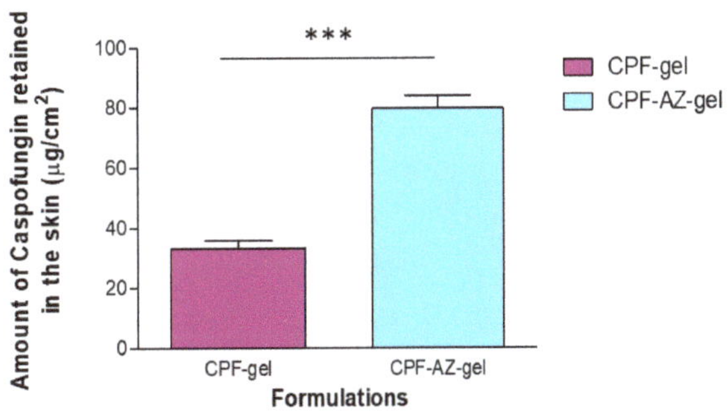

(b)

Figure 11. Permeation of caspofungin from CPF-gel and CPF-AZ-gel: (**a**) permeation profiles of caspofungin depicted as the cumulative amounts of caspofungin permeated through the ex vivo skin over time; (**b**) amount of caspofungin remaining in the skin at the end of the permeation study (28 h). *** significant statistical differences between the two gels ($p < 0.0001$).

The permeation parameters, such as the flux and the cumulative amount of drug permeated at a given time, characterize the rate and extent of the drug through the skin. Evaluating these parameters is important in the development and optimization of topical products. Table 5 reports typical permeation parameters.

Table 5. Permeation parameters estimated for CPF-gel and CPF-AZ-gel.

Parameter	CPF-Gel	CPF-AZ-Gel
A_{28h} (µg)	1854.4 ± 193.2	1623.9 ± 173.2
J (µg/h)	129.2 ± 14.7	86.9 ± 1.3 ***
J/cm^2 (µg/h/cm^2)	50.8 ± 5.8	34.2 ± 0.5 ***
Tl (h)	13.9 ± 1.2	9.2 ± 1.4 ***
K_P (10^{-3} cm/h)	2.54 ± 0.29	1.71 ± 0.02 ***
P_2 (10^{-2} 1/h)	1.20 ± 0.14	1.79 ± 0.11 ***
P_1 (10^{-2} cm)	21.23 ± 0.21	9.53 ± 0.02 ***
Css (ng/mL)	26.99 ± 3.08	18.16 ± 2.65 **

A_{28h}: cumulative amount of caspofungin permeated at 28 h. J: flux; Tl: lag-time; K_P: permeability coefficient; P_1: partition coefficient vehicle-skin; P_2: diffusion coefficient and Css: predicted plasma concentration at the steady state. ** statistical significance $p < 0.01$ and *** statistical significance $p < 0.0001$.

The amount of caspofungin that permeated through the skin over 28 h was slightly higher from the CPF-gel; however, no statistical differences were observed between the formulations for this parameter. The flux corresponds to the permeation rate of the drug through the skin, which is obtained as the slope of the linear regression of the linear part of the permeation profile; this parameter can also be expressed per unit area. The flux was 1.5-fold higher for CPF-gel, indicating that caspofungin crosses the skin faster when applied formulated in CPF-gel than CPF-AZ-gel. Apparently, the inclusion of Azone and Transcutol-P restricts the permeation of caspofungin and retains the drug in the skin. This is an interesting result for local therapy. The same pattern was observed for parameters Tl and K_P, which were 1.5-fold higher for CPF-gel with respect to CPF-AZ-gel. A shorter lag time predicts an earlier onset of action; hence, it is expected that CPF-AZ-gel will initiate its therapeutical action more rapidly than CPF-gel. The permeability coefficient is greater for CPF-gel, which is logical because K_P is estimated from the flux, and CPF-gel exhibited a higher flux. The higher the permeability coefficient, the higher the permeation of the drug into the receptor fluid.

The permeation of drugs depends on both the partition and diffusion processes. When analyzing the partition and diffusion coefficients, it is evident that the partition has a greater impact on the permeation of caspofungin since it shows higher values than the diffusion coefficient. Interestingly, the addition of Azone and Transcutol-P increases the partition coefficient of caspofungin between the formulation and the skin with regard to the gel without the enhancers while decreasing the diffusion of caspofungin through the skin, resulting in a higher retention of caspofungin in the skin. This is also observed when analyzing the predicted plasma concentration at the steady state (C_{ss}); CPF-AZ-gel would reach a lower C_{ss} than CPF-gel. C_{ss} was estimated by taking into account the human plasma clearance for caspofungin (9.42 L/h) [19] and considering an application surface of 5 cm^2.

2.9. Histological Analysis of the Ex Vivo Human Skin after Permeation of Caspofungin

After the ex vivo permeation test, in which we evaluated the permeability capacity of caspofungin from the two gels, an histological analysis was carried out to assess whether any structural change on the skin occurred. Figure 12 shows the histological images of the formulations, as well as the positive and negative controls. The ethanol solution induced the loss of stratum corneum. CPF-gel and CPF-AZ-gel did not change the stratum corneum or the epithelium.

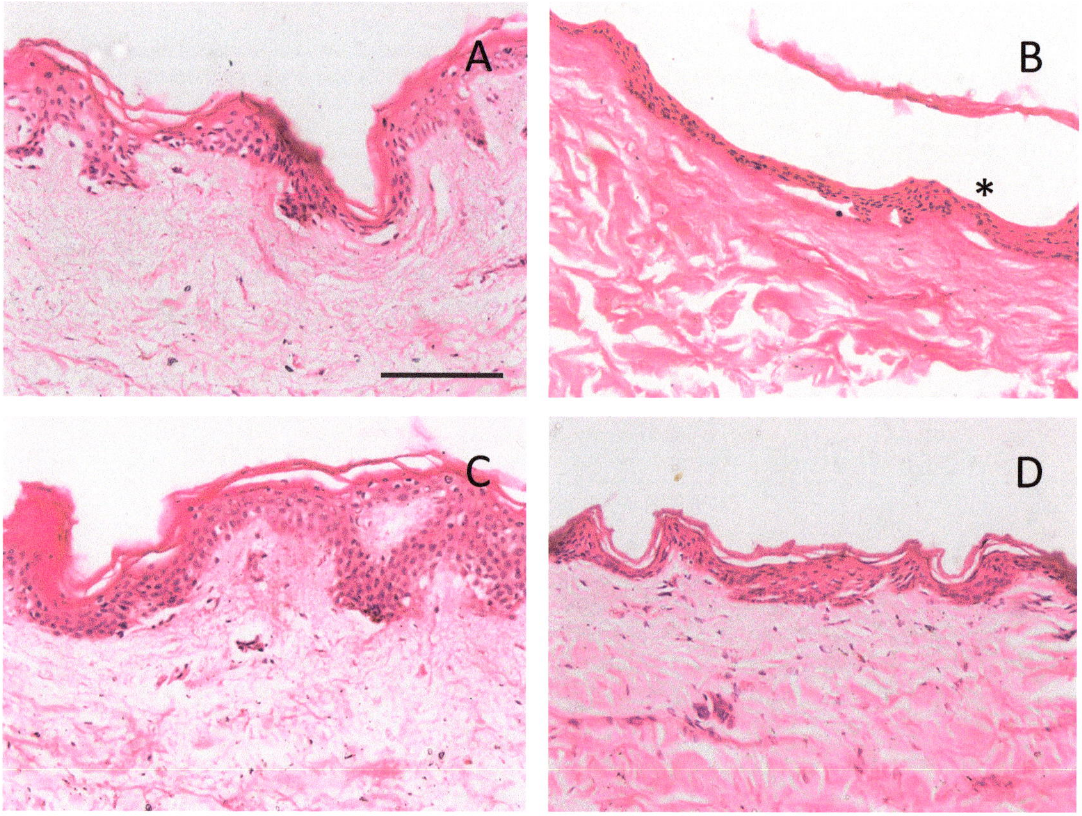

Figure 12. Skin sections stained with hematoxylin and eosin after the ex vivo permeation study, stained and observed under the microscope at 200X. Scale bar = 100 μm. (**A**) PBS as negative control; (**B**) ethanol as positive control; (**C**) CPF-gel and (**D**) CPF-AZ-gel. Asterisk indicates loss of stratum corneum.

2.10. Tolerance Studies by Evaluating Biomechanical Skin Properties

The skin integrity indicates the state of the skin as a physical barrier that protects the body from the environment. The integrity of the skin can be evaluated by measuring TEWL, which determines the ability of the skin to prevent water loss. TEWL increases when the skin barrier is compromised, for instance, via a cut, burn, or some skin diseases, including atopic dermatitis, eczema, or psoriasis. Altered skin barrier function, which may lead to increased transepidermal water loss, can result in irritation and skin dryness. Other factors can also impact the TEWL values, such as age and gender. For example, aging skin typically has a higher TEWL than younger skin, and exposure to dry or hot environments can increase TEWL.

Hence, measuring TEWL can be useful in dermatology and cosmetic science to evaluate the effectiveness of skin care products and treatments, as well as in clinical research to assess the skin barrier function in individuals with skin diseases or conditions [47].

Figure 13 shows the progression of the monitored parameters (TEWL and SCH) before and after the application of the formulations for up to 120 min. These parameters are indicative of the effect of the formulations on skin hydration and integrity. The transepidermal water loss (TEWL) values obtained from CPF-gel and CPF-AZ-gel showed a slight increase 5 min after the application and then a decrease at 30 min, and finally, remaining unchanged, being greater than the values for the formulation CPF-AZ-gel (Figure 13A,B,E). The stratum corneum hydration (SCH) values obtained from CPF-AZ-gel presented a slight increase 5 min after the application, with this increase being higher than the one for the CPF-gel formulation, and then this descended quickly; this occurs because the water in the formulas evaporates after 30 min and then both formulations remain unchanged, reaching values lower than the basal value (Figure 13C,D,F). Considering that skin hydration is directly related to skin capacitance, the results suggest that the formulations slightly increase hydration in relation to the normal behavior of the skin. No visible skin irritation was observed after the formulations were applied to the skin of the patients, indicating that both formulations were well tolerated on the skin.

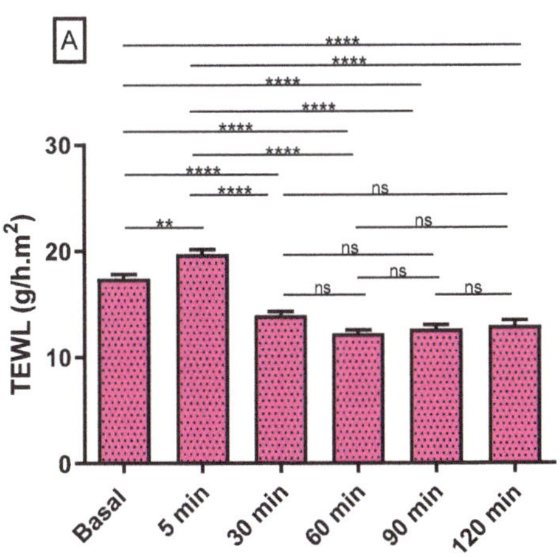

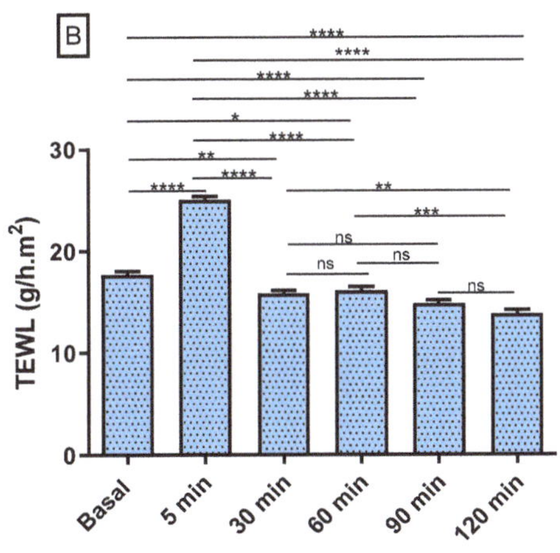

Figure 13. *Cont.*

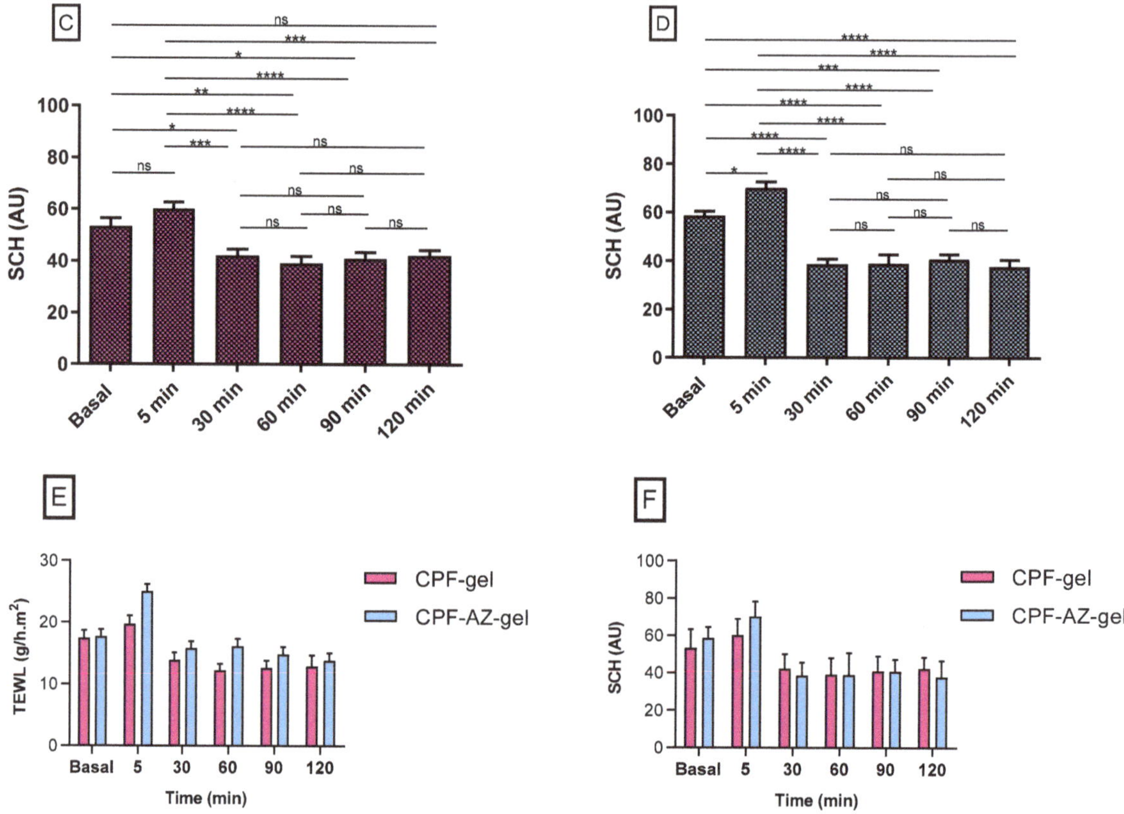

Figure 13. Biomechanical parameters evolution in human volunteers was monitored before the application of the formulations (basal) and 5 min, 30 min, 60 min, 90 min, and 120 min post-application. (**A**,**B**) TEWL of CPF-gel and CPF-AZ-gel, respectively, expressed as $g/h \times m^2$. (**C**,**D**) the SCH of CPF-gel and CPF-AZ-gel respectively expressed as arbitrary units (AU). Significant statistical differences: * $p < 0.05$, ** $p < 0.01$, *** $p < 0.001$, **** $p < 0.0001$, ns = non-significant. (**E**) Comparison of the evolution of the TEWL between both formulas: CPF-gel and CPF-AZ-gel. (**F**) Comparison of the evolution of the SCH between both formulas: CPF-gel and CPF-AZ-gel. Each value represents the mean ± SD (n = 10).

2.11. Antimicrobial Efficacy

Caspofungin belongs to an essential class of antifungals, the echinocandins, which are used to treat invasive candidiasis that is resistant to conventional treatments. Candidiasis is caused mainly by *C. albicans*, *C. glabrata*, *C. parapsilosis*, and *C. tropicalis* [48]. In this study, in vitro tests were carried out to determine if the Candida strains tested were susceptible to the formulations under investigation.

In this study, the CPF-gel and CPF-AZ-gel formulations demonstrated that they are effective at the in vitro level in three of the four Candida strains tested, producing broad areas of inhibition (Figure 14A–C, yellow circles), which indicates a beneficial effect since resistance to these types of drugs is gaining momentum over time, especially among isolates from immunosuppressed patients [49]. In the case of *C. albicans*, the growth of intrahalo colonies was observed within the zone of inhibition, showing resistance (orange arrow). This behavior has been observed in a previous study [39]. The gel without the drug produced a slight inhibitory effect in the area where the formulations were added.

This could be due to the impact caused by chitosan since the latter has been shown to have an antibacterial and antifungal effect (Figure 14A–C, blue circles) [50]. The results are summarized in Table 6.

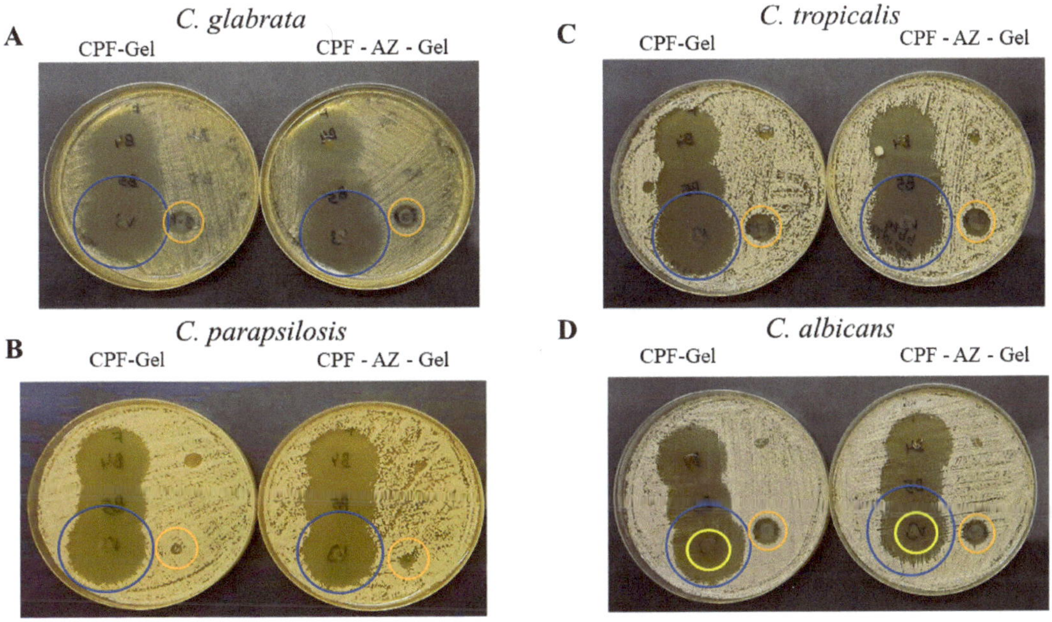

Figure 14. In vitro antifungal activity. (**A**) *C. glabrata* (**B**) *C. parapsilosis* (**C**) *C. tropicalis* and (**D**) *C. albicans*. The blue circles show the inhibition halo for CPF-gel and CPF-AZ-gel, the orange circles show the halo growth around the disc with an excipient (gel), and the yellow circles indicate the intrahalo growth of C. albicans.

Table 6. Growth inhibition on *Candida* species.

Yeast Tested	CPF-Gel	CPF-AZ-Gel	Gel
C. albicans	$R^{(a)}$	$R^{(a)}$	$R^{(b)}$
C. glabrata	S	S	$R^{(b)}$
C. parapsilosis	S	S	$R^{(b)}$
C. tropicalis	S	S	$R^{(b)}$

S: susceptible, $R^{(a)}$: resistant to inhibition of halo and intrahalo colonies, and $R^{(b)}$: resistant, with the formation of a slight halo of inhibition.

3. Conclusions

Two gels containing caspofungin, which is an echinocandin antifungal, were prepared, CPF-gel and CPF-AZ-gel, with the latter containing Azone and Transcutol-P, which are penetration enhancers. The gels were analyzed for their viscosity and rheological behavior. The spreadability of the gels was also investigated, and all characteristics were suitable for topical administration.

The drug release profile from the gels and the capacity of caspofungin to diffuse through the skin were also evaluated. The release of caspofungin followed a one-phase exponential association model, showing CPF-AZ-gel to have a higher drug amount released. The gel containing Azone and Transcutol-P (CPF-AZ-gel) resulted in a higher retention of the caspofungin in the skin while limiting the diffusion of the drug to the receptor fluid. Hence, a higher amount of caspofungin is available in the skin for a local effect.

Both gels were well-tolerated since they did not show any damage to the skin in the histological study, nor was any irritation observed when the formulation was applied to the skin. A decrease in the TEWL values indicated that the gels did not alter the skin function barrier.

Finally, the antifungal activity was assessed on different *Candida* sp. The two gels inhibited the growth of *C. glabrata*, *C. parapsilosis*, and *C. tropicalis*. However, *C. albicans* showed *resistance*.

In summary, the two gels showed satisfactory properties for a cutaneous application. Caspofungin was located in the skin, especially when applied through CPF-AZ-gel, through which it can exert a local effect. The formulations were well-tolerated, and the gels were effective against three of the four strains tested.

4. Materials and Methods

4.1. Materials

Caspofungin acetate salt (molecular weight ~1200 Da) was acquired from SunPharma (Barcelona, Spain). Poloxamer 407 (Pluronic F-127, molecular weight ~12,500 Da) was supplied by BASF (Barcelona, Spain), chitosan medium molecular weight (190–310 KDa and deacetylation degree $\geq$ 75%) was purchased from Sigma Aldrich (Madrid, Spain) and diethylene glycol monoethyl ether (Transcutol-P, molecular weight ~130 Da) was provided by Gattefossé (Saint-Priest, France). Azone (molecular weight ~280 Da) was acquired from Netqem (Durham, NC, USA), and acetic acid and reagents for the analytical method were acquired from Panreac (Barcelona, Spain). Purified and filtered water was obtained using a Milli-Q® Gradient A10 system apparatus (Millipore Iberica SAU.; Madrid, Spain).

4.2. Preparation of Gel Formulations

For CPF-gel preparation, Pluronic F-127 and caspofungin were dissolved using cold purified water at 4 °C under magnetic stirring for 30 min, then chitosan previously dissolved in 0.1 M aqueous acetic acid solution was incorporated, maintaining stirring conditions for 20 min. For CPF-AZ-gel preparation, the water proportions of CPF-gel were adjusted to incorporate Azone previously mixed with Transcutol-P.

4.3. Thermosensitive Properties

In order to observe if there was a phase transition, the gels were acclimatized in vials at 4 °C, 25 °C, and 32 °C. When the desired temperature was reached, the vials were inclined at approximately 45° to determine if the hydrogels flowed or not. The gels were photographed with and without the drug at different temperatures.

4.4. Fourier Transform Infrared

We investigated any possible chemical interaction between caspofungin and the polymers by Fourier Transform Infrared (FTIR). For this, the gels were desiccated in an oven at 37 °C. The FTIR spectra were obtained by a Nicolet iZ10 (Thermo Scientific, Waltham, MA, USA) with a DTGS detector within the range of 4000–525 cm^{-1} with a spectral resolution of 4 cm^{-1} using attenuated total reflectance (ATR) with a diamond crystal. A total of 32 scans per spectrum were obtained.

4.5. Morphological Study and Determination of the Porosity of the Hydrogels

Scanning electron microscopy (SEM) was performed to investigate the gels' structure. To this end, the gels were desiccated in an oven at the temperature of 37 °C, monitoring the desiccation process: the gels were weighed every day until a constant weight was observed. A small amount of the dried gel was coated with a thin film of carbon to obtain a conductive sample suitable for observation by SEM JSM-7001F (JEOL, Inc, Peabody, MA, USA).

The porosity of the hydrogels (CPF-gel and CPF-AZ-gel) was determined by the density method [REF]. Briefly, weighed amounts of dried gels were placed in Eppendorf with 1 mL of pure ethanol (n = 3 for each hydrogel). The experiment was conducted at 32 °C.

At pre-established time points, the Eppendorfs were centrifuged at 3000 rpm for 3 min, the supernatants of ethanol were withdrawn by automated pipette, and the hydrogels were weighed to monitor the increase of weight, which corresponds to the ethanol uptake by the gel. The increase in weight was observed at different time points until a constant weight was obtained. The porosity was calculated according to Equation (2):

$$P = \frac{W_s - W_d}{\rho \times V_s} \times 100 \quad (2)$$

where W_d is the dried hydrogel's weight, W_s is the swollen hydrogel's weight, ρ is ethanol's density, and V_s is the volume of the swollen hydrogel determined by a pycnometer.

4.6. Rheological Behavior

Determining the rheological behaviour in topical products is essential in topical products because theological properties are related to the consistency, texture and spreadability of the product. The viscosity of the formulation will impact the ease of application of the formulation to the skin. The rheological study was carried out with a Haake Rheostress 1® rheometer (Thermo Fisher Scientific, Karlsruhe, Germany) using a cone-cone system (C60/2°Ti: 60 mm diameter, 2° angle). The shear stress (τ) and the viscosity (η) were determined as a function of the shear rate (γ) at 25 ± 0.1 °C. The temperature was set with a thermostatic circulator Thermo Haake Phoenix II + Haake C25P. The Rotational measurements involved a 3-phase program which consisted of a ramp-up shear rate from 0 to 50 s^{-1} for 3 min, followed by a steady shear rate at 50 s^{-1} for 1 min and finally, a ramp-down from 50 to 0 s^{-1} for another period of 3 min. The viscosity was calculated at a steady shear at 100 s^{-1}.

The obtained data were analyzed with Data Manager v. 4.87 software (Haake Rheowin®, Thermo Electron Corporation, Karlsruhe, Germany) Data from the flow curves were modelled to different mathematical models [51]; and the best-fit model was selected on the basis of correlation coefficient and chi-square value.

4.7. Spreadability Analysis

Extensibility testing was carried out by placing a sample of 0.5 g of each formulation (CPF-gel and CPF-AZ-gel) between two glass plates, the one located in the bottom position pre-marked. The standard weight was added onto the upper plate for 60 s forcing the sample to spread. Nine pieces of increasing standard weight (10, 20, 30, 40, 50, 100, 150, 200, and 250 g) were added successively with 60 s between weights. The increase in the spreading area (diameter) was recorded as a function of the weight applied. The spreading area at each applied mass was calculated according to the following equation:

$$S = d^2 \times \frac{\pi}{4} \quad (3)$$

in which S is the spreading area (cm^2) calculated from the applied mass (g), and d is the mean diameter (cm) reached by the sample [39].

The formulations were analyzed in accordance with the best kinetic model, and the extensibility data were fitted to different mathematical equations (hyperbola, Boltzmann) using GraphPad Prism® version 8.0.0 for Windows, GraphPad Software, San Diego, CA, USA. The model fitting appropriateness was confirmed by the r value.

4.8. Evaluation of the Drug Release Kinetics

The rate and extent that caspofungin was released from the formulations was evaluated by Franz cells, which consist of two compartments: the donor and receiver chambers. The latter was filled with saline solution as the receptor medium, and the system was kept at 32 ± 1 °C with constant stirring. These experimental conditions provided the sink conditions. The two compartments are divided by a membrane that acts as an inert support for the formulation. The membrane used in this study was polytetrafluoroethylene (PTFE)

47 mm in diameter and 0.45 μm pore size (Merck, Spain). An amount og 0.3 g of formulation for either CPF-gel or CPF-AZ-gel was applied to the membrane in the donor chamber, and the drug diffused through the membrane into the receiver compartment was assessed over time by collecting samples (200 μL) at the following time points: 1, 2, 4, 6, 22, 26, and 28 h. The samples were analyzed by UPLC. 6 replicates for each formulation were included in this study.

The cumulative amount of caspofungin released was calculated and plotted as a function of time. The release rate was calculated by linear regression analysis as the slope of the linear part in the release profile. Kinetic modeling was performed to describe the behavior of caspofungin release over time. To this end, the data were fitted to several mathematical models, and the best-fitted one was chosen on the basis of the coefficient of determination (R^2).

4.9. Evaluation of the Permeation Capacity of Caspofungin through Ex Vivo Human Skin

To assess the potential of caspofungin to be absorbed into the bloodstream and determine the effectiveness of the gels developed as topical delivery systems, an ex vivo permeation test was conducted by Franz cells. The set-up is similar to the drug release test, with the particularity of using ex vivo human skin as a membrane. The skin was obtained from the abdominal area of donors subjected to aesthetic surgery; the "Docencia e Investigación" Committee of SCIAS Hospital de Barcelona approved the study (approval date 17 January 2020). The skin was cut at the thickness of 0.4 mm by a dermatome (Dermalab GA630 dermatome, Aesculap, Tuttlingen, Germany). This thickness presents the main layers of the skin: stratum corneum, viable epidermis, and a representative part of the dermis, which contains the blood vessels. The integrity of the skin discs was evaluated before the permeation test by measuring the transepidermal water loss (TEWL). Briefly, the sensor was placed on the skin for 2 min to allow it for temperature and humidity equilibration. Afterward, the TEWL values were recorded for 20 s, and only skin discs with TEWL below 13 $g/h{\cdot}cm^2$ were included in the study [43].

The permeation test was performed under an infinite dose approach, and 0.3 g of gel (CPF-gel or CPF-AZ-gel) was applied to the skin on the stratum corneum side, and samples were collected from the receiver medium at the time points 0.5, 1, 2, 4, 6, 22, 26 and 28 h. Samples were analyzed by UPLC to estimate the amount of caspofungin that permeated through the skin over time. Six replicates per formulation were included in the study. Once finished the permeation tests, 3 skin discs per formulation were destined for drug extraction, and the other 3 replicates were used for a histological evaluation.

To determine the amount of caspofungin that remained in the skin after 28 h of permeation, the skin discs were processed for drug extraction as follows: firstly, the excess gel on the skin surface was gently wiped and rinsed with distilled water; then, the diffusional area exposed to the formulations (2.54 cm^2) was cut, punched, and immersed in distilled water. The drug was extracted by ultrasonic technique for 20 min, and samples were analyzed by UPLC.

Data Analysis

The cumulative amounts of permeated caspofungin were plotted for each sampling time point. To evaluate the rate and extent of caspofungin permeation, we calculated the following permeation parameters: flux (J) as the permeation rate, expressed in μg/h; the lag-time (Tl), the permeability coefficient (K_P), and the partition and diffusion coefficients (P1 and P2, respectively) were also estimated, as well as we predicted the plasma concentration at the steady state that caspofungin would achieve after a topical application of the formulations, CPF-gel or CPF-AZ-gel, in a theoretical surface of application of 5 cm^2 [52].

The amount of caspofungin retained in the skin after 28 h of exposure was calculated by extracting the drug from the tissue and applying the factor of drug recovery in the skin of 45.5%.

4.10. Analytical Method for Determining Caspofungin

The samples from the drug release and permeation studies were analyzed by Ultra Performance Liquid Chromatography (UPLC) using an Acquity I-Class UPLC System (Waters, Milford, CT, USA). For the determination of caspofungin, we used a Lichrospher RP-8 column (125 × 4 mm, 5 µm, Phenomenex) in gradient elution, which consisted of 50% B for 0–5 min; 100% B for 5–6 min and finally, 50% B for 6–8 min, being the composition of the mobile phase: A = 0.1 % of trifluoroacetic acid (TFA), and B = methanol. The flow rate was set at 0.8 mL/min, and we injected a sample volume of 10 µL. The detection of caspofungin was carried out by an Acquity Fluorescence Detector at the excitation wavelength of 224 nm and emission wavelength of 304 nm [39].

4.11. Histological Analysis of the Ex Vivo Human Skin after Permeation of Caspofungin

The histological analysis aimed to examine the skin tissues to evaluate whether any alteration in the skin structure had occurred after the permeation study. To this end, 3 skin discs per formulation from the ex vivo permeation test were processed for the histological examination. After treatment with the different formulations, the skin was rinsed in PBS, dehydrated, and finally embedded in paraffin wax. Tissue sections of 5 µm were stained with hematoxylin and eosin to assess the impact of the different formulations on the skin layers. Serum was used as a control condition, and ethanol was used as a positive control. Finally, the skin samples were observed under the microscope (Olympus BX41) and photographed (camera Olympus XC50).

4.12. Tolerance Studies by Evaluating Biomechanical Skin Properties

An in vivo skin tolerance study was conducted to evaluate the biomechanical properties of human skin. The study protocol was approved by the Ethics Committee of the University of Barcelona on 30/01/2019 (IRB00003099). The assessment of the total amount of water loss (TEWL) through the skin was carried out using a Tewameter® TM 300 (Courage-Khazaka Electronics GmbH, Cologne, Germany) to measure the amount of water reaching the surrounding atmosphere through the diffusion and evaporation processes of the epidermal layer of the skin. Ten healthy-skinned participants ranging in age from 25 to 40 years were recruited after medical assessment and notification (written informed consent) of the nature of the study and associated procedures. The subjects were asked not to use skin-care cosmetics on the measurement site (flexor side of the left forearm) during the day before the study. The volunteers stayed in the testing room for at least 20 min before taking the measurements. The measurement site was marked drawing circles around 4 cm in diameter The readings were collected (baseline readings), and then we applied a uniform layer of 0.5 g of the formulations (CPF-gel and CPF-AZ-gel) to the center of the circle using a gentle tool using a circular motion with the thumb to help distribute samples. A total of 20 laps were carried out in a clockwise direction. New measurements were collected just after application and at 30, 60, 90, and 120 min post-application [31]. To measure, the electrode, a small hollow cylinder, was maintained on the different tissue's surface for 1 min. TEWL values ($g/m^2{\cdot}h$) were expressed as the mean ± SD of 10 replicates before and after the application of the formulations for at least 2 h. All measurements were carried out in accordance with published procedures [53,54].

The measurement of the hydration of the stratum corneum (SCH) was determined before application in the basal state and 5, 30, 60, 90, and 120 min post-applications of CPF-gel and CPF-AZ-gel on the treated area using Corneometer® 825 a Multi Probe installed Hydration Adapter® MPA5 (Courage & Khazaka Electronics GmbH, Cologne, Germany). The measurements were performed using the capacitance method, which takes advantage of the relatively high dielectric constant of water compared to other skin substances. Stratum corneum hydration (SCH) values (arbitrary units, AU) are expressed as mean ± SD, n = 10.

4.13. Antimicrobial Efficacy

The in vitro antifungal activity of the prepared formulations was determined using the methodology described in the test protocol [citation from CLSI], similar to the disk diffusion method (also known as the Kirby-Bauer method) but with some modifications.

The *Candida* strains used in this assay were: *Candida albicans* ATCC 10231, *Candida glabrata* ATCC 66032, *Candida parapsilosis* ATCC 22019, and *Candida tropicalis* ATCC 7349 (American Type Culture Collection, Manassas, VA, USA).

For the inoculum preparation, the culture medium was first prepared; Muller Hinton Agar (MH) supplemented with 2% glucose (MH-Glucose 2%) and 500 μg/mL of chloramphenicol to avoid possible bacterial contamination by the excipients and incubating at 30 °C for 48 h. Subsequently, each of the *Candida* strains was seeded. The fungal inoculum was prepared by suspending one to two isolated yeast colonies in Ringer's solution to achieve 0.5 McFarland equivalent density.

The 2% MH-glucose plates were inoculated three times over the entire surface of the agar with the aid of a swab soaked in the yeast with the scratching action, rotating the plate approximately 60° each time to ensure a good distribution of the inoculum.

The following formulations were studied: CPF-gel, CPF-AZ-gel, gel (excipient), and controls of 100 UI/mL nystatin and 250 μg/mL amphotericin B. Approximately 5 μL of these products were placed in the inoculated yeast and incubated at 30 °C for 48 h. The inhibition zone for yeast growth was observed.

Author Contributions: Conceptualization, A.C.C. and B.C.-N.; methodology, A.C.C.; validation, L.S. and M.M.; formal analysis, M.R., M.M. and A.C.C.; investigation, N.P.-G., J.S.-C. and N.B.-d.F.; resources, A.C.C., M.M. and B.C.-N.; data curation, L.C.E., M.R. and L.S.; writing—original draft preparation, N.P.-G., J.S.-C.; writing—review and editing, L.C.E., M.R., L.S. and M.M.; visualization, L.C.E., M.R., L.S. and M.M.; supervision, A.C.C. and B.C.-N.; project administration, A.C.C. and B.C.-N.; funding acquisition, A.C.C., M.M. and B.C.-N. All authors have read and agreed to the published version of the manuscript.

Funding: This research received no external funding.

Institutional Review Board Statement: The study was conducted according to the guidelines of the Declaration of Helsinki and approved by the Institutional Review Board (or Ethics Committee) of the University of Barcelona (IRB00003099, dated 30 January 2019).

Informed Consent Statement: Informed consent was obtained from all subjects involved in the study.

Data Availability Statement: Data is contained within the article.

Acknowledgments: The authors would like to thank T. Huang for her support in the permeation studies; M.J. Rodríguez-Lagunas for her support in the histological analysis, and A. Boix for his assistance in the administrative procedures for the ethical committee approval.

Conflicts of Interest: The authors declare no conflict of interest.

References

1. Sawant, B.; Khan, T. Recent Advances in Delivery of Antifungal Agents for Therapeutic Management of Candidiasis. *Biomed. Pharmacother.* **2017**, *96*, 1478–1490. [CrossRef] [PubMed]
2. Talapko, J.; Juzbasic, M.; Matijevic, T.; Pustijanac, E.; Bekic, S.; Kotris, I.; Skrlec, I. Candida Albicans—The Virulence Factors and Clinical Manifestations of Infection. *J. Fungi* **2021**, *7*, 79. [CrossRef] [PubMed]
3. Espinoza, L.C.; Sosa, L.; Granda, P.C.; Bozal, N.; Diaz-Garrido, N.; Chulca-Torres, B.; Calpena, A.C. Development of a Topical Amphotericin B and Bursera Graveolens Essential Oil-Loaded Gel for the Treatment of Dermal Candidiasis. *Pharmaceuticals* **2021**, *14*, 1033. [CrossRef]
4. Hussein, M.; Wong, L.J.M.; Zhao, J.; Rees, V.E.; Allobawi, R.; Sharma, R.; Rao, G.G.; Baker, M.; Li, J.; Velkov, T. Unique Mechanistic Insights into Pathways Associated with the Synergistic Activity of Polymyxin B and Caspofungin against Multidrug-Resistant Klebsiella Pneumoniae. *Comput. Struct. Biotechnol. J.* **2020**, *20*, 1077–1087. [CrossRef] [PubMed]
5. Saadi Ali, H.; Barzani, H.A.H.; Yardim, Y.; Şentürk, Z. First Electrochemical Study of a Potent Antifungal Drug Caspofungin: Application to Its Enhanced Voltammetric Sensing Based on the Performance of Boron-Doped Diamond Electrode in CTAB-Mediated Measurements. *Diam. Relat. Mater.* **2022**, *125*, 109031. [CrossRef]

6. Wang, Y.; Yan, H.; Li, J.; Zhang, Y.; Wang, Z.; Sun, S. Antifungal Activity and Potential Mechanism of Action of Caspofungin in Combination with Ribavirin against Candida Albicans. *Int. J. Antimicrob. Agents* **2023**, *61*, 106709. [CrossRef]
7. Deresinski, S.; Stevens, D.C. Caspofungin. *Infect. Dis.* **2003**, *36*, 1445–1457. [CrossRef]
8. Patil, A.; Majumdar, S. Echinocandins in Antifungal Pharmacotherapy. *J. Pharm. Pharmacol.* **2017**, *69*, 1635–1660. [CrossRef]
9. Mroczynska, M.; Brillowska-Dabrowska, A. Review on Current Status of Echinocandins Use. *Antibiotics* **2020**, *9*, 227. [CrossRef]
10. Alshehri, A.F.; Almangour, T.A.; Alhifany, A.A.; Alhossan, A. Assessment of Caspofungin Use at a Tertiary Teaching Hospital and Compliance with IDSA Guidelines and FDA Labelings. *Saudi Pharm. J.* **2022**, *30*, 212–216. [CrossRef]
11. Chandel, A.K.S.; Nutan, B.; Raval, I.H.; Jewrajka, S.K. Self-Assembly of Partially Alkylated Dextran-Graft-Poly[(2-Dimethylamino)Ethyl Methacrylate] Copolymer Facilitating Hydrophobic/Hydrophilic Drug Delivery and Improving Conetwork Hydrogel Properties. *Biomacromolecules* **2018**, *19*, 1142–1153. [CrossRef]
12. Vasile, C.; Pamfil, D.; Stoleru, E.; Baican, M. New Developments in Medical Applications of Hybrid Hydrogels Containing Polymers. *Molecules* **2020**, *25*, 1539. [CrossRef]
13. Islam, M.; Park, T.-E.; Reesor, E.; Cherukula, K.; Hasan, A.; Firdous, J.; Singh, B.; Kang, S.-K.; Choi, Y.-J.; Park, I.-K.; et al. Mucoadhesive Chitosan Derivatives as Novel Drug Carriers. *Curr. Pharm. Des.* **2015**, *21*, 4285–4309. [CrossRef]
14. Jiménez-Gómez, C.P.; Cecilia, J.A. Chitosan: A Natural Biopolymer with a Wide and Varied Range of Applications. *Molecules* **2020**, *25*, 3981. [CrossRef] [PubMed]
15. Sahariah, P.; Másson, M. Antimicrobial Chitosan and Chitosan Derivatives: A Review of the Structure—Activity Relationship. *Biomacromolecules* **2017**, *18*, 3846–3868. [CrossRef]
16. Antunes, F.E.; Gentile, L.; Rossi, C.O.; Tavano, L.; Ranieri, G.A. Gels of Pluronic F127 and Nonionic Surfactants from Rheological to Controlled Drug Permeation. *Colloids Surf. B Biointerfaces* **2011**, *87*, 42–48. [CrossRef] [PubMed]
17. Ruan, J.; Liu, C.; Song, H.; Zhong, T.; Quan, P.; Fang, L. A Skin Pharmacokinetics Study of Permeation Enhancers: The Root Cause of Dynamic Enhancement Effect on In Vivo Drug Permeation. *Eur. J. Pharm. Biopharm.* **2023**, *184*, 170–180. [CrossRef] [PubMed]
18. Carvajal-Vidal, P.; Mallandrich, M.; Garcia, M.L.; Calpena, A.C. Effect of Different Skin Penetration Promoters in Halobetasol Propionate Permeation and Retention in Human Skin. *Int. J. Mol. Sci.* **2017**, *18*, 2475. [CrossRef] [PubMed]
19. Chen, Y.; Quan, P.; Liu, X.; Wang, M.; Fang, L. Novel Chemical Permeation Enhancers for Transdermal Drug Delivery. *Asian J. Pharm. Sci.* **2014**, *9*, 51–64. [CrossRef]
20. Williams, A.C.; Barry, B.W. Penetration Enhancers. *Adv. Drug Deliv. Rev.* **2012**, *64*, 128–137. [CrossRef]
21. Karande, P.; Mitragotri, S. Enhancement of Transdermal Drug Delivery via Synergistic Action of Chemicals. *Biochim. Biophys. Acta* **2009**, *1788*, 2362–2373. [CrossRef] [PubMed]
22. Lane, M.E. Skin Penetration Enhancers. *Int. J. Pharm.* **2013**, *447*, 12–21. [CrossRef] [PubMed]
23. Lee, S.G.; Kang, J.; Kim, S.; Kim, C.; Yeom, D.; Yoon, H.; Kwak, S.; Choi, Y. Enhanced Topical Delivery of Tacrolimus by a Carbomer Hydrogel Formulation with Transcutol P. *Drug Dev. Ind. Pharm.* **2016**, *42*, 1636–1642. [CrossRef] [PubMed]
24. Pitzanti, G.; Rosa, A.; Nieddu, M.; Valenti, D.; Pireddu, R.; Lai, F.; Cardia, M.; Fadda, A.; Sinico, C. Transcutol® P Containing SLNs for Improving 8-Methoxypsoralen Skin Delivery. *Pharmaceutics* **2020**, *12*, 973. [CrossRef]
25. Sanz, R.; Calpena, A.; Mallandrich, M.; Gimeno, Á.; Halbaut, L.; Clares, B. Development of a Buccal Doxepin Platform for Pain in Oral Mucositis Derived from Head and Neck Cancer Treatment. *Eur. J. Pharm. Biopharm.* **2017**, *117*, 203211. [CrossRef]
26. Ciurlizza, C.; Fernández, F.; Calpena, A.C.; Lázaro, R.; Parra, A.; Clares, B. Semisolid Formulations Containing Cetirizine: Human Skin Permeation and Topical Antihistaminic Evaluation in a Rabbit Model. *Arch. Dermatol. Res.* **2014**, *306*, 711–717. [CrossRef]
27. Letscher-Bru, V.; Herbrecht, R. Caspofungin: The First Representative of a New Antifungal Class. *J. Antimicrob. Chemother.* **2003**, *51*, 513–521. [CrossRef]
28. Sullivan, D.W.; Gad, S.C.; Julien, M. A Review of the Nonclinical Safety of Transcutol®, a Highly Purified Form of Diethylene Glycol Monoethyl Ether (DEGEE) Used as a Pharmaceutical Excipient. *Food Chem. Toxicol.* **2014**, *72*, 40–50. [CrossRef]
29. Mfoafo, K.; Kwon, Y.; Omidi, Y.; Omidian, H. Contemporary Applications of Thermogelling PEO-PPO-PEO Triblock Copolymers. *J. Drug Deliv. Sci. Technol.* **2022**, *70*, 103182. [CrossRef]
30. Soliman, K.A.; Ullah, K.; Shah, A.; Jones, D.S.; Singh, T.R.R. Poloxamer-Based in Situ Gelling Thermoresponsive Systems for Ocular Drug Delivery Applications. *Drug Discov. Today* **2019**, *24*, 1575–1586. [CrossRef]
31. Zarrintaj, P.; Ramsey, J.D.; Samadi, A.; Atoufi, Z.; Yazdi, M.K.; Ganjali, M.R.; Amirabad, L.M.; Zangene, E.; Farokhi, M.; Formela, K.; et al. Poloxamer: A Versatile Tri-Block Copolymer for Biomedical Applications. *Acta Biomater.* **2020**, *110*, 37–67. [CrossRef]
32. Sacco, P.; Furlani, F.; de Marzo, G.; Marsich, E.; Paoletti, S.; Donati, I. Concepts for Developing Physical Gels of Chitosan and of Chitosan Derivatives. *Gels* **2018**, *4*, 67. [CrossRef]
33. Samani, S.M.; Ahmadi, F.; Oveisi, Z.; Amoozgar, Z. Chitosan Based Hydrogels: Characteristics and Pharmaceutical Applications. *Res. Pharm. Sci.* **2015**, *10*, 1–16.
34. Wang, W.Y.; Hui, P.C.L.; Wat, E.; Ng, F.S.F.; Kan, C.W.; Lau, C.B.S.; Leung, P.C. Enhanced Transdermal Permeability via Constructing the Porous Structure of Poloxamer-Based Hydrogel. *Polymers* **2016**, *8*, 406. [CrossRef] [PubMed]
35. Desai, P.R.; Jain, N.J.; Sharma, R.K.; Bahadur, P. Effect of Additives on the Micellization of PEO/PPO/PEO Block Copolymer F127 in Aqueous Solution. *Colloids Surf. A Physicochem. Eng. Asp.* **2001**, *178*, 57–69. [CrossRef]

36. Raina, N.; Pahwa, R.; Bhattacharya, J.; Paul, A.K.; Nissapatorn, V.; Pereira, M.d.L.; Oliveira, S.M.R.; Dolma, K.G.; Rahmatullah, M.; Wilairatana, P.; et al. Drug Delivery Strategies and Biomedical Significance of Hydrogels: Translational Considerations. *Pharmaceutics* **2022**, *14*, 574. [CrossRef]
37. Espinoza, L.C.; Guaya, D.; Calpena, A.C.; Perotti, R.M.; Halbaut, L.; Sosa, L.; Brito-Llera, A.; Mallandrich, M. Comparative Study of Donepezil-Loaded Formulations for the Treatment of Alzheimer's Disease by Nasal Administration. *Gels* **2022**, *8*, 715. [CrossRef] [PubMed]
38. Berenguer, D.; Alcover, M.M.; Sessa, M.; Halbaut, L.; Guillen, C.; Boix-Montanes, A.; Fisa, R.; Calpena-Campmany, A.C.; Riera, C.; Sosa, L. Topical Amphotericin B Semisolid Dosage Form for Cutaneous Leishmaniasis: Physicochemical Characterization, Ex Vivo Skin Permeation and Biological Activity. *Pharmaceutics* **2020**, *12*, 149. [CrossRef]
39. Perez-Gonzalez, N.; Bozal-de Febrer, N.; Calpena-Campmany, A.C.; Nardi-Ricart, A.; Rodriguez-Lagunas, M.J.; Morales-Molina, J.A.; Soriano-Ruiz, J.L.; Fernandez-Campos, F.; Clares-Naveros, B. New Formulations Loading Caspofungin for Topical Therapy of Vulvovaginal Candidiasis. *Gels* **2021**, *7*, 259. [CrossRef]
40. Brugués, A.P.; Naveros, B.C.; Campmany, A.C.C.; Pastor, P.H.; Saladrigas, R.F.; Lizandra, C.R. Developing Cutaneous Applications of Paromomycin Entrapped in Stimuli-Sensitive Block Copolymer Nanogel Dispersions. *Nanomedicine* **2015**, *10*, 227–240. [CrossRef]
41. European Medicines Agency. *Draft Guideline on Quality and Equivalence of Topical Products*; European Medicines Agency: London, UK, 2018.
42. Ilic, T.; Pantelic, I.; Savic, S. The Implications of Regulatory Framework for Topical Semisolid Drug Products: From Critical Quality and Performance Attributes towards Establishing Bioequivalence. *Pharmaceutics* **2021**, *13*, 710. [CrossRef]
43. Dey, S.; Rothe, H.; Page, L.; O'Connor, R.; Farahmand, S.; Toner, F.; Marsh, R.; Wehmeyer, K.; Zhou, S. An In Vitro Skin Penetration Model for Compromised Skin: Estimating Penetration of Polyethylene Glycol [(1)(4)C]-PEG-7 Phosphate. *Skin Pharmacol. Physiol.* **2015**, *28*, 12–21. [CrossRef]
44. Kulkarni, V.S.; Shaw, C. Use of Polymers and Thickeners in Semisolid and Liquid Formulations. In *Essential Chemistry for Formulators of Semisolid and Liquid Dosages*; Kulkarni, V.S., Shaw, C., Eds.; Elsevier: Amsterdam, The Netherlands, 2016; pp. 43–69.
45. Mohammadi-Meyabadi, R.; Beirampour, N.; Garros, N.; Alvarado, H.L.; Limon, D.; Silva-Abreu, M.; Calpena, A.C.; Mallandrich, M. Assessing the Solubility of Baricitinib and Drug Uptake in Different Tissues Using Absorption and Fluorescence Spectroscopies. *Pharmaceutics* **2022**, *14*, 2714. [CrossRef]
46. Osborne, D.W.; Musakhanian, J. Skin Penetration and Permeation Properties of Transcutol(R)-Neat or Diluted Mixtures. *AAPS PharmSciTech* **2018**, *19*, 3512–3533. [CrossRef]
47. Montero-Vilchez, T.; Segura-Fernandez-Nogueras, M.V.; Perez-Rodriguez, I.; Soler-Gongora, M.; Martinez-Lopez, A.; Fernandez-Gonzalez, A.; Molina-Leyva, A.; Arias-Santiago, S. Skin Barrier Function in Psoriasis and Atopic Dermatitis: Transepidermal Water Loss and Temperature as Useful Tools to Assess Disease Severity. *J. Clin. Med.* **2021**, *10*, 359. [CrossRef] [PubMed]
48. Mori, M.; Imaizumi, M.; Ishiwada, N.; Kaneko, T.; Goto, H.; Kato, K.; Hara, J.; Kosaka, Y.; Koike, K.; Kawamoto, H. Pharmacokinetics, Efficacy, and Safety of Caspofungin in Japanese Pediatric Patients with Invasive Candidiasis and Invasive Aspergillosis. *J. Infect. Chemother.* **2015**, *21*, 421–426. [CrossRef] [PubMed]
49. Farmakiotis, D.; Tarrand, J.J.; Kontoyiannis, D.P. Drug-resistant Candida glabrata infection in cancer patients. *Emerg. Infect. Dis.* **2014**, *20*, 1833–1840. [CrossRef] [PubMed]
50. Raafat, D.; von Bargen, K.; Haas, A.; Sahl, H.G. Insights into the Mode of Action of Chitosan as an Antibacterial Compound. *Appl. Environ. Microbiol.* **2008**, *74*, 3764–3773. [CrossRef]
51. Park, E.K.; Song, K.W. Rheological Evaluation of Petroleum Jelly as a Base Material in Ointment and Cream Formulations: Steady Shear Flow Behavior. *Arch. Pharm. Res.* **2010**, *33*, 141–150. [CrossRef]
52. El Moussaoui, S.; Fernández-Campos, F.; Alonso, C.; Limón, D.; Halbaut, L.; Garduño-Ramirez, M.L.; Calpena, A.C.; Mallandrich, M. Topical Mucoadhesive Alginate-Based Hydrogel Loading Ketorolac for Pain Management after Pharmacotherapy, Ablation, or Surgical Removal in Condyloma Acuminata. *Gels* **2021**, *7*, 8. [CrossRef]
53. Rincon, M.; Silva-Abreu, M.; Espinoza, L.C.; Sosa, L.; Calpena, A.C.; Rodriguez-Lagunas, M.J.; Colom, H. Enhanced Transdermal Delivery of Pranoprofen Using a Thermo-Reversible Hydrogel Loaded with Lipid Nanocarriers for the Treatment of Local Inflammation. *Pharmaceuticals* **2021**, *15*, 22. [CrossRef] [PubMed]
54. El Moussaoui, S.; Abo-horan, I.; Halbaut, L.; Alonso, C.; Coderch, L.; Garduño-ramírez, M.L.; Clares, B.; Soriano, J.L.; Calpena, A.C.; Fernández-campos, F.; et al. Polymeric Nanoparticles and Chitosan Gel Loading Ketorolac Tromethamine to Alleviate Pain Associated with Condyloma Acuminata during the Pre- and Post-ablation. *Pharmaceutics* **2021**, *13*, 1784. [CrossRef] [PubMed]

gels MDPI

Review

Hydrogels with Essential Oils: Recent Advances in Designs and Applications

Mariana Chelu

"Ilie Murgulescu" Institute of Physical Chemistry, 202 Spl. Independentei, 060021 Bucharest, Romania; mchelu@icf.ro

Abstract: The innovative fusion of essential oils with hydrogel engineering offers an optimistic perspective for the design and development of next-generation materials incorporating natural bioactive compounds. This review provides a comprehensive overview of the latest advances in the use of hydrogels containing essential oils for biomedical, dental, cosmetic, food, food packaging, and restoration of cultural heritage applications. Polymeric sources, methods of obtaining, cross-linking techniques, and functional properties of hydrogels are discussed. The unique characteristics of polymer hydrogels containing bioactive agents are highlighted. These include biocompatibility, nontoxicity, effective antibacterial activity, control of the sustained and prolonged release of active substances, optimal porosity, and outstanding cytocompatibility. Additionally, the specific characteristics and distinctive properties of essential oils are explored, along with their extraction and encapsulation methods. The advantages and disadvantages of these methods are also discussed. We have considered limitations due to volatility, solubility, environmental factors, and stability. The importance of loading essential oils in hydrogels, their stability, and biological activity is analyzed. This review highlights through an in-depth analysis, the recent innovations, challenges, and future prospects of hydrogels encapsulated with essential oils and their potential for multiple applications including biomedicine, dentistry, cosmetics, food, food packaging, and cultural heritage conservation.

Keywords: hydrogels; essential oils; encapsulation; biomedical applications; cosmetics; dentistry; active food packaging; restoration of cultural heritage

check for updates

Citation: Chelu, M. Hydrogels with Essential Oils: Recent Advances in Designs and Applications. *Gels* **2024**, *10*, 636. https://doi.org/10.3390/gels10100636

Academic Editor: Xiaoqiu Dou

Received: 9 September 2024
Revised: 27 September 2024
Accepted: 29 September 2024
Published: 30 September 2024

1. Introduction

Hydrogels form a unique category, being among the most modern multifunctional materials that can be applied in numerous fields. As a result, they have captured the interest of many scientists in various research fields. Hydrogels are essentially 3D cross-linked networks formed of hydrophilic polymeric materials that can retain large volumes of water and fluids [1]. They can be formulated from both synthetic polymers and biopolymers. Hydrogels based on natural biodegradable polymers, such as polysaccharides, polypeptides, and proteins have many advantages over synthetic ones and have gained particular importance lately [2]. One of the advantages is the porous macromolecular structure, which can be easily adjusted so that the hydrogels can incorporate different bioactive compounds and then release them in a controlled manner.

Due to their special properties, the so-called intelligent hydrogels have the ability to swell in an aqueous environment, show sensitivity to temperature, light, pH variations and other stimuli, self-healing, and shape memory [3].

Through different strategies, such as molecular design, cross-linking techniques, or the incorporation of different bioactive compounds, the properties and functions of hydrogels can be adapted for a wide variety of applications. They are widely used in the biomedical field for wound healing and tissue engineering, as well as drug delivery systems, pharmaceutical products, the food industry, cosmetics, hygiene products, and dentistry, etc.

Similarly to the remarkable class of hydrogels, essential oils (EOs) have shown an extraordinary increase in scientific interest in recent years, providing huge potential for a diverse range of modern applications, including cutting-edge fields such as nanotechnology, bioengineering, and biomedicine. Extracted from aromatic plants, EOs are hydrophobic products of high concentration and contain molecules with low molecular weight. The extraordinary therapeutic potential of EOs is due to the biological activity of their volatile chemical components (terpenoids, terpenes, and other aromatic compounds) and non-volatile (hydrocarbons, fatty acids, sterols, carotenoids, waxes, and flavonoids) [4].

This review brings a new perspective, scrutinizing the potential to incorporate different essential oils into hydrogels in order to develop efficient delivery systems for bioactive molecules of natural origin. Through this versatile delivery method, new products with improved bioactive activity can find use for many types of applications to provide both cost-effectiveness and high efficiency.

At present, the global production of essential oils is driven by the strong demand for consumption, from natural options to synthetic antioxidants, due to remarkable biological properties, including antimicrobial, antioxidant, anti-inflammatory, antiviral, and antitumor effects [5]. These distinctive qualities pave the way for promising opportunities in utilizing EOs across different fields.

As potent curative agents, EOs offer a viable alternative to synthetic drugs, particularly for their antimicrobial effectiveness against a wide range of pathogenic microorganisms. They are most commonly applied in the biomedical and pharmaceutical sectors.

Since ancient times, the sensory and pharmacological activities of essential oils have been studied for their use in preventive and curative treatments, particularly in cosmetics and countless personal care products.

In the food industry, EOs are used on a large scale, as a beneficial mechanism to combat undesirable microorganisms in food products. The natural bioactive capacities of EOs, such as antimicrobial and antioxidant properties, together with their aroma, flavor, or spicy taste, make them suitable for use as preservatives in the food industry.

EOs can be incorporated into different food systems to increase the shelf life of food while maintaining its quality. They can also be encapsulated in the form of edible coatings or films, to mitigate microbial development on the surface and protect the environment from synthetic packaging.

However, EOs cannot be applied directly, in their raw form, because they are unstable volatile compounds, fragile, and of very high concentration. They require special post-obtaining conditions so that the original chemical profile is not modified by different environmental conditions (light, heat, oxidation) [6].

Consequently, EOs can be encapsulated in different hydrogel matrices to prolong their effective biological activity and ensure their release in a controlled and sustained manner.

Hydrogels are versatile platforms with an adjustable structure, of a non-toxic nature that can be used safely [7]. The high-water content, the sustainable nature of the constituent polymers, and the ability to incorporate EOs give hydrogels favorable biocompatibility and key structural characteristics, making them suitable for a wide range of implementation [8].

This review aims to showcase significant research and findings through a detailed analysis of the main applications of essential oil-embedded hydrogels. Therefore, this review article provides a comprehensive overview of recent studies on obtaining hydrogels, and EOs, as well as their incorporation into hydrogel matrices, for a multitude of uses. These include fields such as biomedicine, dentistry, cosmetics, functional food products, food packaging, and even the preservation of stone cultural heritage.

2. Preparation of Hydrogels

Hydrogels are systems made up of polymers and solvents obtained in the form of 3D cross-linked networks. Cross-linking can be formed (i) physically, (ii) chemically or (iii) by ionizing radiation [9].

The main advantage of hydrogels obtained by the physical method is the absence of crosslinking agents which eliminates the risk of possible residual toxicity [9,10].

However, a notable disadvantage is that physically cross-linked hydrogels are reversible. Between the polymer chains, there are temporary bonds that appear in response to composition, pH, or temperature changes. Moreover, these hydrogels show weak mechanical and viscoelastic properties. Depending on the size of the polymer particles and the nature of the solvent in the 3D network, there may be hydrogen bonds, van der Walls bonds, electrostatic, hydrophobic interactions, or between-polymer chains with local crystallite generation [9].

Chemically obtained hydrogels are created by covalent cross-linking between the existing polymer chains. This leads to a stable or irreversible bond. Chemical cross-linking can be produced using various methods, including addition polymerization, photopolymerization, volume condensation, plasma or electromagnetic radiation, or interpenetration [10].

In general, a cross-linking agent with a multifunctional role is used, which unites the monomeric units and leads to the development of the polymer chain [10]. Chemically cross-linked hydrogels are advantageous due to their resistance to degradation and improved mechanical and viscoelastic properties compared to physically cross-linked ones [9]. On the other hand, hydrogels, obtained by anionic or cationic polymerization have as their main disadvantage the sensitivity to water, respectively, the limitation to non-polar monomers [9].

In the case of hydrogels made by radiation, they can be obtained at ambient temperature and physiological pH, even without a crosslinking agent, which makes them suitable to be applied in a wide range of biomedical, food, or cosmetic applications [11].

Hydrogels can be categorized according to various characteristics, providing deeper insight into their properties and potential uses (Figure 1). These characteristics include [9]:

- According to source: natural, synthetic, or hybrid.
- According to polymer structure: linear, branched, or cross-linked.
- According to physical appearance: macroporous, microporous, or nanoporous.
- According to charge: neutral, anionic (negatively charged), cationic (positively charged), or amphoteric, meaning they contain both positive and negative charges.
- According to responsiveness to stimuli: Hydrogels can also be classified by their sensitivity to external stimuli such as temperature, pH, light, or electric fields. Called "smart hydrogels," these materials can undergo reversible changes in their structure or properties when subjected to these environmental influences.
- According to water content: superabsorbent hydrogels, and less moisture.
- According to degradability: biodegradable, and non-biodegradable.

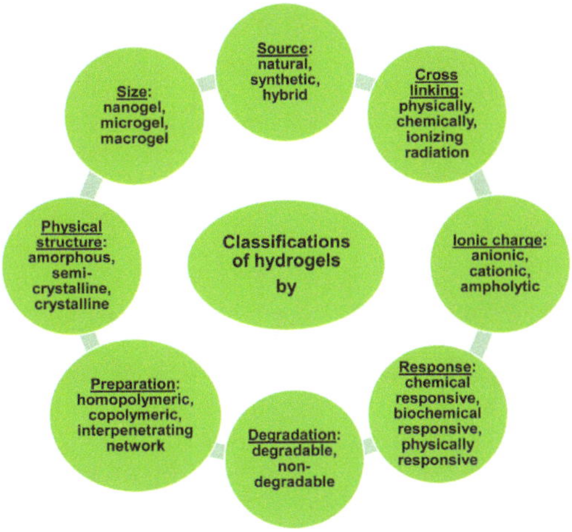

Figure 1. Classifications of hydrogels.

Polymers play an important role in the matrix of hydrogels, greatly influencing their properties. They can be divided into two main categories: natural polymers and synthetic polymers. Biodegradable polymeric materials, both natural and synthetic, have received more interest lately, due to the importance of environmentally friendly products and the possibility of their application in many fields such as biomedicine, pharmaceuticals, or agriculture [12].

The creation of advanced materials formulated from bioavailable and renewable raw materials, including waste, has been increasingly promoted, in agreement with the 12 principles of Green Chemistry and the achievement of the Sustainable Development Goals provided for the UN 2030 Agenda.

Natural polymers are polymer molecules of biological origin that can be obtained from different sources such as animals, bacteria, microorganisms (algae and fungi), or plants. The chemical structures of natural polymers are composed of monomers of amino acids, nucleotides, esters, or monosaccharides, that are covalently coupled to form peptides, polyphenols, polyesters, or polysaccharides [13].

Because they are similar to the components of the extracellular matrix (ECM), natural polymers have a reduced toxicity, with a low risk of causing adverse reactions. This biocompatibility has determined their widespread use in numerous biomedical, pharmaceutical, and cosmetic applications, as additives in textile products, and in food or agriculture [14].

The most used natural polymers include sodium alginate, starch, gelatin, chitosan, collagen, hyaluronic acid, κ-carrageenan, cellulose, gum arabic, silk, fibrin, and bacterial polyesters [15,16].

Synthetic polymers are created artificially in laboratories and can be mainly classified as thermoplastic and thermosetting polymers and elastomers. They are very often found in multiple fields, such as packaging and construction, as plastic materials, fibers, elastomers, or adhesives. As synthetic polymers, we can mention polyvinyl alcohol, poly(lactic acid), polyvinylpyrrolidone, poly(ε-caprolactone), polyurethane, polyethylene glycol, polyethylene oxide, poly(L-lactide-co-caprolactone), carboxymethyl cellulose and poly(vinylidene fluoride) [17]. Among them, some synthetic polymers are biocompatible and biodegradable, such as poly(lactic acid), carboxymethyl cellulose, poly(acrylic acid), poly(vinyl alcohol), or polyethylene glycol. Some of them have shown antitumor, antibiotic, antiviral, or antithrombotic activities and are often used as drug carriers, implants, diagnostic imaging agents, or as bio-ink in 3D printing for various biological scaffolds [18,19].

A new approach in hydrogel engineering is the design of complex systems through which hybrid hydrogels are obtained that incorporate both natural and synthetic polymers, but also other functional components [20]. The hybrid hydrogels that have been developed are capable of integrating nano- or microstructures, allowing for targeted action, controlled transport, and an adjustable release profile.

Hybrid nanogels have the ability to respond faster than macroscopic ones to environmental variations, proving their usefulness especially in biomedical applications such as therapies for tissue engineering, transport and delivery of chemicals in cancer therapy, or in optical detection [21].

3. Methods of Obtaining Essential Oils

Since ancient times, people have used EOs because they were believed to contain essential components that are necessary for healing and prolonging life. Alchemists referred to them as the "quintessence of plants".

Essential oils are aromatic oil liquids in the form of complex natural mixtures of different polar and non-polar compounds, made from natural raw material of plant origin. The main chemical constituents of essential oils are volatile, lipophilic, and odoriferous substances that are commonly found in different parts of plants (leaves, flowers, fruits, or stems), giving them specific properties [22]. Since the beginning, aromatic plants have been used empirically as spices in kitchens, in perfumes, cosmetics, and aromatherapy, for preventive, curative, or therapeutic purposes. With the advent of distillation centers, the methods of obtaining essential oils have advanced and the therapeutic benefits of EO have been scientifically evaluated [23].

The different production techniques involve (i) hot distillation, with water or steam, (ii) cold or dry distillation, and (iii) through various mechanical processes. While the yield of obtaining pure essential oils from aromatic plants is very low, their price is commensurate with their natural bioactivity and implicit pharmaceutical and therapeutic benefits.

Over time, it has been proven that essential oils contain over 300 different aromatic components, showing an extremely varied therapeutic potential [24]. In practice, these capabilities of essential oils have been continuously explored and harnessed on a large scale for a wide range of multiple applications that have been successfully integrated into a variety of industries and contexts. The chemical composition of essential oils can vary, even for the same species, depending on certain parameters such as climatic factors, soil characteristics, harvesting conditions, post-harvest treatment, or the extraction methods used [24].

The uniqueness and importance of EOs are given by their medicinal and bioactive compounds, as well as by the other valuable constituents. EOs show significant variability in their composition, both in terms of quality and quantity. Since this variability is heavily influenced including on the extraction method used, it is crucial to identify optimal, and especially non-toxic extraction techniques.

This chapter will offer a concise summary of the main extraction methods, including both traditional and modern approaches, which are continuously being refined for improvement.

The main compounds discovered in EOs are terpenoid derivatives (80%) and phenylpropanoids (which give the specific spicy smell and aroma) [25].

There are several ways by which EOs are extracted from aromatic plants. The selection of an extraction method is influenced by the plant's texture and characteristics, the specific essential oils being targeted, and the intended application of the final product. Every method demonstrates its own advantages and drawbacks.

Some classical EO extraction methods, also called conventional, have been practiced for hundreds of years and include steam distillation, water distillation, combined water and steam distillation, cohobation (or repeated distillation), maceration, cold pressing, and enfleurage. The limitations of these methods are mainly represented by low extraction yields, thermal degradation, or the need to use high mechanical power [23].

Other more recent methods, also called alternatives, try to demonstrate their efficiency in operation, to be ecological and viable from an economic point of view. Among these can be listed extraction with solvents, supercritical CO_2, or resins, and fractional distillation, percolation, and the phytonic process (uses a new solvent based on hydrofluorocarbon 134) [23]. The main benefits of supercritical extraction with CO_2 are its low cost and non-corrosive nature, which enables the production of thermally unstable EOs at an industrial level with a high yield. In addition, it is an ecological way by which safe EOs can be produced to be used in various applications in the food industry.

New extraction methods use "green concepts" to extract valuable components from aromatic plants. They act in accordance with the United Nations 2030 strategy, pursuing sustainable developments by reducing waste, using discarded by-products, recycling them, and reducing the carbon footprint in processing, which will have a positive impact towards a cleaner environment. The global EO market has been continuously growing, reaching USD 7.51 billion in 2018, and is expected to grow at a CAGR of over 9% between 2019 and 2026 [26].

Innovative techniques that respect these green concepts include ultrasound-assisted extraction of bioactive compounds, microwave-assisted extraction of essential oils, high-pressure liquid extraction, sub- and supercritical fluid extraction, pulsed electric fields, and high-voltage electric discharges [27].

Essential oils are categorized into three main categories which include (i) terpenes, (ii) terpenoids, and (iii) phenylpropanoids in their chemical composition. Terpene constituents can be classified into two primary groups: (i) components that have a hydrocarbon structure (such as monoterpenes diterpenes and sesquiterpenes) and (ii) those that are oxygenated, including acids, aldehydes, alcohols, esters, ketones, lactones, oxides, and phenols [28]. Among the most common terpenes, we can distinguish limonene, sabinene, α-pinene and p-cymene, for example in thyme and oregano, but also in lemon, grapefruit, eucalyptus, and rosemary EOs. Carvacrol, citronellal, carvone, and thymol are all terpenoids that are present in EOs like mint, lavender, tea tree, chamomile, or geranium. Clove, jasmine, rose, or pepper EOs contain phenylpropanoids that can be identified as cinnamaldehyde, eugenol, safrole, and vanillin. EOs also includes other components derived from amino acids, such as alanine, leucine, isoleucine, methionine, and valine [25].

The primary drawbacks of using EOs include their volatility, high sensitivity, poor stability, high sensitivity, and vulnerability to degradation at processing temperatures. To address these challenges, the encapsulation of EO in polymeric matrices improves their bioactivity, stability, and water solubility, and enables long-term sustained delivery across various applications [29].

4. Encapsulation of Essential Oils in Hydrogels

Most essential oils cannot be applied through direct contact with biological systems, because they can be irritating or even toxic in certain cases. Incorporating essential oils directly into hydrophilic matrices is not beneficial due to their hydrophobic nature, which diminishes the inherent bioactivity of their components, leading to the use of high concentrations to be functional. To preserve their biological activity over an extended period, particularly for biomedical applications, EOs should be encapsulated in various systems, such as lipidic nanoparticles, liposomes, films, emulsion gels, oil-in-water emulsions, or spray-dried microparticles [30].

Through encapsulation, risks due to possible toxicity are reduced, ensuring a safe delivery system. Moreover, this method increases the biological activities and efficiency of particularly volatile EOs through better absorption. In addition, encapsulation has often been widely used as a way of protecting essential oils, prolonging their active biocapacity and efficient delivery, which offers the possibility of their implementation in the medical, pharmaceutical, cosmetic, and food fields [31].

These benefits can be achieved by employing different techniques to encapsulate es- essential oils with diverse bioactivities [26,32]: (i) chemical; (ii) physico-chemical; (iii) mechanical; (iv) ultrasound-assisted emulsification; and (v) electrostatic extrusion.

An example is thyme EO, which is often used due to its therapeutic properties. Thyme (*Thymus vulgaris* L.) is a plant native to the Mediterranean area with both dietary and medicinal uses. It contains many polyphenolic compounds of biological interest, such as carvacrol, 5-isopropyl-2-methylphenol, and a p-cymene derivative with a characteristic smell, with antioxidant, antimicrobial, antidiabetic, anti-inflammatory, immunomodulatory, and anticancer bioactivities [33,34]. Thyme EO was encapsulated in the first step in sodium caseinate nanomicelles by a physical method [35]. Then, in the second step, these nanomicelles were introduced into the preparation of a hydrogel. This formulation aimed to improve the stability and protect the bioactivity of thyme EO. Finally, a gelatin nanocomposite hydrogel as a drug delivery platform was obtained, having antibacterial potential for wound healing both in vitro and in vivo [35].

The ionic gelation method, achieved by ionic bonding between alginate and some divalent cations, led to the creation of a biocompatible hydrogel material in the form of alginate microspheres that encapsulated thyme and calendula EO [36]. The study investigated the loading capacity, the encapsulation efficiency of EO, and the dissolution of microspheres under simulated digestion conditions.

Lipid matrices offer a suitable and stable environment for incorporating EOs and ensuring their controlled release [37]. In particular, solid lipid nanoparticles have aroused special interest for the encapsulation of bioactive compounds due to their large surface area and the potential to facilitate the protection of bioactive constituents in ambient conditions.

An interesting method for obtaining lipid matrices is homogenization at high shear followed by the ultrasonication method [37]. Thus, chitosan and polyvinyl alcohol hydrogels containing solid lipid nanoparticles loaded with EOs of *Origanum vulgare* and *Thymus vulgaris* were formulated and investigated as alternatives to synthetic fungicides. The materials made with EOs content have demonstrated abilities to reduce the infestation with phytopathogenic fungi responsible for the degradation of perishable fruits [37].

A new work investigated the capabilities of *Perilla frutescens* (L.), the annual aromatic plant cultivated and used for thousands of years in traditional medicine or as food. In the first step, microcapsule powders of *Perilla frutescens* (L.) [38] EOs were prepared by the spray drying method of a wall material (octenyl succinic anhydride starch). In the next step, they were further encapsulated with sodium alginate and chitosan by the polyelectrolyte complex coacervate method, obtaining stable hydrogel balls for aqueous and acidic food formulations with a complete and prolonged release of the encapsulated EOs [38].

Hydrophobic clove EO was loaded in situ into a hydrophilic chitosan polymer matrix to obtain functional coatings as food packaging [32]. By using this method, bioactive materials were obtained without the need for crosslinking agents.

The electrostatic extrusion technique that was applied to encapsulate fennel EO in an alginate polymer matrix, together with the incorporation of a whey protein followed by freeze-drying, is an original approach to improve the encapsulation efficiency and loading capacity [39]. The encapsulated EO maintained its qualitative appearance by keeping 58.95% of the volatile compounds [39].

5. Applications of Hydrogel Materials Enriched with Essential Oil

The exploration and application of the bioactivities of essential oils as natural phytotherapeutic agents in various biomedical fields arose from the need to develop alternative therapeutic approaches to traditional synthetic treatments.

Hydrogels are ideal host matrices for some limitations of EOs, such as volatility, high sensitivity to environmental factors, and lower stability [26]. Together, EOs and hydrogels are biocompatible and biodegradable materials, which demonstrate remarkable physicochemical properties and antibacterial, antioxidant, anti-inflammatory, and anticancer activities [40]. The porous 3D structure of hydrogels facilitates the incorporation of essential oils through hydrophobic interactions, enabling their sustained and controlled release in response to various stimuli such as hydrolytic and enzymatic activity, pH changes, or temperature variations [41].

The beneficial combination of essential oils with the engineering of hydrogels can be an advanced approach to the design and development of the next generation of hybrid biomedical systems that embed natural therapeutic compounds.

5.1. Biomedical Applications

5.1.1. Topical or Transdermal Delivery Systems

Natural polysaccharides are among the most widely used biopolymers in biomedical applications due to their biocompatibility, bioactivity, biodegradability, and exceptional rheological and biomucoadhesive properties. These attributes make them ideal for developing a wide variety of topical formulations, for wound healing, or as effective and inexpensive drug delivery systems [42]. Additionally, marine polysaccharides enhance hydrogel formation capabilities, making them particularly effective for skin applications in treating various dermatological conditions.

Alginate and fucoidan hydrogels loaded with menthol, L-linalool, bergamot oil, and β-pinene essential oils have been developed to improve skin permeability [43]. The aim of the study was to evaluate the way in which these EOs influence the penetration of the active ingredients through the skin, and the effect of the composition, in order to create effective formulations for topical or transdermal administration [43]. The porous morphology of the prepared hydrogels, presented in Figure 2, could be due to the lyophilized oil droplets, which can lead to these structures. Menthol, a cyclic monoterpene, is widely recognized for its ability to improve skin permeability by disrupting intercellular lipids in the stratum corneum. Bergamot EO (*Citrus bergamia*) is mainly composed of limonene, linalyl acetate, and linalool with anti-inflammatory properties and b-pinene, a bicyclic monoterpene, with antioxidant, anti-inflammatory, and analgesic effects. Combining the activities of EOs like menthol, L-linalool, bergamot oil, and β-pinene can indeed be a powerful strategy for overcoming the skin barrier and treating inflammation. Each of these essential oils has unique properties that, when combined, can work synergistically to enhance skin permeability and provide anti-inflammatory benefits [43].

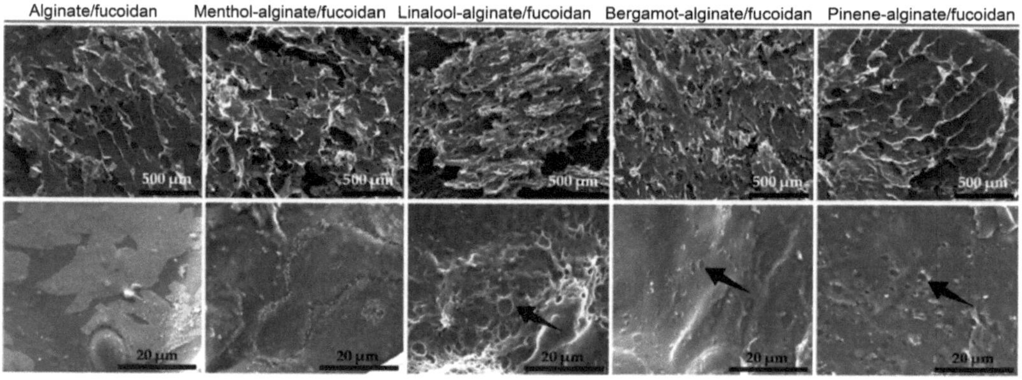

Figure 2. SEM images of the hydrogel samples [43].

5.1.2. Antimicrobial and Anti-Inflammatory Activity

Oral candidiasis is a fungal infection primarily produced by *Candida* species for which there is a rather limited antifungal treatment. This condition is particularly challenging to manage due to the limited availability of effective antifungal treatments and the potential for these treatments to cause adverse effects and contribute to the development of antifungal resistance.

Encapsulating biocides within hydrogels is an effective strategy for targeted delivery, offering controlled release and enhanced therapeutic effects. Specifically, using methylcellulose-based hydrogels incorporated with *Melissa officinalis* EO can enhance antimicrobial efficacy while maintaining biocompatibility with biological tissues [44]. The hydrogel formulation based on methylcellulose with *Melissa officinalis* EO demonstrated both antimicrobial activity and antifungal potential, making it an effective treatment for inhibiting oral candidiasis [44].

A complex study focused on the development of hydrogel films made from a combination of polyvinyl alcohol (PVA), corn starch, patchouli oil, and silver nanoparticles (Figure 3) [45]. These materials were chosen for their bioactive properties, particularly their effectiveness against *Staphylococcus aureus* and *Staphylococcus epidermidis*, both of which are common bacteria responsible for various infections, including skin and soft tissue infections. The nanoparticles were prepared by green synthesis, in the presence of both aqueous and methanolic extracts from patchouli plants (*Pogostemon cablin* Benth). The use of cross-linked polymeric hydrogel films with glutaraldehyde and containing biosynthesized silver nanoparticles with phytochemicals presents an advanced approach to developing antimicrobial materials [45].

Figure 4 shows SEM and photo images during the swelling experiments.

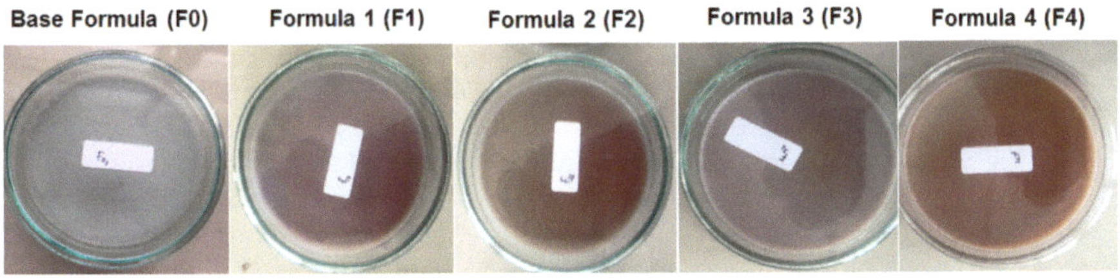

Figure 3. Photo of hydrogel films based on polyvinyl alcohol–cornstarch–patchouli oil and Ag nanoparticles [45].

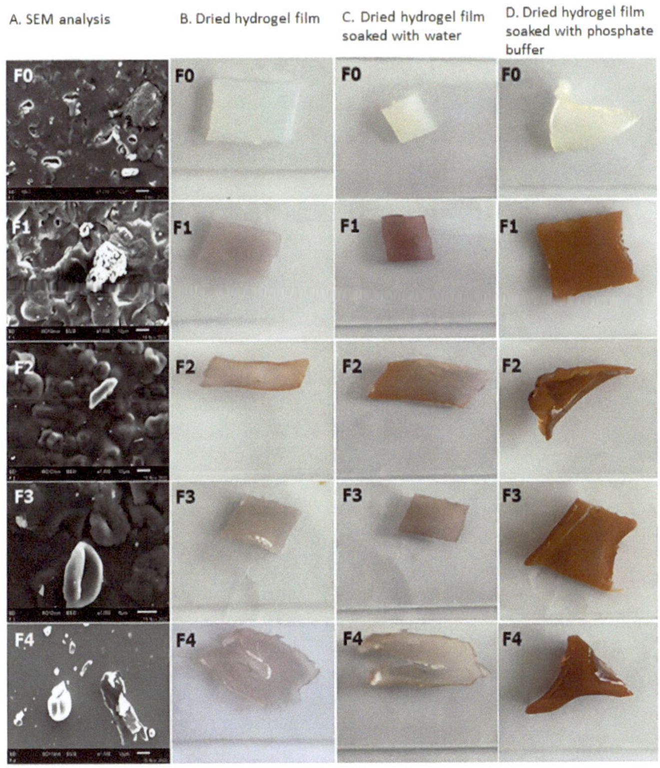

Figure 4. SEM images of hydrogel films based on polyvinyl alcohol/corn starch/patchouli oil (**A**) with Ag nanoparticles (samples F0–F4); (**B**) Dry samples; (**C**) Dry samples soaked in water; (**D**) Dry samples soaked with phosphate buffer [45].

Acne Vulgaris is a common inflammatory skin condition that affects many young individuals and often persists into adulthood. Traditional acne treatments, which mainly rely on antibiotics, have shown limited effectiveness and frequently disrupt the balance of the skin microbiome. Recent research suggests that essential oils and herbs could offer promising benefits for treating acne, a long-lasting inflammatory condition that can lead to scarring [46].

Thyme EO has excellent antibacterial and antioxidant properties that are suitable for inflammatory skin conditions such as acne [47]. Obtained by steam distillation of the flowering stems, Thyme EO contains thymol (37–55%) and 0.5-carvacrol (0.5–5.5%). These biocomponents have antibacterial activity, easily penetrating the lipid layer. Although Thyme EO is recognized for its insolubility in water, high volatility, and tendency to degrade rapidly when exposed to air, light, or high temperatures for long periods, the extraordinary potential of this plant has been explored in numerous studies.

In a recent research, Thyme EO was encapsulated in biodegradable nanoparticles of poly-(D,L)-(lactic-co-glycolic acid) for skin and pharmaceutical applications [47]. Through functionalization, the nanosystems remained stable for a period of 6 months, by cold storage. In vitro, ex vivo, and in vivo evaluations on human volunteers indicated that Thyme EO demonstrated excellent antioxidant activity and healing of skin inflammation without leaving acne scars (Figure 5A,B) [47].

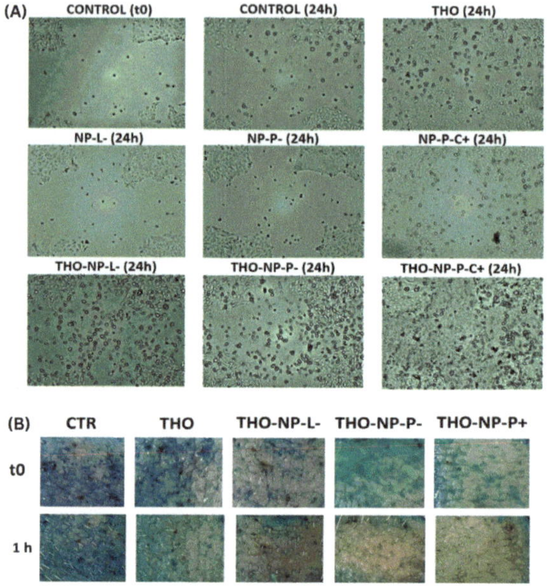

Figure 5. Wound healing activity: (**A**) for in vitro scratch assay in HaCaT cell lines. Images were taken before (Control t0) and 24 h after incubation: untreated samples (control and 24 h) and treated samples, respectively thyme oil (THO) and functionalized hydrogels and their corresponding empty NPs. (**B**) Skin surface showed a reduction in methylene blue following ex vivo antioxidant activity of the samples [47].

5.1.3. Wound Dressing Applications

Numerous formulations of composite hydrogels have been studied as intricate systems composed of biopolymers, incorporating various bioactive elements from essential oils. These platforms, exhibiting synergistic properties, are being explored for use as advanced wound dressings with enhanced therapeutic potential [48].

Cedarwood EO obtained from several types of conifers (e.g., *Cedrus* sp. and *Juniperus* sp.) is a mixture of safe organic chemicals with pesticidal and preservative properties. In order to develop effective hemostatic and antibacterial dressings for treating wounds, composite porous sponges were designed [49]. Polyvinyl alcohol was physically cross-linked with kaolin and incorporated cedar essential oil, through a freeze–thaw approach, yielding sponge hydrogels with distinct lamellar architectures. The addition of cedar and kaolin in the formulation improved the pore sizes and structure of the resulting sponges (Figure 6). Studies have shown the biocompatibility of these sponges, improved antibacterial activity against *Bacillus cereus* and *Escherichia coli*, and high free radical scavenging capacity and hemostatic performance [49].

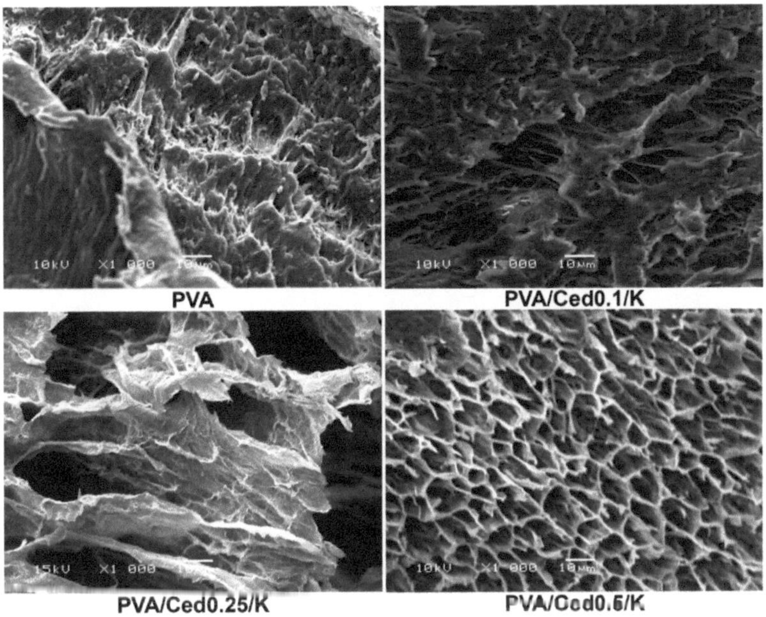

Figure 6. SEM micrographs at 1000× magnifications showing the microstructure in cross-sectional of composite sponges [49].

Clove (*Syzygium aromaticum* L. Myrtaceae) EO possesses significant biological activities beneficial to human health, such as antimicrobial, antioxidant, and insecticidal properties. Consequently, it has attracted considerable attention for its widespread use in the medical world, perfume, cosmetic, flavoring, and food industries [50]. It can be extracted by (i) hydrodistillation, (ii) steam distillation, (iii) ultrasound-assisted extraction, (iv) microwave-assisted extraction, (v) cold pressing, or (vi) supercritical fluid extraction. The extraction methods used determine the concentration of primary volatile compounds in clove essential oil and organic clove extracts. It contains mostly eugenol (at least 50%), respectively, eugenyl acetate, β-caryophyllene, and α-humulene (10–40%).

The development of different materials for biomedical applications has been in continuous growth lately [51]. Thus, the biological capabilities of clove EO were used in a very interesting recent study for the generation of hydrogels as wound dressings. The hydrogels were loaded with cloves EO by combining covalent and physical cross-linking methods. In the first step, EO was emulsified and stabilized in a chitosan-based solution, which was further strengthened by covalent cross-linking of the Schiff base with another polysaccharide, namely oxidized pullulan (Figure 7). In the next step, several freeze–thaw cycles were performed to stabilize the cloves EO in the physically cross-linked polymer walls. The hydrogels formed with a sponge-like porous structure (Figure 8) exhibited outstanding elasticity [52].

The antibacterial activity of hydrogels containing clove essential oil was evaluated by the time-kill method, for different incubation time intervals, against three bacterial strains and demonstrated antibacterial and antifungal effectiveness against *S. aureus* and *E. coli* (Figure 9) [52].

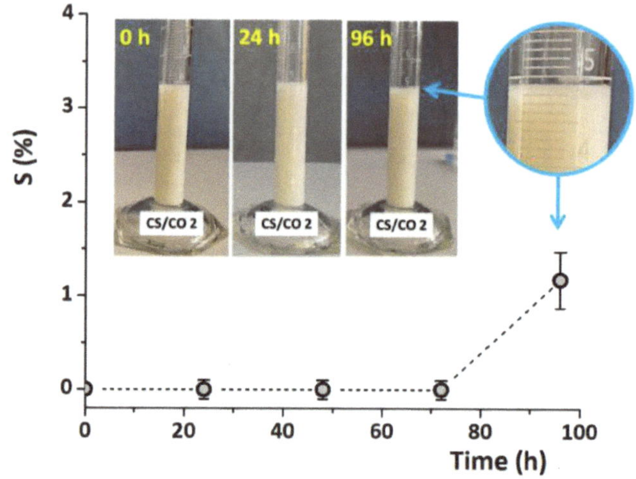

Figure 7. Photos of the stable chitosan emulsion with clove EO content (after 0–4 days) kept in the dark and ambient conditions [52].

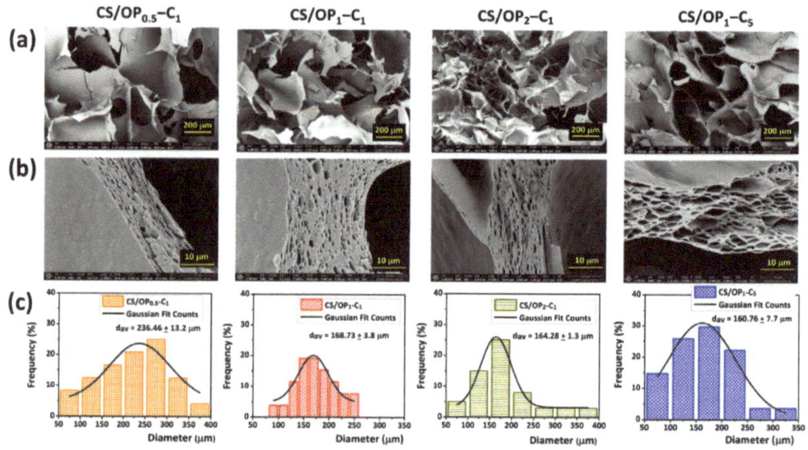

Figure 8. Cross-sectional SEM images of chitosan and oxidized pullulan-based hydrogels loaded with clove EO: overview (**a**), wall detail (**b**), and pore size distribution with corresponding diagrams (**c**) [52].

A novel gelatin nanocomposite hydrogel formulation encapsulated thyme essential oil in sodium caseinate nanomicelles formulated as a gelatin nanocomposite hydrogel which has been investigated as a drug delivery platform for in vitro antibacterial and in vivo wound healing potential [35]. The evaluation tests of the biocompatible and hemocompatible hydrogel showed a sustained in vitro release profile of EOs, with a strong antibacterial effect. In addition, the wound-healing potential of the nanocomposite was investigated in vivo, demonstrating a significant wound reduction in the group of animals it was tested on, after only 18 days. Antibacterial hydrogel may be a promising active and biocompatible platform for sustained delivery of thyme essential oil [35].

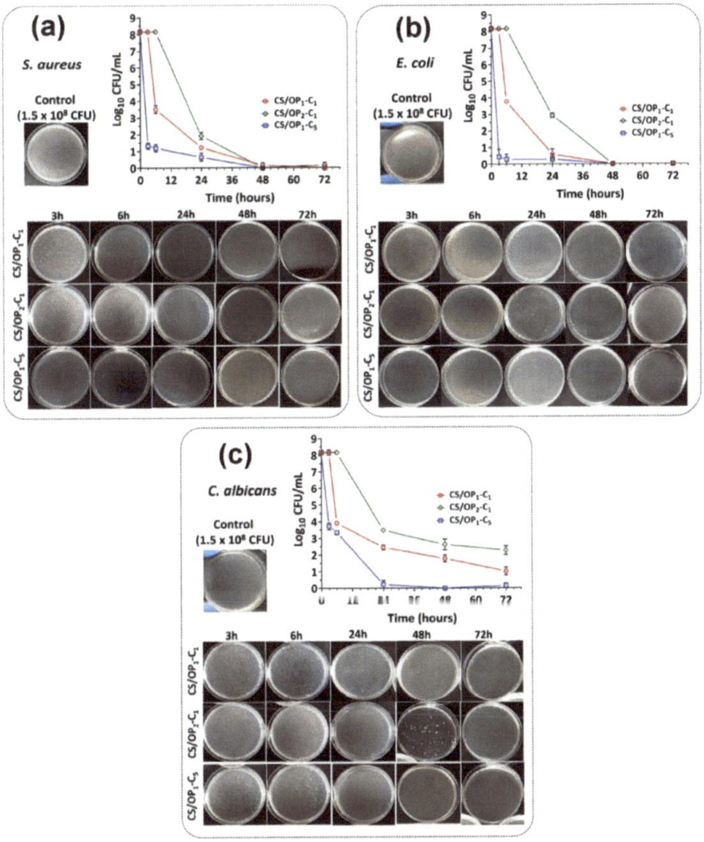

Figure 9. Antibacterial effect of clove oil-loaded hydrogels evaluated by the time-kill method, after 3–72 h of incubation with *S. aureus* (**a**), *E. coli* (**b**), and *C. albicans* (**c**) [52].

Origanum vulgare L. (oregano) has been used since ancient times all over the world, as a culinary ingredient, spice, or preservative and in curative treatments, being carminative, tonic, stimulant, and diaphoretic. Numerous studies have reported the main characteristics of this common plant, which demonstrate good analgesic, antimicrobial, antifungal, antiviral, antioxidant, and anti-inflammatory activities. In addition, it helps to easily penetrate the skin for transdermal drug administration [53]. In local applications, it is useful in anti-aging treatments, due to its antioxidant and anti-inflammatory properties which provides protection against free radicals of various reactive oxygen species [54,55]. It has a wealth of volatile and non-volatile components such as flavonoids, phenolic acids, and tannins, mainly phenolic monoterpenes such as carvacrol and thymol, with a variable chemical profile depending on the species and the geographical area [56].

Recent research used oregano EO in an innovative hydrogel formulation based on polymeric micelles (Figure 10) [57]. The release and permeation profile of the EO, the in vivo effects on biocompatibility, and the impact of the hydrogel on in ovo-angiogenesis were evaluated. It should be noted that the study avoided animal testing and a chick chorioallantoic membrane was used. The results showed a sustained release of EO, having a potential anti-angiogenic effect. This hydrogel with oregano EO content could be a natural therapeutic alternative in skin pathologies, such as fibroepithelial polyps [57].

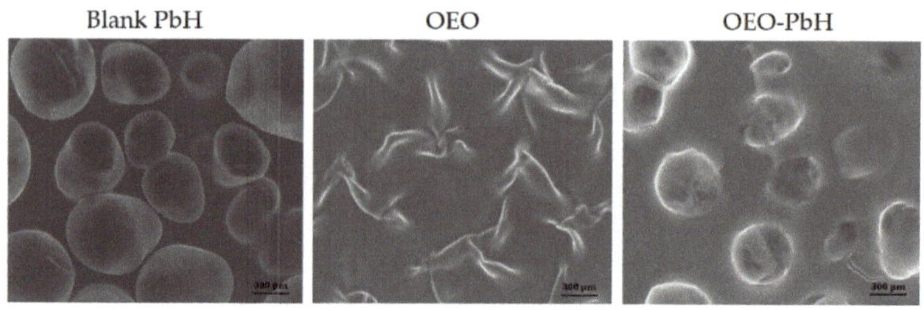

Figure 10. SEM micrographs of polymeric hydrogel (control), oregano essential oil, and micelle-based hydrogel sample containing oregano EO [57].

The bioactive and curative phytotherapeutic potential of essential oils has been exploited in a multitude of applications for wound healing [58,59].

Peppermint (*Mentha* × *piperita*) EO is widely used in the cosmetic industry for its aromatic fragrance. Its main constituents are menthol, menthone, and menthol acetate, and as secondary components, it contains bitter substances, caffeic acid, flavonoids, tannins, 1,8-eucalyptol, and propanone [60]. The pine needles EO mainly contains α-terpineol, linalool, and limonene, but also anethole, caryophyllene, and eugenol [61]. Fennel EO predominantly contains (E)-anethole, but also α-phellandrene and fenchone, methyl chavicol, p-cymene, and β-phellandrene [62].

The healing capacities of four types of EOs have been used advantageously by encapsulating them in microcapsules in the first step, and then by incorporating them in polymer matrices in the form of films to develop dressings for wounds [63]. Polyvinyl alcohol, polyvinyl pyrrolidone, and hydroxypropyl methylcellulose were selected as polymeric materials. Poly(ethylene glycol) and glycerol were used as plasticizers, together with Zn stearate as a stabilizer, and vitamins A and E for the antioxidant effect. EOs of mint, thyme, pine, and fennel were loaded into the polymer matrices as active substances with antimicrobial effects. The different types and compositions of EOs and polymer components affect the shape and aspect of the microcapsules, which can be visibly observed (Figure 11). The results of the investigations showed that the samples made with EO content presented good inhibitory activity and antimicrobial properties against *Staphylococcus aureus*, *Enterococcus faecalis*, *Escherichia coli*, *Pseudomonas aeruginosa*, and *Candida albicans* [63].

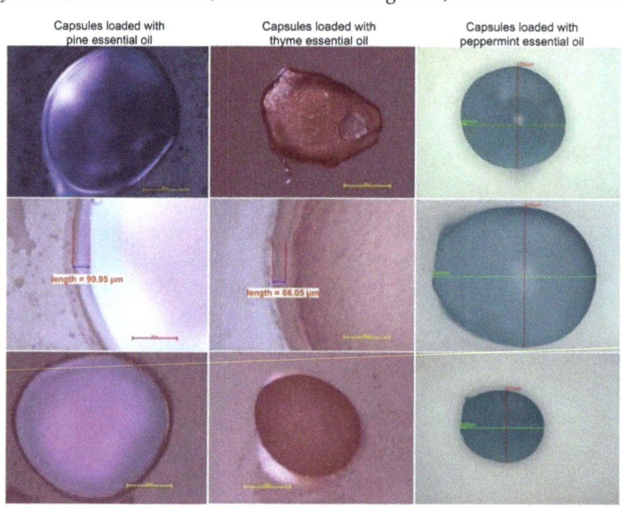

Figure 11. Optical microscopy images of polymer capsules of different shapes loaded with pine, thyme, and mint EO [63].

5.1.4. Chemotherapeutic

The alarming increase in the number of cancer cases in recent years highlights the pressing need for intensified efforts to improve therapeutic protocols [64,65]. Breast cancer, brain cancer, or invasive skin cancer affect millions of people every year and cause suffering and death all over the world [66,67]. With the multitude of different cancer treatment protocols, both surgical interventions and targeted therapies (chemo-, radio-, hormone-), the increase in the survival of cancer patients has been almost constant in recent years [68]. From here it is obvious the major urgency with which new therapies, combinatory, and targeted strategies are approached for a synergistic effect that will prolong the survival time and decrease the mortality rate [69,70].

Owing to some serious side effects of currently used anticancer chemotherapeutic methods or agents, there is a growing trend to use herbal medicine and its phytocompound derivatives [71]. It is important to use them both as ideal therapeutic alternatives and alongside chemotherapy treatments for many types of cancer [72–74].

Research is in a continuous dynamic and is actively focused on the discovery of new "green" pharmacological components for chemotherapies that offer potent potential activity with minimal side effects. A study aimed at obtaining new synergistic therapeutic agents (antimicrobial, antioxidant, and anticancer) was carried out by nanoencapsulation of clove essential oil in a nanogel based on squid chitosan and another phytochemical component, namely ρ-coumaric acid [75]. The in vitro evaluation of the nanogel encapsulated with clove essential oil indicated chemotherapeutic effects and potential for the prevention or therapy of pathologies induced by oxidative stress, microbial infection, or breast and skin cancer [75].

5.1.5. Carrier for Drug Delivery

Hydrogel delivery systems are excellent therapeutic tools for multiple clinical uses [76]. The adjustable 3D structure of hydrogels allows the inclusion of small molecules, macromolecules, or growth factors and they have the ability to protect drugs susceptible to degradation. It also ensures precise spatial and temporal control over the release of therapeutic factors and degradability [77].

Bioactive molecule delivery systems are designed and developed in the form of films, pearls, and nanogels. In order to create a smart drug carrier with intestinal release activity, alginate hydrogel beads containing essential oils were made [78]. Glycyrrhizic acid, licorice root extract, and Thymus EO were loaded into ß-cyclodextrin. By co-encapsulating them with alginate, active alginate hydrogel beads were obtained. Studying the release of EO from alginate beads in simulated gastric fluid and simulated intestinal fluids indicated a high release rate of both EOs [78].

The characteristics of a natural hydrogel nanoliposome hybrid system were evaluated for the controllable release of thyme essential oil in the gastrointestinal tract [79]. Hydrogels based on pea protein and gum Arabic indicated the need for intermediates such as maltodextrin for stabilization. Figure 12 shows different photo and SEM images of different formulations, with and without nanoliposomes and nanoliposome–maltodextrin complexes [79].

5.1.6. Burn Healing

Burns or scalds to the skin are particularly serious injuries, sometimes life-threatening, as they can disrupt the body's essential functions. This disruption is primarily due to the loss of water, electrolytes, and proteins, because of the wounds [80].

Burns need emergency medical care along with very strict infection control and surveillance measures to increase the rate of healing and survival. The appearance of multi-resistant organisms to antibiotics or some treatments, such as dressings or ointments inappropriate for the degree of burn, can lead to invasive infections [81].

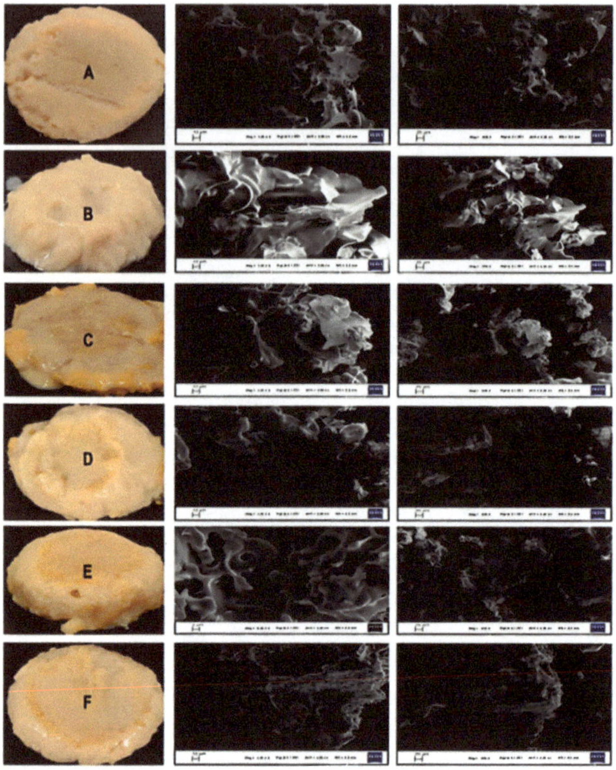

Figure 12. Photo and SEM images of different hydrogel formulations, with and without EO, nanoliposomes, and nanoliposome/maltodextrin complexes. (**A**): control hydrogel; (**B**): hydrogel with EO; (**C**): hydrogel with lecithin and EO (14.23%); (**D**): hydrogel with lecithin, maltodextrin and EO (20%); (**E**): hydrogel with lecithin, maltodextrin and EO (25%); (**F**): hydrogel with lecithin, maltodextrin, and EO (33%) [79].

Polysaccharide-based hydrogel dressings are more advantageous materials for the treatment of burns, compared to traditional textile dressings, due to the easy application and removal and rapid coverage of the wounds together with the surrounding areas, the good capacity to absorb exudate, and the comfort given by the improvement quickness of pain. In addition, the transparency of most hydrogels allows for easier management of lesions.

Numerous studies have reported the results of the use of polymeric hydrogels with different EOs content in the care of wounds and burns [82–84]. New formulations of hydrogel materials based on polyvinyl alcohol and gelatin enriched with ginger extract have been proposed as dressings for burn wound healing. The hydrogels demonstrated comparable wound healing efficacy to the commercial dressing on rabbit back burn wounds in vivo. In addition, they showed significantly higher wound healing activity than the control group, as evidenced by intensive collagen development observed in histopathological analysis [85]. A new study obtained and tested materials for the treatment of burns by designing dressings based on physically cross-linked carboxymethyl chitosan and carbomer 940 hydrogels. EOs of eucalyptus, ginger, and cumin were selected and loaded into them [86]. The hydrogel containing eucalyptus EO showed favorable antibacterial activities against *S. aureus* and *E. coli*. Moreover, experiments performed in vivo on mice demonstrated that hydrogel with eucalyptus EO improved wound healing in burn models and considerably promoted the regeneration of the dermis and epidermis. The histological analysis highlighted the decrease in the values of IL-6, TNF-α, and the increase in the values of the factors TGF-β, VEGF, and EGF, specific to the burn wound tissue area [86].

5.2. Dental Applications

Traditionally, plants, herbal extracts, and essential oils have been successfully used in dentistry to clean teeth and dental caries [87]. People traditionally crafted toothbrushes using natural bristles from twigs selected from medicinal plants, which were rich in oils. Fir, clove, bay, eucalyptus, juniper, neem, or oak were used, with a rich content of volatile oils that acted to stimulate blood circulation, and with tannins for contraction and cleaning of the gums [88]. They also used poppies or cranberries, rich in vitamins, to keep their gums healthy. It has been observed that aloe vera plants, marigolds, and grapefruit seeds have beneficial and anti-inflammatory effects in the oral cavity [89]. They inhibit the growth of aerobic or anaerobic bacteria and act to reduce gingival bleeding and gingivitis [90].

Phytochemicals provide a potential strategy in the prevention and treatment of dental caries, inflammation, and other oral infections and could be a powerful substitute for antibiotics [91,92].

A promising strategy for the prevention and treatment of dental caries, inflammation, and other oral infections is the use of phytochemicals both in current care products and in oral treatments [87]. These natural compounds could serve as a powerful alternative to antibiotics [93].

Extensive recent research has developed hydrogels with incorporated essential oils for the therapy of periodontitis [94]. These materials are described as dental drugs that could be used as photosensitizers in photodynamic therapy for the treatment of periodontitis. Oregano®, Frankincense®, and the Thieves® blend were incorporated as EOs, with a content of cloves, lemon, cinnamon bark, eucalyptus radiata, and rosemary extract. The main constituents identified from the mixture of selected and used oils included eugenol, pinene, limonene, carvacrol, and cymene [94].

5.3. Cosmetics Applications

Essential oils are integral to the formulation of care products and cosmetics, offering a wide range of benefits thanks to their rich and diverse composition of biocompounds [95,96]. Moreover, hydrogels combined with various chemical compounds can be incorporated into cosmetic formulations, offering multiple topical applications for both skin and hair [97,98].

Hydrogels for cosmetic preparations can be obtained from biopolymers of natural origin, such as alginate, collagen, gelatin, hyaluronic acid, chitosan, xanthan gum, pectin, starch, or cellulose [99,100]. These biopolymers themselves possess bioactivities advantageous to cosmetics. Thus, new cosmetic products were designed and made in the form of gels, microcapsules, or masks, both for skin and hair, with excellent hydration, softening, and elasticity performances, supporting and actively promoting anti-aging. Also, superabsorbent hydrogels have been developed in comfortable hygiene products, capable of absorbing fluids.

The combination of hydrogels with essential oils is a successful mixture, particularly useful and advantageous as cosmetic preparations or beauty and care products [101–103].

Recently, a study was reported that aimed at the design and creation of new cosmeceutical materials based on hydrogels with improved biological properties [104]. In the first step, *Camellia oleifera* EO was loaded into chitosan nanoparticles by emulsification and then ionic gelation. Then, hydrogels based on poly(vinyl alcohol), silk sericin, and gelatin were prepared, in which chitosan nanoparticles were embedded. Materials that showed tyrosinase inhibition and antioxidant activity could be useful in cosmeceutical applications, such as facial masks [104].

5.4. Food Applications

Hydrogel-based formulations with incorporated EOs have numerous applications in the food industry [105].

Hydrogels for food application should be categorized as follows: (i) Delivery; (ii) Packaging; (iii) Coating; (iv) Fat replacer; and (v) Texturizing. A growing field of research centers on hydrogel beads, which act as carriers for nano- or microparticles. These systems are highly effective for targeted drug delivery and can also be used as food supplements, including dietary additives, probiotics, or food components for special medical purposes [105,106]. These types of materials, which can be administered orally, have exploited the biological origin of natural polymers, especially polysaccharides and proteins, their specific biodegradability, and pH sensitivity [107]. A considerable amount of research has been devoted to the development of hydrogels for the encapsulation of food-grade components, such as vitamins, natural extracts, and essential oils [108].

Hydrogel beads show great potential for improving the bioavailability and performance of some compounds from the range of nutraceuticals, including EO, offering them protection against chemical degradation [109]. The granules are made through an accessible technique, in two steps: (i) obtaining the particles enriched with biopolymeric materials and nutraceutical content, and (ii) crosslinking the biopolymeric materials. The first step can be achieved by injection, phase separation, shearing, or templating, and the second step can be achieved by degree changes in the solvent quality, the incorporation of counter-ions or enzymes, or by heating–cooling cycles [110].

Recently, substantial amounts of waste from coffee pulp, generated during the extraction of essential oils, were analyzed [111]. These wastes were used to extract two different pectin fractions (highly methoxylated and low methoxylated). Pectins have been studied for their performance as EO carrier systems. The pectin fractions formed two systems of hydrogel beads, with or without chitosan, to encapsulate the EO of roasted coffee or green coffee. The two systems were analyzed in terms of their antioxidant activity and EO release profile for potential food applications. On the one hand, the highly methoxylated pectin obtained from *Coffea arabica* presented better EO encapsulation performances. On the other hand, surprisingly, the EOs obtained from roasted coffee showed superior antioxidant activity compared to that obtained from green coffee [111].

Food packaging serves as a passive barrier, shielding products from environmental factors, extending their shelf life by preventing contamination, and ensuring safe transportation and storage. Active packaging incorporated with essential oils allows interaction between food and the external environment, helping to regulate temperature, moisture levels, and microbial control, which ultimately enhances the quality and extends the shelf life of the food [112]. The upcoming chapter will focus on the key concerns and challenges faced by the food packaging industry. It will explore how essential oils influence the microstructure of packaging materials and examine their specific properties.

A new direction regarding the applications of hydrogels with EO content is represented by edible coatings that extend the shelf life of some perishable foods, by delaying oxidation and reducing the amount of packaging. EOs are known to be excellent natural antimicrobial and antibacterial agents [113]. Some contaminants such as gram-negative bacteria (*E. coli*) can cause serious diseases by contaminating food such as milk and meat or gram-positive ones (*S. aureus* and *B. cereus*), which cause the contamination of fruits or food products with starch content.

The richness of volatile compounds contained in EO, such as phenolics, determines the use of oils in edible films or coatings for flavoring, packaging, or preservation of food products. Different gels incorporating EO such as basil leaves, clove, cypress, fennel, lavender, oregano, pine, rosemary, thyme, and verbena have been used to inhibit lipid oxidation and microorganism growth in coatings for fish fillets, cheese, fruits, or vegetables [114–118].

Studies on two types of gelatin hydrogels containing rosemary and orange EO microdroplets prepared by simple emulsification in the presence of Tween®80 surfactant showed interesting conclusions [119]. The mechanical and antibacterial properties of these gels against some food contaminants such as *E. coli*, *S. aureus*, and *B. cereus*. indicated adequate characteristics as edible coatings of perishable foods, in order to preserve foods such as meat [119].

A novel area of research that has gained attention recently is the replacement of animal fats through the immobilization of oils within hydrogels [120]. Healthier meat products are a direction imposed both by worldwide recommendations and by consumer demands. Traditional products should be adapted to the nutritional characteristics recommended by specialists, by reformulating them. Therefore, a recent approach to improving the health of meat products is the use of healthy oils (vegetable or marine) as fat substitutes. Also, it is important to develop food products that are low in fat, but which retain their functional qualities, such as mayonnaise or ice cream [121]. However, it is a great challenge for researchers to keep the specific bioactivities of hydrogels with incorporated essential oils and use them as fat substitutes or as materials with specific textures. Future scientific discoveries based on nanotechnology will also develop such products.

5.5. Food Packaging Applications

Active packaging is representing intelligent materials that improve the preservation of food, especially perishable ones, extend the shelf life, and ensure safety by interacting with the food product through its various components [122].

Currently, there is a multitude of advantageous active packaging for applications in different fields. In the food industry, but also in the beverage industry, there is the most active packaging, due to the very high demand for increased shelf life, freshness, and safety. The pharmaceutical industry, medical technology, agriculture, and courier and delivery services are just a few other areas where there is a demand for these types of modern packaging.

Active packaging is created based on the active components of biopolymers or with different biocompounds incorporated into them [123].

For instance, biopolymeric hydrogels with essential oils incorporated as antimicrobial substances are advantageous systems for obtaining active food packaging. In addition to monitoring the condition and ensuring food safety, minimizing the risk of contamination, increasing the shelf life, or obtaining more durable packaging, smart packaging can help reduce food waste. In this context, adding EOs to packaging is a natural alternative that can replace chemical additives [124,125].

Cinnamon EO is a natural bacteriostatic agent, with potential applications in the field of food preservation [126]. In general, it has found many uses in culinary and medicinal applications. It contains numerous chemical constituents, of which, depending on the different species of *Cinnamomum* trees or shrubs, the most important are the compounds (E)-cinnamaldehyde, linalool, β-caryophyllene, eucalyptol, and eugenol [126]. Apart from the specific spicy taste and cinnamon flavor that is due to the cinnamaldehyde compound, the wide variety of components of cinnamon EO have antimicrobial, antioxidant, antifungal, and antidiabetic biological properties [127,128].

Sodium alginate and acacia gum hydrogels loaded with cinnamon EO were prepared as edible films and analyzed in terms of view of physicochemical characteristics [129]. The antioxidant capacity of the films was improved with increasing cinnamon EO concentration, making them promising candidates for use as active food packaging materials. Figure 13 shows the composite films prepared with different concentrations of cinnamon EO between 0–30 μL EO [129].

Recently, the phytochemical components of the essential oil obtained from the *Artemisia dracunculus* plant, widely distributed geographically, were identified and evaluated [130]. *Artemisia dracunculus* EO was valorized by incorporating different amounts into hydrogel matrices based on polyvinyl alcohol and agar. The results of the antimicrobial tests indicated sustained antimicrobial activity against nine pathogenic strains (four Gram-positive and five Gram-negative). The incorporation of *Artemisa dracunculus* EO in these hydrogel models can lead to practical applications in the area of food technology, as an active and biodegradable alternative to classic packaging [130].

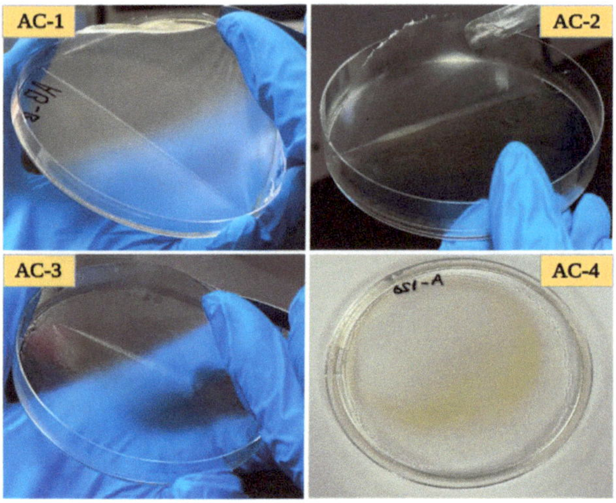

Figure 13. Photo of the composite films based on sodium alginate and acacia gum with different concentrations of cinnamon EO: (AC1—without EO) (AC2—15 μL EO) (AC3—20 μL EO) (AC4—30 μL EO) [129].

In another study, a quantity of powdered starch was obtained in the first stage from the use of residual biomass, then in the second stage, it was introduced into a formulation, to prepare cryogels and hydrogels [131]. The materials prepared by absorption or cross-linking by the Schiff-base reaction were loaded with diacetyl and mint EO. The prepared materials showed a good ability to adsorb water and deliver antimicrobial substances, being advantageous for possible fresh food packaging applications [131].

Furthermore, new antibacterial hydrogels were prepared by the method of freeze–thaw cycles [127]. Inclusion complexes methyl-β-cyclodextrin and thyme oil were incorporated into a polyvinyl alcohol matrix with polysaccharide content, respectively, dendrobium and guar gum, in various ratios. These materials presented very good mechanical performance, as well as antimicrobial and antioxidant activities favorable for the preservation of chicken breast, extending the shelf life by four days. These results indicate the potential of the materials for possible active packaging applications [132].

5.6. Restoration of Stone Cultural Heritage

Stone monuments in the sphere of cultural heritage suffer from biological damage. The variation in environmental conditions determines the growth of phototrophic microorganisms on stone surfaces, in the form of biofilms [133–135]. These microorganisms are made of microbial aggregates that act mechanically and produce micro-decohesion of the substrates [136]. In addition, biofilms, together with atmospheric pollutants, promote chemical corrosion, pigmentation, or discoloration of stone surfaces [137].

The classic restoration of stone surfaces uses both physical and chemical methods [138]. However, mechanical brushing can damage the surface of the monument, and chemical treatment can lead to a selection of resistant microbial species or can be harmful to the environment or the operators [139,140].

An innovative and eco-sustainable restoration technique is the use of essential oils with natural biocidal action, embedded in hydrogels, as alternatives to chemical treatments for the restoration of cultural heritage [141–144].

EOs of lavender and thyme were encapsulated in alginate hydrogel in order to create an easy-to-use and non-invasive restoration method [136]. The vitality of cyanobacterial biofilms was discouraged by applying hydrogel for different periods of time. The results of the tests indicated that the best inhibitory effect on the photosynthetic activity of microorganisms was shown by thyme oil rich in thymol, for a concentration of 0.1% (v/v) it was shown by thyme oil rich in thymol. It retained an effective antimicrobial action against cyanobacteria. Notably, the developed protocol allowed the use of a very small amount of essential oil as a green biocide [136].

In another study, thyme EO was also used in the preparation of poly(vinyl)alcohol and borax-based hydrogels, together with a double-layered hydroxide of ZnAl intercalated with sodium alginate, and silver nanoparticles or a mix of silver–silver chloride nanoparticles [145]. The hydrogels were thus formulated to mechanically remove the biopatina from two types of biodamaged stones: Carrara marble and St. Margarethen. The hydrogel with thyme EO content worked effectively for cleaning stones with porous structures and different compositions, damaged by the natural environment [145].

A real case study reported the results of in situ application of a sodium alginate hydrogel containing thyme EO. The experiment followed the restoration of three selected parts of Fortunato Depero's mosaic located in a neighborhood in Rome (Italy) [146]. The material was prepared by a simple method and easily applied on large and vertical surfaces. The images taken before and after application demonstrated that a single treatment was enough to completely eliminate the microbial patina. The hydrogel loaded with thyme EO as a natural biocide showed a very good biocide performance [146].

In summary, Table 1 outlines the key characteristics of various essential oils discussed in this review, highlighting the main bioactive compounds, the extraction methods used, pharmacological features and potential applications. The brief summary of essential oil-enriched hydrogel applications introduced in this manuscript can be found in Table 2.

Table 1. General characteristics of essential oils incorporated in hydrogels.

Plants	Essential Oils	Main Constituents	Extraction Procedure	Pharmacological Properties	Applications	Reference
Cinnamomum zeylanicum	Cinnamon oil	cinnamaldehyde	Steam distillation and Soxhlex extraction	Antimicrobial, antibiotic, antioxidant	Food packaging materials, food preservation	[129,147–149]
Lavandula angustifolia	Lavandin essential oils	Terpenes (e.g., linalool, linalyl acetate, terpinen-4-ol) and terpenoids (e.g., eucalyptol)	Steam distillation	Antioxidants, antibacterial, anxiolytics, analgesics, and anti-inflammatories	Wound healing, Microparticles as delivery system	[48,150–153]
Cymbopogon (spp.)	Lemongrass essential oils	Terpenes and Terpenoids (Terpinen-4-ol, α-Terpineol (neral, isoneral, geranial, isogeranial, geraniol, geranyl acetate, citronellal, citronellol, germacrene-D, and elemol)	Steam distillation	Antifungal, antibacterial, antiviral, anticancer, and antioxidant	Pharmaceutical, cosmetics, and food preservations industries	[154,155]
Melaleuca alternifolia	Tea tree essential oils	Terpenes (e.g., terpinen-4-ol, 1,8-cineole)	Steam distillation	Antimicrobial and anti-inflammatory	Beads for food preservation	[156–158]

Table 1. *Cont.*

Plants	Essential Oils	Main Constituents	Extraction Procedure	Pharmacological Properties	Applications	Reference
Mentha piperita	Peppermint essential oils	Menthol, menthone, neomenthol and iso-menthone	Steam distillation, hydrodistillation, microwave-assisted extraction, supercritical fluid extraction, ultrasonic-assisted extraction and countercurrent extraction	Anti-inflammatory, antibacterial, antiviral, scolicidal, immunomodulatory, antitumor, neuroprotective, antifatigue and antioxidant; hypoglycemic and hypolipidemic effects, gastrointestinal and dermatological diseases	Patches, wound dressing	[63,159–162]
Ocimum basilicum (L.)	Basil essential oils	Eugenol, e α-Pinene, β-Pinene, Methyl chavicol, 1,8 cineole, L-linalool, Ocimene, Borneol, Geraneol, B-Caryphyllone, and n-Cinnamate	Hydrodistillation	Carminative, galactogogue, stomachic and antispasmodic tonic, vermifuge,	Food packaging, antiperspirant in agriculture	[163–166]
Thymus vulgaris (L.)	Thyme essential oils	Carvacrol, 5-isopropyl-2-methylphenol, and a p-cymene	Hydrodistillation, steam distillation	Antioxidant, antimicrobial, antidiabetic, anti-inflammatory, immunomodulatory and anticancer bioactivities	Wound healing, wound dressing; beads as delivery systems	[33,35,36,47,63,79,167]

Table 2. Applications of essential oils incorporated in hydrogels.

	Method of Preparations	Materials	Encapsulated Essential Oils	Applications	References
Biomedical applications	Physical crosslinking	Sodium alginate/Fucoidan	Menthol, L-linalool, bergamot oil, and β-pinene	Topical or transdermal administration	[43]
	Physical crosslinking	Methylcellulose (10% (*w*/*v*))	*Melissa officinalis* EO	Treatment of oral candidiasis.	[44]
	Chemical crosslinking	Polyvinyl Alcohol/Corn Starch Hydrogel Films loaded with Silver Nanoparticles	Patchouli EO	Antimicrobial materials (against *Staphylococcus aureus* and *Staphylococcus epidermidis*)	[45]
	Solvent displacement method	Poly-(D,L)-(lactic-co-glycolic acid)	Thyme EO	Inflammatory skin disorders	[47]
	Physical crosslinking	Polyvinyl alcohol/kaolin	Cedar EO	Wound dressing	[49]
	Covalent and physical crosslinking	Chitosan/oxidized pullulan	Clove EO	Wound dressings	[52]
	Cold gelation process	Polymeric-Micelles-Based Hydrogels (Pluronic F127–20%*w*/*w*; and Pluronic L 31—1%*w*/*w*)	Oregano EO	Cutaneous application	[57]

Table 2. *Cont.*

	Method of Preparations	Materials	Encapsulated Essential Oils	Applications	References
	Chemical crosslinking; casting method	Polyvinyl alcohol/polyvinyl pyrrolidone; hydroxypropyl methyl cellulose; sodium alginate; polyethylene glycol; glycerol; Zn stearate; vitamin A and E	Fennel, pine, mint and thyme EO	Wound dressings	[63]
	Chemical crosslinking	Ultrasound-assisted deacetylated chitosan/ρ-coumaric acid	Clove EO	Chemotherapeutic/ chemopreventive agent	[75]
	Thin-film dispersion technique/heat-induced gelation.	Pea protein (30%) and gum Arabic (1.5%); Soybean lecithin; maltodextrin and gum Arabic	Thyme EO	Delivery of bioactive compounds (food packing material, tissue engineering or drug delivery)	[79]
	Physical crosslinking	Carboxymethyl chitosan/carbomer 940	Eucalyptus, ginger, and cumin EO	Burn dressing material for skin burn repair	[86]
Dental applications	Physical crosslinking	Xanthan gum/Glycerin/Lyophilized Whey/Polyvinylpyrrolidone/ PEG 400	Oregano®, Frankincense®, Thieves®, Frankincense® EO	Therapy of periodontitis	[94]
Cosmetics applications	Ionic gelation	Poly(vinyl alcohol), silk sericin, and gelatin/chitosan nanoparticles	*Camellia oleifera* EO	Cosmetic product (facial masks)	[104]
Food applications	Emulsification/ionic crosslinking	Pectin and pectin/chitosan hydrogel beads	Green and roasted coffee EO	Systems for the delivery and controlled release of essential oils; food applications	[111]
Food packaging applications	Gelation/casting	Sodium alginate/ acacia gum	Cinnamon EO	Hydrogel-based films as active food packaging materials	[129]
Restorations of the stone cultural heritage	Preparation directly in situ	Sodium alginate	Thyme EO	Biocides for restoration in a real case of study, i.e., the mosaic Le Professioni e le Arti of Fortunato Depero	[146]

6. Challenges and Perspectives

The advancement of polymeric hydrogels presents emerging opportunities for their use across various fields, thanks to their biocompatibility, simple gelation process, ease of application, and the potential for functionalization. Hydrogels have the ability to alter their volume, phase, and structure when exposed to specific external stimuli, making them versatile for use in a wide range of sectors. However, the limited mechanical rigidity commonly found in some biocompatible hydrogels presents a significant challenge that needs to be addressed, particularly when dealing with rapid dynamic changes or when considering structural uniformity and long-term stability.

Integrating EO into the hydrogel matrix can enhance their biological activities, shield them from degradation, and serve as a platform for creating innovative biotechnological products. Furthermore, encapsulation seeks to address certain limitations of EOs, such as their volatility, reduced stability, and high sensitivity to environmental conditions. The controlled release of bioactive compounds from EOs encapsulated within hydrogels is crucial for effectively delivering these compounds to their target. On the other hand, a huge difficulty, especially in medical applications, is the very high qualitative and quantitative variability of the EO composition. This is determined by intrinsic factors that influence each other and are due to the varieties and age of the plants, the type of soil, the climate, or the time of harvesting, but also by some extrinsic factors such as the extraction methods used.

However, to enhance their effectiveness as targeted delivery systems of EOs, further research is required to assess their safety across various applications, ranging from biomedical to food industries.

The limitations related to the insufficient data on the stability, safety, and long-term bioactivity of these materials are emphasized. Moreover, the limited number of in vivo studies, particularly in the medical field, could delay their commercialization in the pharmaceutical and biomedical sectors.

7. Conclusions

This review aims to provide an overview of polymeric hydrogels containing essential oils, emphasizing their vast potential for applications in various fields. Hydrogels are valuable as delivery systems because of their ability to be biocompatible, biodegradable, and provide controlled release of plant-derived bioactive ingredients. Hydrogel structures, known for their remarkable swelling, gelling, and bioactive loading capabilities, play a vital role in the creation of functional materials. Most essential oils are accepted, credited, and appreciated as valuable bioactive ingredients, capable of performing multiple pharmacological functions, such as anticancer, antiseptic, antiviral, and antioxidant activities. Results from the cited literature suggest that hydrogels containing essential oils are ecological, sustainable materials, with improved biological properties, demonstrating effective antibacterial, antifungal, anticancer, and anti-inflammatory activities. Nevertheless, challenges remain in this field, including the need for standardization and the absence of cost-effective methods for scaling up production on a larger scale. Additionally, issues concerning stability and toxicity require thorough investigation. The advancement of essential oil-enriched hydrogel materials presents a growing opportunity to be applied in a wide range of fields, including biomedicine, cosmetics, dentistry, the food industry, and even heritage conservation.

Funding: This research received no external funding.

Institutional Review Board Statement: Not applicable.

Informed Consent Statement: Not applicable.

Data Availability Statement: Not applicable.

Conflicts of Interest: The author declares no conflicts of interest.

References

1. Bayles, A.V.; Pleij, T.; Hofmann, M.; Hauf, F.; Tervoort, T.; Vermant, J. Structuring hydrogel cross-link density using hierarchical filament 3D printing. *ACS Appl. Mater. Interfaces* **2022**, *14*, 15667–15677. [CrossRef] [PubMed]
2. Vasile, C.; Pamfil, D.; Stoleru, E.; Baican, M. New Developments in Medical Applications of Hybrid Hydrogels Containing Natural Polymers. *Molecules* **2020**, *25*, 1539. [CrossRef] [PubMed]
3. Löwenberg, C.; Balk, M.; Wischke, C.; Behl, M.; Lendlein, A. Shape-memory hydrogels: Evolution of structural principles to enable shape switching of hydrophilic polymer networks. *Acc. Chem. Res.* **2017**, *50*, 723–732. [CrossRef] [PubMed]
4. Unalan, I.; Boccaccini, A.R. Essential oils in biomedical applications: Recent progress and future opportunities. *Curr. Opin. Biomed. Eng.* **2021**, *17*, 100261. [CrossRef]
5. Yan, M.-R.; Wang, C.-H.; Cruz Flores, N.H.; Su, Y.-Y. Targeting Open Market with Strategic Business Innovations: A Case Study of Growth Dynamics in Essential Oil and Aromatherapy Industry. *J. Open Innov. Technol. Mark. Complex.* **2019**, *5*, 7. [CrossRef]
6. Mkaddem Mounira, G. Essential Oils and Their Bioactive Molecules. In *Recent Advances and New Applications*; IntechOpen: Rijeka, Croatia, 2024. [CrossRef]
7. Chelu, M.; Moreno, J.C.; Atkinson, I.; Cusu, J.P.; Rusu, A.; Bratan, V.; Aricov, L.; Anastasescu, M.; Seciu-Grama, A.M.; Musuc, A.M. Green synthesis of bioinspired chitosan-ZnO-based polysaccharide gums hydrogels with propolis extract as novel func-tional natural biomaterials. *Int. J. Biol. Macromol.* **2022**, *211*, 410–424. [CrossRef]

8. Majcher, M.J.; Hoare, T. Applications of Hydrogels. In *Functional Biopolymers*; Jafar Mazumder, M., Sheardown, H., Al-Ahmed, A., Eds.; Polymers and Polymeric Composites: A Reference Series; Springer: Cham, Switzerland, 2019. [CrossRef]
9. Chelu, M.; Musuc, A.M. Polymer Gels: Classification and Recent Developments in Biomedical Applications. *Gels* **2023**, *9*, 161. [CrossRef]
10. Sapuła, P.; Bialik-Wąs, K.; Malarz, K. Are Natural Compounds a Promising Alternative to Synthetic Cross-Linking Agents in the Preparation of Hydrogels? *Pharmaceutics* **2023**, *15*, 253. [CrossRef]
11. Ahmed, M.S.; Islam, M.; Hasan, M.K.; Nam, K.-W. A Comprehensive Review of Radiation-Induced Hydrogels: Synthesis, Properties, and Multidimensional Applications. *Gels* **2024**, *10*, 381. [CrossRef]
12. Chelu, M.; Musuc, A.M. Advanced Biomedical Applications of Multifunctional Natural and Synthetic Biomaterials. *Processes* **2023**, *11*, 2696. [CrossRef]
13. Yosri, N.; Khalifa, S.A.; Attia, N.F.; Du, M.; Yin, L.; Abolibda, T.Z.; Zhai, K.; Guo, Z.; El-Seedi, H.R. Advancing sustainability in the green engineering of nanocomposites based on marine-derived polymers and their applications: A comprehensive review. *Int. J. Biol. Macromol.* **2024**, *274*, 133249. [CrossRef] [PubMed]
14. Rodrigues, D.B.; Reis, R.L.; Pirraco, R.P. How are natural-based polymers shaping the future of cancer immunotherapy—A review. *Polym. Rev.* **2024**, *64*, 371–406. [CrossRef]
15. Monia, T. Sustainable natural biopolymers for biomedical applications. *J. Thermoplast. Compos. Mater.* **2024**, *37*, 2505–2524. [CrossRef]
16. Manivannan, R.K.; Sharma, N.; Kumar, V.; Jayaraj, I.; Vimal, S.; Umesh, M. A comprehensive review on natural macro-molecular biopolymers for biomedical applications: Recent advancements, current challenges, and future outlooks. *Carbohydr. Polym. Technol. Appl.* **2024**, *8*, 100536. [CrossRef]
17. Buriti, B.M.A.d.B.; Figueiredo, P.L.B.; Passos, M.F.; da Silva, J.K.R. Polymer-Based Wound Dressings Loaded with Essential Oil for the Treatment of Wounds: A Review. *Pharmaceuticals* **2024**, *17*, 897. [CrossRef] [PubMed]
18. Satchanska, G.; Davidova, S.; Petrov, P.D. Natural and Synthetic Polymers for Biomedical and Environmental Applications. *Polymers* **2024**, *16*, 1159. [CrossRef]
19. Harini, A.; Sofini, S.P.S.; Balasubramanian, D.; Girigoswami, A.; Girigoswami, K. Biomedical applications of natural and synthetic polymer-based nanocomposites. *J. Biomater. Sci. Polym. Ed.* **2023**, *35*, 269–294. [CrossRef]
20. Cai, M.-H.; Chen, X.-Y.; Fu, L.-Q.; Du, W.-L.; Yang, X.; Mou, X.-Z.; Hu, P.-Y. Design and Development of Hybrid Hydrogels for Biomedical Applications: Recent Trends in Anti-cancer Drug Delivery and Tissue Engineering. *Front. Bioeng. Biotechnol.* **2021**, *9*, 630943. [CrossRef]
21. Katopodi, T.; Petanidis, S.; Floros, G.; Porpodis, K.; Kosmidis, C. Hybrid Nanogel Drug Delivery Systems: Transforming the Tumor Microenvironment through Tumor Tissue Editing. *Cells* **2024**, *13*, 908. [CrossRef]
22. Sadgrove, N.J.; Padilla-González, G.F.; Phumthum, M. Fundamental Chemistry of Essen-tial Oils and Volatile Organic Compounds, Methods of Analysis and Authentication. *Plants* **2022**, *11*, 789. [CrossRef]
23. Stratakos, A.C.; Koidis, A. Chapter 4—Methods for Extracting Essential Oils. In *Essential Oils in Food Preservation, Flavor and Safety*; Preedy, V.R., Ed.; Academic Press: Cambridge, MA, USA, 2016; pp. 31–38. [CrossRef]
24. Dhifi, W.; Bellili, S.; Jazi, S.; Bahloul, N.; Mnif, W. Essential Oils' Chemical Characterization and Investigation of Some Biological Activities: A Critical Review. *Medicines* **2016**, *3*, 25. [CrossRef] [PubMed]
25. Masyita, A.; Mustika Sari, R.; Dwi Astuti, A.; Yasir, B.; Rahma Rumata, N.; Emran, T.B.; Nainu, F.; Simal-Gandara, J. Terpenes and terpenoids as main bioactive compounds of essential oils, their roles in human health and potential application as natural food preservatives. *Food Chem. X* **2022**, *13*, 100217. [CrossRef] [PubMed]
26. Yammine, J.; Chihib, N.E.; Gharsallaoui, A.; Ismail, A.; Karam, L. Advances in essential oils encapsulation: Development, characterization and release mechanisms. *Polym. Bull.* **2024**, *81*, 3837–3882. [CrossRef]
27. Sousa, V.I.; Parente, J.F.; Marques, J.F.; Forte, M.A.; Tavares, C.J. Microencapsulation of Essential Oils: A Review. *Polymers* **2022**, *14*, 1730. [CrossRef]
28. Moghaddam, M.; Mehdizadeh, L. Chapter 13—Chemistry of Essential Oils and Factors Influencing Their Constituents. In *Soft Chemistry and Food Fermentation*; Academic Press: Cambridge, MA, USA, 2017; pp. 379–419. [CrossRef]
29. Lammari, N.; Louaer, O.; Meniai, A.H.; Elaissari, A. Encapsulation of Essential Oils via Nanoprecipitation Process: Overview, Progress, Challenges and Prospects. *Pharmaceutics* **2020**, *12*, 431. [CrossRef]
30. Syed, I.; Garg, S.; Sarkar, P. Entrapment of essential oils in hydrogels for biomedical applications. In *Polymeric Gels*; Woodhead Publishing: Cambridge, UK, 2018; pp. 125–141. [CrossRef]
31. El Asbahani, A.; Miladi, K.; Badri, W.; Sala, M.; Aït Addi, E.H.; Casabianca, H.; El Mousadik, A.; Hartmann, D.; Jilale, A.; Renaud, F.N.R.; et al. Essential oils: From extraction to encapsulation. *Int. J. Pharm.* **2015**, *483*, 220–243. [CrossRef]
32. Stoleru, E.; Dumitriu, R.P.; Ailiesei, G.-L.; Yilmaz, C.; Brebu, M. Synthesis of Bioactive Materials by In Situ One-Step Direct Loading of *Syzygium aromaticum* Essential Oil into Chitosan-Based Hydrogels. *Gels* **2022**, *8*, 225. [CrossRef] [PubMed]
33. Spyrou, S.; Bellou, M.G.; Papanikolaou, A.; Nakou, K.; Kontogianni, V.G.; Chatzikonstan-tinou, A.V.; Stamatis, H. Evaluation of Antioxidant, Antibacterial and Enzyme-Inhibitory Properties of Dittany and Thyme Extracts and Their Application in Hydrogel Preparation. *BioChem* **2024**, *4*, 166–188. [CrossRef]
34. Mostaghimi, M.; Majdinasab, M.; Hosseini, S.M.H. Characterization of alginate hydrogel beads loaded with thyme and clove essential oils nanoemulsions. *J. Polym. Environ.* **2022**, *30*, 1647–1661. [CrossRef]

35. Alsakhawy, S.A.; Baghdadi, H.H.; El-Shenawy, M.A.; Sabra, S.A.; El-Hosseiny, L.S. Encapsulation of *thymus vulgaris* essential oil in caseinate/gelatin nano-composite hydrogel: In vitro antibacterial activity and in vivo wound healing potential. *Int. J. Pharm.* **2022**, *628*, 122280. [CrossRef]
36. Çakır, C.; Gürkan, E.H. Enhancing therapeutic effects alginate microencapsulation of thyme and calendula oils using ionic gelation for controlled drug delivery. *J. Biomater. Sci. Polym. Ed.* **2024**, 1–29. [CrossRef] [PubMed]
37. Fincheira, P.; Espinoza, J.; Levío-Raimán, M.; Vera, J.; Tortella, G.; Brito, A.M.M.; Seabra, A.B.; Diez, M.D.; Quiroz, A.; Rubilar, O. Formulation of essential oils-loaded solid lipid nanoparticles-based chitosan/PVA hydrogels to control the growth of *Botrytis cinerea* and *Penicillium expansum*. *Int. J. Biol. Macromol.* **2024**, *270*, 132218. [CrossRef] [PubMed]
38. Zhao, Y.; Li, H.; Wang, Y.; Zhang, Z.; Wang, Q. Preparation, characterization and release kinetics of a multilayer encapsulated *Perilla frutescens* L. essential oil hydrogel bead. *Int. J. Biol. Macromol.* **2023**, *249*, 124776. [CrossRef] [PubMed]
39. Dobroslavić, E.; Cegledi, E.; Robić, K.; Elez Garofulić, I.; Dragović-Uzelac, V.; Repajić, M. Encapsulation of Fennel Essential Oil in Calcium Alginate Microbeads via Electrostatic Extrusion. *Appl. Sci.* **2024**, *14*, 3522. [CrossRef]
40. Chelu, M.; Musuc, A.M. Biomaterials-Based Hydrogels for Therapeutic Applications. In *Biomaterials in Microencapsulation*; IntechOpen: London, UK, 2024. [CrossRef]
41. De France, K.J.; Xu, F.; Hoare, T. Structured Macroporous Hydrogels: Progress, Challenges, and Opportunities. *Adv. Healthc. Mater.* **2018**, *7*, 1700927. [CrossRef]
42. Chelu, M.; Calderon Moreno, J.M.; Musuc, A.M.; Popa, M. Natural Regenerative Hydrogels for Wound Healing. *Gels* **2024**, *10*, 547. [CrossRef] [PubMed]
43. Barbosa, A.I.; Lima, S.A.C.; Yousef, I.; Reis, S. Evaluating the Skin Interactions and Permeation of Alginate/Fucoidan Hydrogels Per Se and Associated with Different Essential Oils. *Pharmaceutics* **2023**, *15*, 190. [CrossRef]
44. Serra, E.; Saubade, F.; Ligorio, C.; Whitehead, K.; Sloan, A.; Williams, D.W.; Hidalgo-Bastida, A.; Verran, J.; Malic, S. Methylcellulose Hydrogel with *Melissa officinalis* Essential Oil as a Potential Treatment for Oral Candidiasis. *Microorganisms* **2020**, *8*, 215. [CrossRef]
45. Khairan, K.; Hasan, M.; Idroes, R.; Diah, M. Fabrication and Evaluation of Polyvinyl Alcohol/Corn Starch/Patchouli Oil Hydrogel Films Loaded with Silver Nanoparticles Biosynthesized in *Pogostemon cablin* Benth Leaves' Extract. *Molecules* **2023**, *28*, 2020. [CrossRef] [PubMed]
46. Chelu, M.; Musuc, A.M.; Aricov, L.; Ozon, E.A.; Iosageanu, A.; Stefan, L.M.; Prelipcean, A.-M.; Popa, M.; Moreno, J.C. Antibacterial Aloe vera Based Biocompatible Hydrogel for Use in Dermatological Applications. *Int. J. Mol. Sci.* **2023**, *24*, 3893. [CrossRef]
47. Folle, C.; Díaz-Garrido, N.; Mallandrich, M.; Suñer-Carbó, J.; Sánchez-López, E.; Halbaut, L.; Marqués, A.M.; Espina, M.; Badia, J.; Baldoma, L.; et al. Hydrogel of Thyme-Oil-PLGA Nanoparticles Designed for Skin Inflammation Treatment. *Gels* **2024**, *10*, 149. [CrossRef]
48. Rusu, A.G.; Niță, L.E.; Roșca, I.; Croitoriu, A.; Ghilan, A.; Mititelu-Tarțău, L.; Grigoraș, A.V.; Crețu, B.-E.-B.; Chiriac, A.P. Alginate-Based Hydrogels Enriched with Lavender Essential Oil: Evaluation of Physicochemical Properties, Antimicrobial Activity, and In Vivo Biocompatibility. *Pharmaceutics* **2023**, *15*, 2608. [CrossRef]
49. Tamer, T.M.; Sabet, M.M.; Alhalili, Z.A.H.; Ismail, A.M.; Mohy-Eldin, M.S.; Hassan, M.A. Influence of Cedar Essential Oil on Physical and Biological Properties of Hemostatic, Antibacterial, and Antioxidant Polyvinyl Alcohol/Cedar Oil/Kaolin Composite Hydrogels. *Pharmaceutics* **2022**, *14*, 2649. [CrossRef]
50. Cortés-Rojas, D.F.; de Souza, C.R.F.; Oliveira, W.P. Clove (*Syzygium aromaticum*): A precious spice. *Asian Pac. J. Trop. Biomed.* **2014**, *4*, 90–96. [CrossRef] [PubMed]
51. Chelu, M.; Popa, M.; Ozon, E.A.; Pandele Cusu, J.; Anastasescu, M.; Surdu, V.A.; Calderon Moreno, J.; Musuc, A.M. High-Content Aloe vera Based Hydrogels: Physicochemical and Pharmaceutical Properties. *Polymers* **2023**, *15*, 1312. [CrossRef] [PubMed]
52. Suflet, D.M.; Constantin, M.; Pelin, I.M.; Popescu, I.; Rimbu, C.M.; Horhogea, C.E.; Fundueanu, G. Chitosan–Oxidized Pullulan Hydrogels Loaded with Essential Clove Oil: Synthesis, Characterization, Antioxidant and Antimicrobial Properties. *Gels* **2024**, *10*, 227. [CrossRef]
53. Edris, A.E. Pharmaceutical and therapeutic potentials of essential oils and their individual volatile constituents: A review. *Phytother. Res.* **2007**, *21*, 308–323. [CrossRef]
54. Laothaweerungsawat, N.; Sirithunyalug, J.; Chaiyana, W. Chemical Compositions and Anti-Skin-Ageing Activities of *Origanum vulgare* L. Essential Oil from Tropical and Mediterranean Region. *Molecules* **2020**, *25*, 1101. [CrossRef]
55. Jianu, C.; Lukinich-Gruia, A.T.; Rădulescu, M.; Mioc, M.; Mioc, A.; Șoica, C.; Constantin, A.T.; David, I.; Bujancă, G.; Radu, R.G. Essential Oil of *Origanum vulgare* var. *aureum* L. from Western Romania: Chemical Analysis, In Vitro and In Silico Screening of Its Antioxidant Activity. *Appl. Sci.* **2023**, *13*, 5076. [CrossRef]
56. Alekseeva, M.; Zagorcheva, T.; Atanassov, I.; Rusanov, K. *Origanum vulgare* L.—A Review on Genetic Diversity, Cultivation, Biological Activities and Perspectives for Molecular Breeding. *Bulg. J. Agric. Sci.* **2020**, *26*, 1183–1197.
57. Avram, Ș.; Bora, L.; Vlaia, L.L.; Muț, A.M.; Olteanu, G.-E.; Olariu, I.; Magyari-Pavel, I.Z.; Minda, D.; Diaconeasa, Z.; Sfirloaga, P.; et al. Cutaneous Polymeric-Micelles-Based Hydrogel Containing *Origanum vulgare* L. Essential Oil: In Vitro Release and Permeation, Angiogenesis, and Safety Profile In Ovo. *Pharmaceuticals* **2023**, *16*, 940. [CrossRef] [PubMed]
58. Cruz Sánchez, E.; García, M.T.; Pereira, J.; Oliveira, F.; Craveiro, R.; Paiva, A.; Gracia, I.; García-Vargas, J.M.; Duarte, A.R.C. Alginate–Chitosan Membranes for the Encapsulation of Lavender Essential Oil and Development of Biomedical Applications Related to Wound Healing. *Molecules* **2023**, *28*, 3689. [CrossRef] [PubMed]

59. Nawaz, A.; Farid, A.; Safdar, M.; Latif, M.S.; Ghazanfar, S.; Akhtar, N.; Al Jaouni, S.K.; Selim, S.; Khan, M.W. Formulation Development and Ex-Vivo Permeability of Curcumin Hydrogels under the Influence of Natural Chemical Enhancers. *Gels* **2022**, *8*, 384. [CrossRef]
60. Liang, J.; Zhang, Y.; Chi, P.; Liu, H.; Jing, Z.; Cao, H.; Du, Y.; Zhao, Y.; Qin, X.; Zhang, W.; et al. Essential oils: Chemical constituents, potential neuropharmacological effects and aromatherapy—A review. *Pharmacol. Res. Mod. Chin. Med.* **2023**, *6*, 100210. [CrossRef]
61. Zeng, W.C.; Zhang, Z.; Gao, H.; Jia, L.R.; He, Q. Chemical composition, antioxidant, and antimicrobial activities of essential oil from pine needle (*Cedrus deodara*). *J. Food Sci.* **2012**, *77*, C824–C829. [CrossRef]
62. Šunić, L.; Ilić, Z.S.; Stanojević, L.; Milenković, L.; Stanojević, J.; Kovač, R.; Milenković, A.; Cvetković, D. Comparison of the Essential Oil Content, Constituents and Antioxidant Activity from Different Plant Parts during Development Stages of Wild Fennel (*Foeniculum vulgare* Mill.). *Horticulturae* **2023**, *9*, 364. [CrossRef]
63. Gheorghita, D.; Grosu, E.; Robu, A.; Ditu, L.M.; Deleanu, I.M.; Gradisteanu Pircalabioru, G.; Raiciu, A.-D.; Bita, A.-I.; Antoniac, A.; Antoniac, V.I. Essential Oils as Antimicrobial Active Substances in Wound Dressings. *Materials* **2022**, *15*, 6923. [CrossRef]
64. Agelaki, S.; Boukovinas, I.; Athanasiadis, I.; Trimis, G.; Dimitriadis, I.; Poughias, L.; Morais, E.; Sabale, U.; Bencina, G.; Athanasopoulos, C. A systematic literature review of the human papillomavirus prevalence in locally and regionally advanced and recur-rent/metastatic head and neck cancers through the last decade: The "ALARM" study. *Cancer Med.* **2024**, *13*, e6916. [CrossRef]
65. Siegel, R.L.; Giaquinto, A.N.; Jemal, A. Cancer statistics. *CA A Cancer J. Clin.* **2024**, *74*(2), 203. [CrossRef]
66. Alsabbagh, R.; Ahmed, M.; Alqudah, M.A.Y.; Hamoudi, R.; Harati, R. Insights into the Molecular Mechanisms Mediating Extravasation in Brain Metastasis of Breast Cancer, Melanoma, and Lung Cancer. *Cancers* **2023**, *15*, 2258. [CrossRef]
67. Waks, A.G.; Winer, E.P. Breast cancer treatment: A review. *JAMA* **2019**, *321*, 288–300. [CrossRef]
68. Ali, S.; Li, J.; Pei, Y.; Khurram, R.; Rehman, K.u.; Rasool, A.B. State-of-the-Art Challenges and Perspectives in Multi-Organ Cancer Diagnosis via Deep Learning-Based Methods. *Cancers* **2021**, *13*, 5546. [CrossRef] [PubMed]
69. Burz, C.; Pop, V.; Silaghi, C.; Lupan, I.; Samasca, G. Prognosis and Treatment of Gastric Cancer: A 2024 Update. *Cancers* **2024**, *16*, 1708. [CrossRef] [PubMed]
70. Chunarkar-Patil, P.; Kaleem, M.; Mishra, R.; Ray, S.; Ahmad, A.; Verma, D.; Bhayye, S.; Dubey, R.; Singh, H.N.; Kumar, S. Anticancer Drug Discovery Based on Natural Products: From Computational Approaches to Clinical Studies. *Biomedicines* **2024**, *12*, 201. [CrossRef]
71. Tauro, S.; Dhokchawle, B.; Mohite, P.; Nahar, D.; Nadar, S.; Coutinho, E. Natural Anticancer agents: Their therapeutic potential, challenges and Promising outcomes. *Curr. Med. Chem.* **2024**, *31*, 848–870. [CrossRef]
72. Bajpai, P.; Usmani, S.; Kumar, R.; Prakash, O. Recent advances in anticancer ap-proach of traditional medicinal plants: A novel strategy for cancer chemotherapy. *Intell. Pharm.* **2024**, *2*, 291–304. [CrossRef]
73. Jampilek, J.; Kralova, K. Anticancer Applications of Essential Oils Formulated into Lipid-Based Delivery Nanosystems. *Pharmaceutics* **2022**, *14*, 2681. [CrossRef] [PubMed]
74. Angelini, P.; Tirillini, B.; Akhtar, M.S.; Dimitriu, L.; Bricchi, E.; Bertuzzi, G.; Venanzoni, R. Essential oil with anticancer activity: An overview. In *Anticancer Plants: Natural Products and Biotechnological Implements*; Springer: Singapore, 2018; Volume 2, pp. 207–231. [CrossRef]
75. Kamal, I.; Khedr, A.I.; Alfaifi, M.Y.; Elbehairi, S.E.I.; Elshaarawy, R.F.; Saad, A.S. Chemo-therapeutic and chemopreventive potentials of ρ-coumaric acid–Squid chitosan nanogel loaded with *Syzygium aromaticum* essential oil. *Int. J. Biol. Macromol.* **2021**, *188*, 523–533. [CrossRef] [PubMed]
76. Liu, B.; Chen, K. Advances in Hydrogel-Based Drug Delivery Systems. *Gels* **2024**, *10*, 262. [CrossRef]
77. Idumah, C.I.; Nwuzor, I.C.; Odera, S.R.; Timothy, U.J.; Ngenegbo, U.; Tanjung, F.A. Recent advances in polymeric hydrogel nanoarchitectures for drug delivery applications. *Int. J. Polym. Mater. Polym. Biomater.* **2024**, *73*, 1–32. [CrossRef]
78. Shabkhiz, M.A.; Pirouzifard, M.K.; Pirsa, S.; Mahdavinia, G.R. Alginate hydrogel beads containing Thymus daenensis essential oils/Glycyrrhizic acid loaded in β-cyclodextrin. Investigation of structural, antioxidant/antimicrobial properties and release assessment. *J. Mol. Liq.* **2021**, *344*, 117738. [CrossRef]
79. Basyigit, B. Designing Nanoliposome-in-Natural Hydrogel Hybrid System for Controlla-ble Release of Essential Oil in Gastrointestinal Tract: A Novel Vehicle. *Foods* **2023**, *12*, 2242. [CrossRef] [PubMed]
80. Mogosanu, G.D.; Grumezescu, A.M. Natural and synthetic polymers for wounds and burn dressing. *Int. J. Pharm.* **2014**, *463*, 127–136. [CrossRef] [PubMed]
81. George, B.; Bhatia, N.; Suchithra, T.V. Burgeoning hydrogel technology in burn wound care: A comprehensive meta-analysis. *Eur. Polym. J.* **2021**, *157*, 110640. [CrossRef]
82. Alven, S.; Peter, S.; Aderibigbe, B.A. Polymer-Based Hydrogels Enriched with Essential Oils: A Promising Approach for the Treatment of Infected Wounds. *Polymers* **2022**, *14*, 3772. [CrossRef] [PubMed]
83. Goh, M.; Du, M.; Peng, W.R.; Saw, P.E.; Chen, Z. Advancing burn wound treatment: Exploring hydrogel as a transdermal drug delivery system. *Drug Deliv.* **2024**, *31*, 2300945. [CrossRef]
84. Jiji, S.; Udhayakumar, S.; Rose, C.; Muralidharan, C.; Kadirvelu, K. Thymol enriched bacterial cellulose hydrogel as effective material for third degree burn wound repair. *Int. J. Biol. Macromol.* **2019**, *122*, 452–460. [CrossRef] [PubMed]

85. Khan, B.A.; Ullah, S.; Khan, M.K.; Uzair, B.; Menaa, F.; Braga, V.A. Fabrication, Physical Characterizations, and In Vitro, In Vivo Evaluation of Ginger Extract-Loaded Gelatin/Poly(Vinyl Alcohol) Hydrogel Films against Burn Wound Healing in Animal Model. *AAPS PharmSciTech* **2020**, *21*, 323. [CrossRef]
86. Wang, H.; Liu, Y.; Cai, K.; Zhang, B.; Tang, S.; Zhang, W.; Liu, W. Antibacterial polysaccharide-based hydrogel dressing containing plant essential oil for burn wound healing. *Burn. Trauma* **2021**, *9*, tkab041. [CrossRef]
87. Dalir Abdolahinia, E.; Hajisadeghi, S.; Moayedi Banan, Z.; Dadgar, E.; Delaramifar, A.; Izadian, S.; Simin Sharifi, S.; Maleki Dizaj, S. Potential applications of medicinal herbs and phytochemicals in oral and dental health: Status quo and future perspectives. *Oral Dis.* **2023**, *29*, 2468–2482. [CrossRef]
88. Kumar, G.; Jalaluddin, M.; Rout, P.; Mohanty, R.; Dileep, C.L. Emerging trends of herbal care in dentistry. *J. Clin. Diagn. Res.* **2013**, *7*, 1827–1829. [CrossRef]
89. Chelu, M.; Musuc, A.M.; Popa, M.; Calderon Moreno, J. Aloe vera-Based Hydrogels for Wound Healing: Properties and Therapeutic Effects. *Gels* **2023**, *9*, 539. [CrossRef] [PubMed]
90. Jo, Y.H.; Cho, J.H.; Park, D.H.; Yoon, H.I.; Han, S.H.; Yilmaz, B. Antimicrobial activity, surface properties, and cytotoxicity of microencapsulated phytochemicals incorporated into three-dimensionally printable dental polymers. *J. Dent.* **2024**, *141*, 104820. [CrossRef] [PubMed]
91. Wong, Y.Y.; Chow, Y.L. Exploring the potential of spice-derived phytochemicals as alternative antimicrobial agents. *eFood* **2024**, *5*, e126. [CrossRef]
92. Tzimas, K.; Antoniadou, M.; Varzakas, T.; Voidarou, C. Plant-Derived Compounds: A Promising Tool for Dental Caries Prevention. *Curr. Issues Mol. Biol.* **2024**, *46*, 5257–5290. [CrossRef] [PubMed]
93. Pandhi, S.; Kumar, A. Odontonutraceuticals: Phytochemicals for Oral Health Care. In *Innovations in Food Technology*; Mishra, P., Mishra, R.R., Adetunji, C.O., Eds.; Springer: Singapore, 2020. [CrossRef]
94. Muresan, S.M.C.; Dreanca, A.; Repciuc, C.; Dejescu, C.; Rotar, O.; Pop, R.A.; Pantea, S.; Pall, E.; Ciotlaus, I.; Sarosi, C.; et al. Dental Hydrogels with Essential Oils with Potential Activity in Periodontitis. *Appl. Sci.* **2023**, *13*, 1787. [CrossRef]
95. Achagar, R.; Ait-Touchente, Z.; El Ati, R.; Boujdi, K.; Thoume, A.; Abdou, A.; Touzani, R. A Comprehensive Review of Essential Oil–Nanotechnology Synergy for Advanced Dermo-cosmetic Delivery. *Cosmetics* **2024**, *11*, 48. [CrossRef]
96. Rubio, L.; Pita, A.; Garcia-Jares, C.; Lores, M. Natural Extracts and Essential Oils as Ingredients in Cosmetics: Search for Potential Phytomarkers and Allergen Survey. *Cosmetics* **2024**, *11*, 84. [CrossRef]
97. Tocai, A.-C.; Memete, A.R.; Ganea, M.; Vicaș, L.G.; Gligor, O.D.; Vicas, S.I. The Formulation of Dermato-Cosmetic Products Using Sanguisorba minor Scop. Extract with Powerful Antioxidant Capacities. *Cosmetics* **2024**, *11*, 8. [CrossRef]
98. Verger, A.; Kichou, H.; Huang, N.; Perse, X.; Ardeza, I.M.; Pradel, C.; Goncalves Martins Da Conceicao, R.; Atanasova, B.; Legrand, F.-X.; Munnier, E.; et al. Effects of Hydrophilic Natural Deep Eutectic Solvents on the Rheological, Textural, and Sensory Properties of Carboxymethylcellulose-Based Cosmetic Hydrogels. *ACS Sustain. Chem. Eng.* **2024**, *12*, 7187–7199. [CrossRef]
99. Chelu, M.; Musuc, A.M. Natural Biological Macromolecules for Designing Hydrogels as Health Care and Anti-aging Solutions. *Eng. Proc.* **2023**, *56*, 158. [CrossRef]
100. Zagórska-Dziok, M.; Sobczak, M. Hydrogel-Based Active Substance Release Systems for Cosmetology and Dermatology Application: A Review. *Pharmaceutics* **2020**, *12*, 396. [CrossRef] [PubMed]
101. Morais, F.P.; Simões, R.M.S.; Curto, J.M.R. Biopolymeric Delivery Systems for Cosmetic Applications Using *Chlorella vulgaris* Algae and Tea Tree Essential Oil. *Polymers* **2020**, *12*, 2689. [CrossRef] [PubMed]
102. Sosnowska, K.; Tomczykowa, M.; Winnicka, K.; Kalemba, D.; Tomczyk, M. In vivo evaluation of the antipsoriatic effect of hydrogel with lavandin essential oil and its main components after topical application. *Acta Pol. Pharm.* **2022**, *79*, 757–770. [CrossRef]
103. Yap, X.F.; Saw, S.H.; Lim, V.; Tan, C.X. Plant Essential Oil Nanoemulgel as a Cosmeceutical Ingredient: A Review. *Cosmetics* **2024**, *11*, 116. [CrossRef]
104. Kaolaor, A.; Kiti, K.; Pankongadisak, P.; Suwantong, O. Camellia Oleifera oil-loaded chitosan nanoparticles embedded in hydrogels as cosmeceutical products with improved biological properties and sustained drug release. *Int. J. Biol. Macromol.* **2024**, *275*, 133560. [CrossRef]
105. de Vos, P.; Faas, M.M.; Spasojevic, M.; Sikkema, J. Encapsulation for preservation of functionality and targeted delivery of bioactive food components. *Int. Dairy J.* **2010**, *20*, 292–302. [CrossRef]
106. Črnivec, I.G.O.; Ulrih, N.P. Nano-hydrogels of alginate for encapsulation of food ingredients. In *Biopolymer Nanostructures for Food Encapsulation Purposes*; Academic Press: Cambridge, MA, USA, 2019; pp. 335–380. [CrossRef]
107. Chambin, O.; Dupuis, G.; Champion, D.; Voilley, A.; Pourcelot, Y. Colon-specific drug delivery: Influence of solution reticulation properties upon pectin beads performance. *Int. J. Pharm.* **2006**, *321*, 86–93. [CrossRef]
108. Tavares, L.; Noreña, C.P.Z.; Barros, H.L.; Smaoui, S.; Lima, P.S.; Marques de Oliveira, M. Rheological and structural trends on encapsulation of bioactive compounds of essential oils: A global systematic review of recent research. *Food Hydrocoll.* **2022**, *129*, 107628. [CrossRef]
109. Popa, E.G.; Gomes, M.E.; Reis, R.L. Cell delivery systems using alginate–carrageenan hy-drogel beads and fibers for regenerative medicine applications. *Biomacromolecules* **2011**, *12*, 3952–3961. [CrossRef]
110. McClements, D.J. Recent progress in hydrogel delivery systems for improving nutraceutical bioavailability. *Food Hydrocoll.* **2017**, *68*, 238–245. [CrossRef]

111. Reichembach, L.H.; de Oliveira Petkowicz, C.L.; Guerrero, P.; de la Caba, K. Pectin and pectin/chitosan hydrogel beads as coffee essential oils carrier systems. *Food Hydrocoll.* **2024**, *151*, 109814. [CrossRef]
112. Carpena, M.; Nuñez-Estevez, B.; Soria-Lopez, A.; Garcia-Oliveira, P.; Prieto, M.A. Essential Oils and Their Application on Active Packaging Systems: A Review. *Resources* **2021**, *10*, 7. [CrossRef]
113. Gómez-Estaca, J.; De Lacey, A.L.; López-Caballero, M.E.; Gómez-Guillén, M.D.C.; Montero, P. Biodegradable gelatin–chitosan films incorporated with essential oils as antimicrobial agents for fish preservation. *Food Microbiol.* **2010**, *27*, 889–896. [CrossRef] [PubMed]
114. Ahmad, M.; Benjakul, S.; Sumpavapol, P.; Nirmal, N.P. Quality changes of sea bass slices wrapped with gelatin film incorporated with lemongrass essential oil. *Int. J. Food Microbiol.* **2012**, *155*, 171–178. [CrossRef] [PubMed]
115. Kanelaki, A.; Zampouni, K.; Mourtzinos, I.; Katsanidis, E. Hydrogels, Oleogels and Bigels as Edible Coatings of Sardine Fillets and Delivery Systems of Rosemary Extract. *Gels* **2022**, *8*, 660. [CrossRef] [PubMed]
116. Iacovino, S.; Cofelice, M.; Sorrentino, E.; Cuomo, F.; Messia, M.C.; Lopez, F. Alginate-Based Emulsions and Hydrogels for Extending the Shelf Life of Banana Fruit. *Gels* **2024**, *10*, 245. [CrossRef]
117. Bandyopadhyay, S.; Saha, N.; Zandraa, O.; Pummerová, M.; Sáha, P. Essential Oil Based PVP-CMC-BC-GG Functional Hydrogel Sachet for 'Cheese': Its Shelf Life Confirmed with Anthocyanin (Isolated from Red Cabbage) Bio Stickers. *Foods* **2020**, *9*, 307. [CrossRef]
118. Erceg, T.; Šovljanski, O.; Stupar, A.; Ugarković, J.; Aćimović, M.; Pezo, L.; Tomić, A.; Todosijević, M. A comprehensive approach to chi-tosan-gelatine edible coating with β-cyclodextrin/lemongrass essential oil inclusion complex—Characterization and food application. *Int. J. Biol. Macromol.* **2023**, *228*, 400–410. [CrossRef]
119. Goudoulas, T.B.; Vanderhaeghen, S.; Germann, N. Micro-dispersed essential oils loaded gelatin hydrogels with antibacterial activity. *LWT* **2022**, *154*, 112797. [CrossRef]
120. Stefanowska, K.; Bucher, M.; Reichert, C.L.; Sip, A.; Woźniak, M.; Schmid, M.; Dobrucka, R.; Ratajczak, I. Chitosan-based films with nanocellulose and propolis as active packaging materials. *Ind. Crops Prod.* **2024**, *219*, 119112. [CrossRef]
121. Domínguez, R.; Munekata, P.E.S.; Pateiro, M.; López-Fernández, O.; Manuel Lorenzo, J. Immobilization of oils using hydrogels as strategy to replace animal fats and improve the healthiness of meat products. *Curr. Opin. Food Sci.* **2021**, *37*, 135–144. [CrossRef]
122. Nath, P.C.; Debnath, S.; Sridhar, K.; Inbaraj, B.S.; Nayak, P.K.; Sharma, M. A Comprehensive Review of Food Hydrogels: Principles, Formation Mechanisms, Microstructure, and Its Applications. *Gels* **2023**, *9*, 1. [CrossRef] [PubMed]
123. Pascuta, M.S.; Vodnar, D.C. Nanocarriers for Sustainable Active Packaging: An Overview during and Post COVID-19. *Coatings* **2022**, *12*, 102. [CrossRef]
124. Tomić, A.; Šovljanski, O.; Erceg, T. Insight on Incorporation of Essential Oils as Antimicrobial Substances in Biopolymer-Based Active Packaging. *Antibiotics* **2023**, *12*, 1473. [CrossRef] [PubMed]
125. Alonso, P.; Fernández-Pastor, S.; Guerrero, A. Application of Cinnamon Essential Oil in Active Food Packaging: A Review. *Appl. Sci.* **2024**, *14*, 6554. [CrossRef]
126. Behbahani, A.B.; Falah, F.; Lavi Arab, F.; Vasiee, M.; Tabatabaee Yazdi, F. Chemical composition and antioxidant, antimicrobial, and antiproliferative activities of Cinnamomum zeylanicum bark essential oil. *Evid. Based Complement. Altern. Med.* **2020**, *2020*, 5190603. [CrossRef] [PubMed]
127. Abdelwahab, S.I.; Mariod, A.A.; Taha, M.M.E.; Zaman, F.Q.; Abdelmageed, A.H.A.; Khamis, S.; Sivasothy, Y.; Awang, K. Chemical composition and antioxidant properties of the essential oil of Cinnamomum altissimum Kosterm (Lauraceae). *Arab. J. Chem.* **2017**, *10*, 131–135. [CrossRef]
128. Hou, F.; Chen, X.; Yi, F.; Song, L.; Zhan, S.; Han, X.; Zhang, L.; Li, F.; Wang, X.; Liu, Z. Antibacterial and antibiofilm properties of cinnamon essential oil on Pseudomonas tolaasii and application of potato starch/CEO active pads in preservation of mushroom (*Agaricus bisporus*). *Food Control* **2024**, *165*, 110705. [CrossRef]
129. Bhatia, S.; Al-Harrasi, A.; Shah, Y.A.; Altoubi, H.W.K.; Kotta, S.; Sharma, P.; Anwer, M.K.; Kaithavalappil, D.S.; Koca, E.; Aydemir, L.Y. Fabrication, Characterization, and Antioxidant Potential of Sodium Alginate/Acacia Gum Hydrogel-Based Films Loaded with Cinnamon Essential Oil. *Gels* **2023**, *9*, 337. [CrossRef]
130. Rîmbu, C.M.; Serbezeanu, D.; Vlad-Bubulac, T.; Suflet, D.M.; Motrescu, I.; Lungoci, C.; Robu, T.; Vrînceanu, N.; Grecu, M.; Cozma, A.P.; et al. Antimicrobial Activity of Artemisia dracunculus Oil-Loaded Agarose/Poly(Vinyl Alcohol) Hydrogel for Bio-Applications. *Gels* **2024**, *10*, 26. [CrossRef] [PubMed]
131. Boccia, A.C.; Pulvirenti, A.; Cerruti, P.; Silvetti, T.; Brasca, M. Antimicrobial starch-based cryogels and hydrogels for dual-active food packaging applications. *Carbohydr. Polym.* **2024**, *342*, 122340. [CrossRef] [PubMed]
132. Chen, M.; Hu, Z.; Zheng, H.; Wang, J.; Xu, X. Antimicrobial polysaccharide hydrogels embedded with methyl-β-cyclodextrin/thyme oil inclusion complexes for exceptional mechanical performance and chilled chicken breast preservation. *Int. J. Biol. Macromol.* **2024**, *267*, 131586. [CrossRef]
133. Li, Q.; Zhang, B.; Yang, X.; Ge, Q. Deterioration-associated microbiome of stone monuments: Structure, variation, and assembly. *Appl. Environ. Microbiol.* **2018**, *84*, e02680-17. [CrossRef]
134. Vázquez-Nion, D.; Silva, B.; Prieto, B. Influence of the properties of granitic rocks on their bioreceptivity to subaerial phototrophic biofilms. *Sci. Total Environ.* **2018**, *610*, 44–54. [CrossRef] [PubMed]
135. Gaylarde, C. Influence of Environment on Microbial Colonization of Historic Stone Buildings with Emphasis on Cyanobacteria. *Heritage* **2020**, *3*, 1469–1482. [CrossRef]

136. Ranaldi, R.; Rugnini, L.; Gabriele, F.; Spreti, N.; Casieri, C.; Di Marco, G.; Gismondi, A.; Bruno, L. Plant essential oils suspended into hydrogel: Development of an easy-to-use protocol for the restoration of stone cultural heritage. *Int. Biodeterior. Biodegrad.* **2022**, *172*, 105436. [CrossRef]
137. Pinna, D. Microbial Growth and its Effects on Inorganic Heritage Materials. In *Microorganisms in the Deterioration and Preservation of Cultural Heritage*; Joseph, E., Ed.; Springer: Cham, Switzerland, 2021. [CrossRef]
138. Toreno, G.; Isola, D.; Meloni, P.; Carcangiu, G.; Selbmann, L.; Onofri, S.; Caneva, G.; Zucconi, L. Biological colonization on stone monuments: A new low impact cleaning method. *J. Cult. Herit.* **2018**, *30*, 100–109. [CrossRef]
139. Cappitelli, F.; Cattò, C.; Villa, F. The Control of Cultural Heritage Microbial Deterioration. *Microorganisms* **2020**, *8*, 1542. [CrossRef]
140. Sanmartín, P.; Rodríguez, A.; Aguiar, U. Medium-term field evaluation of several widely used cleaning-restoration techniques applied to algal biofilm formed on a granite-built historical monument. *Int. Biodeterior. Biodegrad.* **2020**, *147*, 104870. [CrossRef]
141. Genova, C.; Fuentes, E.; Favero, G.; Prieto, B. Evaluation of the Cleaning Effect of Natural-Based Biocides: Application on Different Phototropic Biofilms Colonizing the Same Granite Wall. *Coatings* **2023**, *13*, 520. [CrossRef]
142. Gabriele, F.; Ranaldi, R.; Bruno, L.; Casieri, C.; Rugnini, L.; Spreti, N. Biodeterioration of stone monuments: Studies on the influence of bioreceptivity on cyanobacterial biofilm growth and on the biocidal efficacy of essential oils in natural hydrogel. *Sci. Total Environ.* **2023**, *870*, 161901. [CrossRef]
143. Russo, R.; Palla, F. Plant Essential Oils as Biocides in Sustainable Strategies for the Conservation of Cultural Heritage. *Sustainability* **2023**, *15*, 8522. [CrossRef]
144. Privitera, A.; Tuti, S.; Laverdura, U.P.; Duranti, L.; Di Bartolomeo, E.; Taddei, A.R.; Sodo, A. One-step nanoencapsulation of essential oils and their application in hybrid coatings: A sustainable long-lasting treatment of stone materials against biodeterioration. *Prog. Org. Coat.* **2024**, *196*, 108759. [CrossRef]
145. Boccalon, E.; Nocchetti, M.; Pica, M.; Romani, A.; Sterflinger, K. Hydrogels: A 'stepping stone' towards new cleaning strategies for biodeteriorated surfaces. *J. Cult. Herit.* **2021**, *47*, 1–11. [CrossRef]
146. Bruno, L.; Casieri, C.; Francesco Gabriele, F.; Ranaldi, R.; Rugnini, L.; Spreti, N. In situ application of alginate hydrogels containing oxidant or natural biocides on Fortunato Depero's mosaic (Rome, Italy). *Int. Biodeterior. Biodegrad.* **2023**, *183*, 105641. [CrossRef]
147. Wong, Y.C.; Ahmad-Mudzaqqir, M.Y.; Wan-Nurdiyana, W.A. Extraction of essential oil from cinnamon (*Cinnamomum zeylanicum*). *Orient. J. Chem.* **2014**, *30*, 37. [CrossRef]
148. Liu, X.; Jia, J.; Duan, S.; Zhou, X.; Xiang, A.; Lian, Z.; Ge, F. Zein/MCM-41 nanocomposite film incorporated with cinnamon essential oil loaded by modified supercritical CO_2 impregnation for long-term antibacterial packaging. *Pharmaceutics* **2020**, *12*, 169. [CrossRef]
149. Wang, Y.; Yuan, C.; Liu, Y.; Cui, B. Fabrication of kappa–carrageenan hydrogels with cinnamon essential oil/hydroxypropyl–β–cyclodextrin composite: Evaluation of physicochemical properties, release kinetics and antimicrobial activity. *Int. J. Biol. Macromol.* **2021**, *170*, 593–601. [CrossRef]
150. Lesage-Meessen, L.; Bou, M.; Sigoillot, J.C.; Faulds, C.B.; Lomascolo, A. Essential oils and distilled straws of lavender and lavandin: A review of current use and potential application in white biotechnology. *Appl. Microbiol. Biotechnol.* **2015**, *99*, 3375–3385. [CrossRef]
151. Pokajewicz, K.; Biało'n, M.; Svydenko, L.; Fedin, R.; Hudz, N. Chemical composition of the essential oil of the new cultivars of Lavandula angustifolia Mill. Bred in Ukraine. *Molecules* **2021**, *26*, 5681. [CrossRef] [PubMed]
152. Cardia, G.F.E.; de Souza Silva-Comar, F.M.; da Rocha, E.M.T.; Silva-Filho, S.E.; Zagotto, M.; Uchida, N.S.; do Amaral, V.; Bersani-Amado, C.A.; Cuman, R.K.N. Pharmacological, medicinal and toxicological properties of lavender essential oil: A review. Research. *Soc. Dev.* **2021**, *10*, e23310514933. [CrossRef]
153. Deng, X.; Chen, J.; Chen, W. Hydrogel particles as a controlled release delivery system for lavender essential oil using pH triggers. *Colloids Surf. A Physicochem. Eng. Asp.* **2020**, *603*, 125134. [CrossRef]
154. Schaneberg, B.T.; Khan, I.A. Comparison of extraction methods for marker compounds in the essential oil of lemon grass by GC. *J. Agric. Food Chem.* **2002**, *50*, 1345–1349. [CrossRef]
155. Mukarram, M.; Choudhary, S.; Khan, M.A.; Poltronieri, P.; Khan, M.M.A.; Ali, J.; Kurjak, D.; Shahid, M. Lemongrass Essential Oil Components with Antimicrobial and Anticancer Activities. *Antioxidants* **2022**, *11*, 20. [CrossRef]
156. Carson, C.F.; Hammer, K.A.; Riley, T.V. Melaleuca alternifolia (Tea Tree) Oil: A Review of Antimicrobial and Other Medicinal Properties. *Clin. Microbiol. Rev.* **2006**, *19*, 50–62. [CrossRef] [PubMed]
157. Castro, J.I.; Valencia-Llano, C.H.; Zapata, M.E.V.; Restrepo, Y.J.; Hernandez, J.H.M.; Navia-Porras, D.P.; Valencia, Y.; Valencia, C.; Grande-Tovar, C.D. Chitosan/polyvinyl alcohol/tea tree essential oil composite films for biomedical applications. *Polymers* **2021**, *13*, 3753. [CrossRef] [PubMed]
158. Li, Z.; Jiang, L.; Wang, Y.; Li, M.; Liu, T.; Liu, Y. Chitosan–gellan gum polyelectrolyte hydrogel beads containing tea tree oil microcapsules: Preparation, characterization and application. *Food Hydrocoll.* **2024**, *157*, 110464. [CrossRef]
159. Gomes, D.S.; Costa, A.; Pereira, A.M.; Casal, M.; Machado, R. Biocomposites of silk-elastin and essential oil from Mentha piperita display antibacterial activity. *ACS Omega* **2022**, *7*, 6568–6578. [CrossRef]
160. Messaoudi, M.; Rebiai, A.; Sawicka, B.; Atanassova, M.; Ouakouak, H.; Larkem, I.; Egbuna, C.; Awuchi, C.G.; Boubekeur, S.; Ferhat, M.A.; et al. Effect of Extraction Methods on Polyphenols, Flavonoids, Mineral Elements, and Biological Activities of Essential Oil and Extracts of *Mentha pulegium* L. *Molecules* **2022**, *27*, 11. [CrossRef]

161. Zhao, H.; Ren, S.; Yang, H.; Tang, S.; Guo, C.; Liu, M.; Tao, Q.; Xu, H. Peppermint essential oil: Its phytochemistry, biological activity, pharmacological effect and application. *Biomed. Pharmacother.* **2022**, *154*, 113559. [CrossRef] [PubMed]
162. Monfared-Hajishirkiaee, R.; Ehtesabi, H.; Latifi, H. Peppermint essential oil and ZnO nanoparticles: A green and effective combination for a cooling bilayer patch with antibacterial activity. *J. Environ. Chem. Eng.* **2024**, *12*, 112833. [CrossRef]
163. Mahcene, Z.; Khelil, A.; Hasni, S.; Akman, P.K.; Bozkurt, F.; Birech, K.; Goudjil, M.B.; Tornuk, F. Development and characterization of sodium alginate-based active edible films incorporated with essential oils of some medicinal plants. *Int. J. Biol. Macromol.* **2020**, *145*, 124–132. [CrossRef] [PubMed]
164. Torpol, K.; Sriwattana, S.; Sangsuwan, J.; Wiriyacharee, P.; Prinyawiwatkul, W. Optimising chitosan–pectin hydrogel beads containing combined garlic and holy basil essential oils and their application as antimicrobial inhibitor. *Int. J. Food Sci. Technol.* **2019**, *54*, 2064–2074. [CrossRef]
165. Khater, R.M. Effect of hydrogel and antitranspirants treatments on the productivity of sweet basil (*Ocimum basilicum* L.) plant. *Egypt. J. Desert Res.* **2015**, *65*, 193–214. [CrossRef]
166. Shahrajabian, M.H.; Sun, W.; Cheng, Q. Chemical components and pharmacological benefits of Basil (*Ocimum basilicum*): A review. *Int. J. Food Prop.* **2020**, *23*, 1961–1970. [CrossRef]
167. Kowalonek, J.; Stachowiak, N.; Bolczak, K.; Richert, A. Physicochemical and antibacterial properties of alginate films containing tansy (*Tanacetum vulgare* L.) essential oil. *Polymers* **2023**, *15*, 260. [CrossRef]

gels

Review

NaDES Application in Cosmetic and Pharmaceutical Fields: An Overview

Carla Villa [1], Debora Caviglia [1], Francesco Saverio Robustelli della Cuna [2], Guendalina Zuccari [1] and Eleonora Russo [1,*]

[1] Department of Pharmacy, Section of Drug and Cosmetic Chemistry, Viale Benedetto XV 3, 16132 Genoa, Italy; carla.villa@unige.it (C.V.); debora.caviglia@edu.unige.it (D.C.); guendalina.zuccari@unige.it (G.Z.)
[2] Environmental Research Center, ICS Maugeri SPA SB, Institute of Pavia, IRCCS, Via Maugeri 2, 27100 Pavia, Italy; fsaveriorobustelli@unipv.it
* Correspondence: eleonora.russo@unige.it

Abstract: Natural deep eutectic solvents (NaDES) represent a new generation of green, non-flammable solvents, useful as an efficient alternative to the well-known ionic liquids. They can be easily prepared and exhibit unexpected solubilizing power for lipophilic molecules, although those of a hydrophilic nature are mostly used. For their unique properties, they can be recommend for different cosmetic and pharmaceutical applications, ranging from sustainable extraction, obtaining ready-to-use ingredients, to the development of biocompatible drug delivery responsive systems. In the biomedical field, NaDES can be used as biopolymer modifiers, acting as delivery compounds also known as "therapeutic deep eutectic systems", being able to solubilize and stabilize different chemical and galenical formulations. The aim of this review is to give an overview of the current knowledge regarding natural deep eutectic solvents specifically applied in the cosmetic and pharmaceutical fields. The work could help to disclose new opportunities and challenges for their implementation not only as green alternative solvents but also as potential useful pathways to deliver bioactive ingredients in innovative formulations.

Keywords: natural deep eutectic solvents; NaDES and formulations; NaDES and drug delivery; NaDES and cosmetic; bioactive compound extraction

Citation: Villa, C.; Caviglia, D.; Robustelli della Cuna, F.S.; Zuccari, G.; Russo, E. NaDES Application in Cosmetic and Pharmaceutical Fields: An Overview. *Gels* **2024**, *10*, 107. https://doi.org/10.3390/gels10020107

Academic Editors: Adina Magdalena Musuc, Magdalena Mititelu and Mariana Chelu

Received: 29 December 2023
Revised: 25 January 2024
Accepted: 26 January 2024
Published: 28 January 2024

1. Introduction

Currently, interest in the development of sustainable processes and green bioactive compounds from renewable sources is steadily increasing in the cosmetic and pharmaceutical fields. From an extractive point of view, avoiding unfriendly solvents, saving sources and energy, and recycling waste have become primary objectives for the pharmaceutic and cosmetic context, according to the green extraction principles [1]. Conventional organic solvents are commonly used for extracting aromas, perfumes, medicines, and dyes from plants, but they are often not sustainable due to toxicity, high environmental impact and flammability [2]. For this reason, in recent years, research in the green extraction context has focused its attention on new non-toxic, biodegradable green solvents [3].

In this context, ionic liquids (ILs) and deep eutectic solvents (DES) can represent an excellent alternative to conventional hazardous organic solvents [2]. ILs are defined as salts deriving from the combination of an organic cation and an anion, characterized by a melting point below 100 °C, being in most cases liquids at room temperature [4]. DESs are defined as homogeneous eutectic mixtures obtained by mixing two or more pure components (liquids or solids, ions or neutral molecules) acting as hydrogen bond acceptors (HBA) and hydrogen bond donors (HBD) [5]. High thermal stability, low volatility, and wide ranges of viscosity and polarity are some of the most interesting properties belonging to both ILs and DES [6,7]. In particular, the class of IL organic salts is characterized by low melting point and minimal vapor pressure, and they can be modified in terms of polarity and selectivity for different applications such as chemical or enzymatic reactions [8,9]. Unfortunately, their use is restricted due to their high toxicity and high production costs, including those for synthesis, purification and disposal [10]. These limitations can be overcome by deep eutectic solvents with comparable or better physical properties and phase behaviors than ILs [11]. First introduced by Abbott et al. [12], DES represent a great and successful alternative to ILs, characterized by easy preparation, purity and low costs [13]. The process to obtain DES involves the simple mixing of a hydrogen-bond acceptor (HBA) (like a quaternary ammonium salt) and a hydrogen-bond donor (HBD) at a suitable molar ratio [14]. Their interaction gives rise to supramolecular compounds, with peculiar chemical-physical characteristics [15,16], with a charge delocalization that is responsible for the lowering of the mixture melting temperature with respect to the individual components (generally from room temperature to 70 °C) [7]. Unfortunately, the use of DES, like green solvents at room temperature, can be hindered by their melting points being too high [17].

In this regard, a new generation of greener DES of natural origin has emerged over the past decade. In nature, it has been hypothesized that in plants, different metabolites can form eutectic mixtures which play different biological roles. They can act as an alternatives to water and lipids, with the ability to transport water-insoluble compounds inside the cells, explaining the co-presence of water soluble and insoluble compounds in the botanical matrix. For this reason, when these metabolites (i.e., sugars, alcohols, amino acids, organic acids) form DES, they are called "Natural DES" (NaDES) [18]. Their green properties and behaviors were first described by Choi et al. in 2011 [17–19].

Synthetic NaDES can reproduce this natural behavior, and they are considered promising new green solvents to be applied in several fields, such as in the cosmetic, pharmaceuticals and food areas. NADES were used successfully to extract phenolic compounds from plant material. For example, in recent years, research has been carried out on the NaDES extraction of phlorotannins from the brown alga *Fucus vesiculosus* L. [20,21]. Phlorotannins are polyphenols with antioxidant, anti-inflammatory, antiallergic, antibacterial, and anti-tumor properties. They have a wide range of cosmetic applications, e.g., in sunscreens as anti-aging and UV-protective agents, and in in food packaging films as preservatives. Some authors have suggested the use of NaDES as solvents to stabilize proteins (lysozyme, amylase, photosynthetic enzymes) and DNA [18,22,23]. This opportunity led to an increasing interest in NaDES as drug delivery systems for active, but poorly soluble, ingredients [24–26]. NaDES show a wide polarity range and high solubilization strength for a variety of compounds. They present several advantages over classical solvents, ILs, and DES, such as natural origin, low cost, biodegradability, absence of toxicity, sustainability, and simple preparation [22]. Although NaDES are recognized as being slightly toxic and with a low environmental impact, it must be mentioned that they show the phenomenon of eutrophication [27]. As extractive alternative solvents, NaDES allow the achievement of efficient extractions when compared to conventional solvents [28–31]. Moreover, they often improve the stability and storage of the extracted compounds of interest, such as phenols, β-carotene, and α-tocopherol [22,23,29,32,33]. Despite the myriad of research fields in which NaDES are involved for diverse types of applications, the dermocosmetic and pharmaceutical topics are relatively unexplored, as can be seen from Figure 1.

In the last 5 years (2019–2023), more than 1600 papers have been published on this subject. Most of them deal with the use of these eutectic systems for the extraction of bioactive compounds from botanical matrices and/or agrifood waste, for the most varied applications. In order to narrow and detail the object of the work, this review considers and reports the latest research and results limited to cosmetic and pharmaceutical applications, where NaDES are explicitly included in the formulations. Papers in which the cosmetic and/or pharmaceutical potential is only mentioned and NaDES are proposed as alternative extractive solvents are summarized and cited in Table 1.

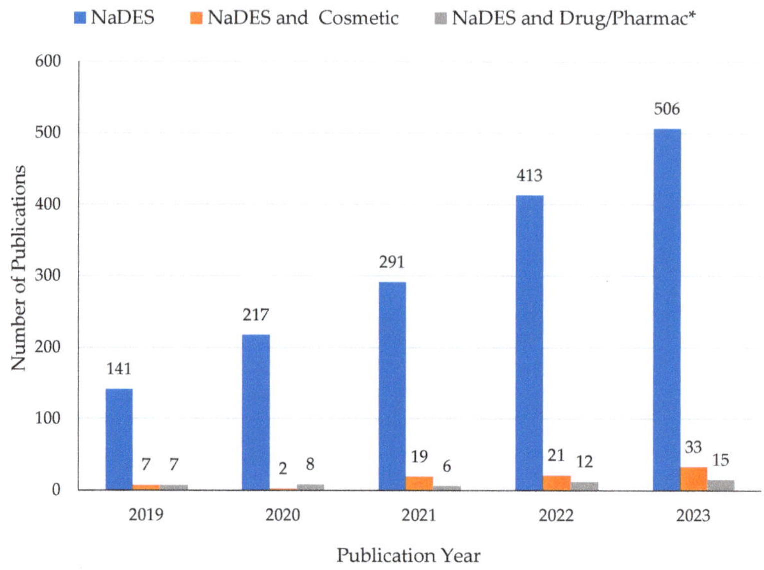

Figure 1. Histogram showing the increase in publications in the past 5 years (2019–2023) regarding the keywords "NaDES or natural deep eutectic solvent" and the corresponding small number related to "NaDES and Cosmetic" and "NaDES and Drug or Pharmac*" (data available on Scopus accessed on 20 December 2023).

Table 1. Extraction of target compounds from natural sources and agri-food byproducts using NADES as alternatives to conventional solvents, and related references.

Target Compound	Natural Matrix	NaDES System	Conventional Solvent	Reference
Anthocyanins	Grape skin	Citric acid/D-(+)-maltose	water or organic solvents such as methanol and ethanol	[34]
Anthocyanins	Mulberry	Choline chloride/citric acid/glucose	methanol, ethanol, acetic acid modified water or hydrochloric acid modified ethanol	[35]
Anthocyanins	Grape pomace	Choline chloride/citric acid Choline chloride/proline/malic acid	methanol, acetone and hydrochloric acid	[36]
Anthocyanins	Sour cherry pomace	Choline chloride/malic acid	acidified ethanol	[37]
Anthocyanins	Blueberry peel	Choline chloride/malic acid Choline chloride/citric acid	acidified ethanol	[38]
Caffeine	Chinese dark tea	Choline chloride/lactic acid	chloroform, dichloromethane, acetone and ethyl acetate.	[39]
Curcumin	Standard solubility tests	Choline chloride/glycine	ethanol, methanol, acetone and ethyl acetate	[40]
Hydroxytyrosol	Olive leaves	Citric acid/glycine/water	ethanol and water	[41]
Isoflavones	Soybeans	Choline chloride/citric acid	acetonitrile acetone, ethanol and methanol	[42]
Pectins	Mango peel	Betaine/citric acid choline chloride/malic acid	alkaline, acidic aqueous solutions and enzyme	[43]
Phenolic acids	Orange peel	Choline chloride/ D-(+)-glucose/water	acetonitrile, methanol and acetone	[44]
Phenolic compounds	Bitter melon	Choline chloride/acetic acid	ethanol, methanol, acetone, ethyl acetate and chloroform	[45]

Table 1. *Cont.*

Target Compound	Natural Matrix	NaDES System	Conventional Solvent	Reference
Phenolic compounds	*Olea europaea*	Water/Choline chloride/fructose	dimethyl sulfoxide, hexane, ethanol and methanol	[46]
Phenolic compounds	Olive pomace	Choline chloride/citric acid	petroleum ether, acetone, ethyl acetate and methanol	[47]
Phenolic compounds	Hazelnut skin	Choline chloride/lactic acid	methanol, ethanol, and methanol/water mixtures	[48]
Phenolic compounds	Cocoa beans	Betaine/glucose	Hexane, petroleum ether, methanol, ethanol, ethyl acetate and acetone	[49]
Phenolic compounds	Waste mango peel	Lactic acid/glucose	methanol, ethanol, acetone and ethyl acetate	[50]
Rosmarinic acid, carnosol, carnosic acid	*Rosmarinus officinalis*	Lactic acid-glucose/menthol-lauric acid (biphasic system)	dichloromethane, ethanol and methanol	[51]
Solenesol	Tobacco leaves	Choline chloride/urea	petroleum ether, acetone, *n*-hexane, ethyl acetate and methanol	[52]
Tryptanthrin, indirubin, and indigo	*Baphicacanthus cusia*	Lactic acid/L-menthol	methanol, ethanol and methanol/dichloromethane	[53]
Triterpenic acids, Ursolic acid	*Eucalyptus globulus*	Choline chloride/D-(+)-glucose	dichloromethane *n*-hexane ethanol or chloroform	[54]

1.1. NaDES Preparation

Many NaDES mixtures are biodegradable and have low toxicity [55–57], partially due to their natural origin. Most of their components present an intrinsic cosmetic or pharmaceutical activity, being well-known and used ingredients (organic acids, sugars, alcohols and polyols, amino acids and quaternary ammonium salts).

Particularly from a cosmetic point of view, this aspect presents many advantages: increasing the naturality of the compositions and the concentration of active ingredients, stabilizing them without adding preservatives, reducing the number of ingredients and allowing a synergistic effect to improve the biological activity of the formulation.

The preparation of NaDES yields easy results with high purity and no waste formation [58] according to the fundamental principles of green chemistry [59].

As mentioned above, NaDES can be prepared by mixing an HBA (i.e., choline chloride, choline acetate or betaine) with an HBD (glycerol, urea, glucose, sorbitol, fructose, etc.), with or without water [17], mainly applying these most common and different physical methods:

1. Heating and stirring method [17], where two components are mixed with a magnetic stirring bar, in a 50 °C water bath until a clear viscous liquid is formed, about 30–90 min later [17,22,60,61]. Otherwise, it is possible to follow the conditions stated by Abbot et al. 2003 [12], or heating at 80 °C under continuous stirring [60,62,63].
2. Freeze-drying method [64], which is the least used and based on freeze-drying by sublimation of both the NaDES aqueous portion and the individual components of the NaDES. This method makes it possible to achieve pure NaDES.
3. Evaporation method [17], which involves the use of rotary evaporator to allow the components' evaporation and dissolution in water at 50 °C. The liquid that is obtained is transferred to a silica gel desiccator until it reaches a constant weight.
4. Grinding method, where the component mixture is ground in a mortar with a pestle, at room temperature, until formation of a homogeneous liquid [65].
5. Ultrasound-assisted heating method, where the component mixture is exposed to ultrasonication until a homogeneous liquid is formed [66].
6. Microwave irradiation technique, where the mixture is irradiated in a microwave oven at low power emission and for a few seconds [67].

The methods mentioned above are shown in Figure 2. The microwave-assisted preparation of NaDES represents a promising green technique, due to its advantages such as higher yields, lower energy consumption and shorter reaction times [68].

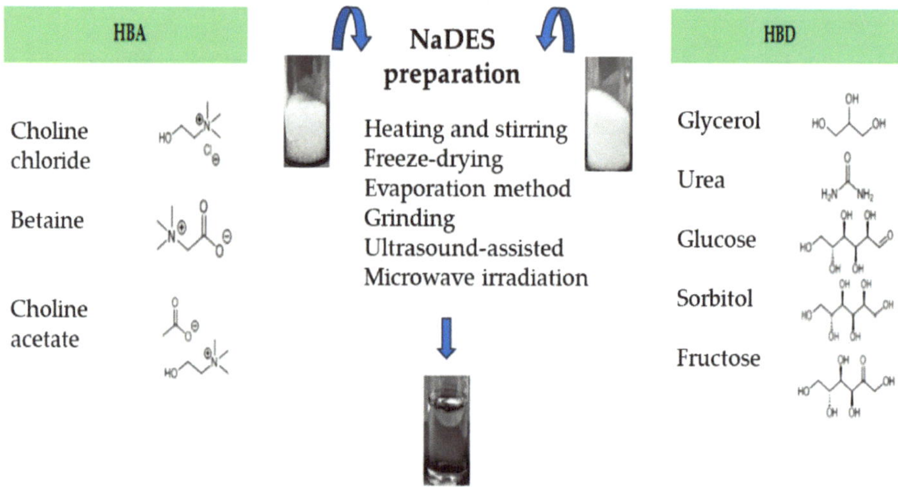

Figure 2. Different preparation methods to obtain NaDES.

1.2. NaDES Structure

The structure and properties of NaDES are conferred by the type and ratio of components and also by the hydrogen (H) bonds established between the metabolites themselves [17,69,70]. The H bonds' strength is related to the phase-transition temperature, stability and solvent properties of the mixture [64]; their key roles in important NaDES features and behaviors (such as stability and formation) depend on their number and location [17].

The lowering of the mixture melting temperature, with respect to the single components, is due to the formation of a charge delocalization between the HBA and HBD and to the van der Waals forces that allow blocking the crystallization of the compounds [29]. Usually, a low freezing point can be determined by a higher binding capacity between the HBD and HBA [70].

The NaDES structures have been evaluated through nuclear magnetic resonance spectroscopy (NMR) studies, crystallographic data, fast atom bombardment-mass spectrometry (FAB)-MS and Fourier transform infrared spectroscopy (FT-IR) [17,60,71]. Thanks to the nuclear Overhauser effect spectroscopy (NOESY) spectra obtained, it has been seen that NaDES are characterized by a supramolecular structure mainly due to bonds established between HBAs and HBDs [17]. This supramolecular structure changes after water dilution [59]; in fact, it was observed that the presence or absence of water plays a significant role. This behavior occurs because the H bond systems that NaDES are able to form, will gradually fade when diluted with water, until disappearing when the water amount exceeds 50% v/v. In this regard, it has been observed that the degradation of concentrated NaDES was slower than that of diluted ones [72].

The amount of added water tolerated by the eutectic system should be determined for each NaDES. Moreover, the types of components used to form NaDES can influence their physicochemical properties, such as viscosity, conductivity, density, and polarity [60].

Craveiro et al. have demonstrated that water can increase polarity, which affects the solubility of NaDES [73]. Simultaneously, dilution with water can result in a decrease in viscosity. This rheological behavior is one of the main problems that NaDES present [17]. A high viscosity interferes with the flow of substances and decreases the extraction efficiency [74]; this problem can be overcome by heating. The high temperature and thermal expansion lead to increased molecular force and to structural damage, respectively [75]. Another way to reduce the viscosity is dilution with water, since, as already mentioned, water leads to the breaking of the hydrogen bonds and consequently to a lower viscosity [60].

Several works in the literature describe the use of NaDES to obtain, from natural sources, bioactive compounds that can be used in cosmetic and pharmaceutical formulations (Figure 3). The main advantage consists of the possibility of directly adding the NaDES-based extract itself to all types of topical formulations, both in the cosmetic and pharmaceutical fields, without dramatic changes in the rheological properties or sensorial profile [76]. However, only a few can actually be used for cosmetic applications because of safety or regulatory issues [75].

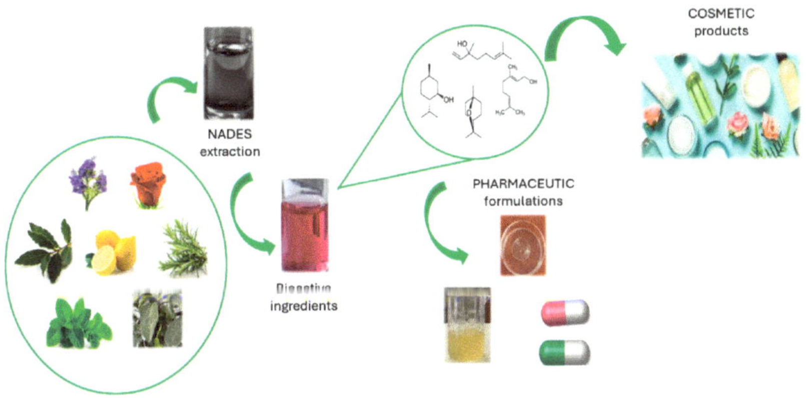

Figure 3. Use of NaDES in cosmetic and pharmaceutical fields.

In the section below, some papers dealing with the use of NaDES to obtain active principles with excellent properties that can be exploited in the future have been taken into consideration.

2. NaDES in the Cosmetic Field

Jeong et al. [30] developed an extractive procedure to obtain catechins from *Camellia sinensis* (*C. sinensis*) green tea leaves, including epigallocatechin-3-gallate (EGCG), a powerful antioxidant. Their optimized extraction method involved the use of a ternary DES mixture, suitable for both pharmaceutical topical preparations and cosmetic formulations. The authors prepared NaDES by both the heating [17] and freeze-drying [64] methods, selecting glycerol, xylitol, citric acid, betaine, D-(+)-glucose, D-sorbitol, D-(+)-maltose, maltitol, urea, D-(−)-fructose, D-(+)-galactose, and sucrose in an adequate molar ratio. They tested several extractive methods, including ultrasound-assisted extraction (UAE), agitation, heating, and heating with agitation.

All of the green solvents assessed by the authors allowed for very efficient extractions, but taking into account the limits of use in cosmetic formulations and the production costs, the final choice was a NaDES composed of betaine, glycerol and D-(+)-glucose, 4:20:1 (BGG-4). Compared to conventional solvent extractions (water and organic solvents), this mixture allowed for a better extraction of EGCG and improved stability. The best extraction conditions, identified by the response surface methodology, involved the application of UAE at room temperature for 6.5 min, using 81% BGG-4. In conclusion, it was highlighted that BGG-4 is an excellent extractive solvent and stabilizer for catechins of *C. sinensis*, useful in topical formulations.

Also, Vasyliev et al. [77] indicated NaDES as promising solvents to extract antioxidant bioactive compounds for use in cosmetic formulations. In their work, they applied the UAE method with one NaDES based on choline chloride (as the HBA) to extract polyphenols from waste tomato pomace. To enhance extraction efficiency, they prepared the NaDES in the presence of water. In particular, they tested a mixture composed of choline chloride: 1,2-propanediol (1:2 *v*/*w*): water (10% *w*/*w*) (DESs-1) and another one containing choline chloride: lactic acid (1:2 *v*/*w*): water (10% *w*/*w*) (DESs-2).

The tomato pomace extracts obtained were then characterized and used as antioxidant agents (being rich in phenolic acids and flavanols) to develop a natural cream formulation (oil-in-water emulsion). The main phenolic compounds extracted with DESs-1 and DESs-2 were gallic acid, chlorogenic acid, caffeic acid, trans-cinnamic acid, p-coumaric acid and ferulic acid. The extract obtained with DESs-2 afforded higher amounts of quercetin, caffeic and ferulic acid, displaying an enhanced antioxidant power when compared to DESs-1. The cosmetic formulation also containing DESs-2 showed greater antioxidant activity. Both emulsions, stabilized by DES, demonstrated antifungal activity against *Candida albicans*. In conclusion, this study showed that DES can extract polyphenols from agri-food waste, such as tomato pomace, to be used as antioxidant additives in the cosmetics industry.

The research team of Petkov et al. [78] also investigated NaDES as possible solvents for extracting bioactive compounds from natural sources. The authors evaluated the antioxidant activity of extracts from *Plantago major* (*P. major*), *Sideritis scardica* (*S. scardica*) and propolis obtained by UAE, assessing the extraction efficiency in terms of total phenols and flavonoids content, using 10 different NaDES. Betaine-malic acid-water 1:1:6 (BMAH), citric acid-1,2-propanediol 1:4 (CAPD), lactic acid-fructose 5:1 (LAFr), lactic acid-1,2-propanediol 1:1 (LAPD), choline chloride-glucose-water 5: 2:25 (XXGlH), choline chloride-glycerol 1:1 (XXGly), choline chloride-1,2-propanediol 1:3 (XXPD), and choline chloride1,2-propanediol-water 1:1:1 (XXPDH) were prepared by the heating and stirring method. In contrast, choline chloride-urea 1:1 (XXU) and choline chloride-xylitol 4:1 (XXXy) were obtained by the vacuum evaporation method The most effective phenolic extraction for both *P. major* and *S. scardica* was achieved using XXGlH as the solvent. When compared to EtOH 70%, used as a reference, XXPDH, XXPD and LAPD extracted a greater phenolic amount and the same quantity of flavonoids. Moreover, in contrast to *P. major* and *S. scardica*, propolis showed a strong correlation between phenolic concentration and antioxidant capacity. In conclusion, Petkov et al. asserted that NaDES extracts can be directly incorporated into formulations considering their intrinsic properties, such as biocompatibility, low toxicity, and excellent antioxidant activity.

Another natural bioactive compound, namely Naringerin (NA), a flavonoid already utilized in various formulations, was extracted by El Maaiden et al. [79] from dried aerial parts of *Searsia tripartita* (*ST*), using NaDES. This study focused on six eutectic solvents, prepared by the heating technique [80], composed by choline chloride in a 1:2 ratio with formic acid (DES-1), ethylene glycol (DES-2), lactic acid (DES-3), urea (DES-4), and glycerol (DES-5), and in a 2:1 ratio with citric acid (DES-6). These solvents were, therefore, used for the NA extraction by UAE from *ST* powder. After characterization, DES-1 was chosen for further analyses as the best-performing solvent with the highest NA concentration, while DES-6 showed the lowest yields. El Maaiden et al. presented the best operative conditions based on their achievements, involving an extraction time of 10 min at 50 °C, with an ultrasound amplitude of 75 W and a solid–liquid ratio of 1/60 g/mL. These extracts proved to be excellent antiaging agents for their antioxidant activity and as enzyme inhibitors of tyrosinase, collagenase, elastase, and hyaluronidase, which are responsible for skin aging.

Further, Jamaleddine et al. [81] conducted an extractive study using NaDES on tomato pomace (TP), rich in bioactive compounds. Specifically, they prepared and used four kinds of NaDES to extract TP by UAE. Jamaleddine and coworkers proposed a novel strategy for sustainable formulations by incorporating their extraction medium directly into the formulations. They selected four different methods, present in the literature, for preparing their NaDES. DES 1, composed of glycerol:glucose (1:3) and water (30%), was prepared using the method developed by Wils et al. [82]; DES 2 (DL-menthol-lactic acid 8:1) was obtained by the method of Silva et al. [83]; DES 3, composed of lactic acid-glucose (5:1) and 15% water, was prepared following the method of Fernandez et al. [84], and finally, DES

4 (L-proline-glycerol 1:2.5 and 30% water) was prepared according to Wu et al. [85] with some modifications. TP extractions required a matrix maceration in DES 1 for 2 h at 40 °C (ratio NaDES/TP 32:2 *w*/*w*) and for 30 min at 50 °C with DES (NaDES/TP 20:3.4 *w*/*w*). For DES 3 and DES 4, the UAE procedure was carried out for 1 h at 40 °C, with a solid–liquid ratio of 40/2 *v*/*w* and 20/2 *v*/*w*, respectively. The results showed that DES 1 demonstrated great suitability for the extraction of phenolic acids, flavones, flavonols and tannins. DES 2 could extract carotenoids, lipids and tocopherol. DES 3 demonstrated greater efficiency for phenols, while flavones, flavanols and flavanones were ultimately extracted by DES 4. Finally, the extracts obtained were employed to develop four cosmetic formulations: a peel-off mask (containing DES 1); a lip balm (DES 2); a water-soluble mask (DES 3); and a moisturizing cream (DES 4).

Another research team, Jin et al. [86], produced extracts with excellent skin properties using a mixture of dried and ground leaves of *Ginkgo biloba* L (GB), *Cinnamomum camphora* (L) *J. Presl* (CC), and *Cryptomeria japonica* (L.f.) *D. Don* (CJ). Using safe, stable, and cost-effective substances approved by the European Commission (2006) [87], Jin et al. prepared and assessed 15 different cosmetics-compliant NaDES using the heating method [17,88]: glycerol-xylitol 2:1 (DES 1), glycerol-maltose 3:1 (DES 2), glycerol-sorbitol 2:1 (DES 3), glycerol-fructose 3:1 (DES 4), glycerol-sucrose 3:1 (DES 5),glycerol-glucose 3:1 (DES 6), glycerol-maltitol 3:1 (DES 7), glycerol-malic acid 1:1 and 1:2 (DES 8 and DES 9), lactic acid-glucose 1:2 (DES 10), fructose-sucrose 1:1 (DES 11), fructose-sucrose-glucose 1:1:1 (DES 12), betaine-sucrose 1:1 and 1:2 (DES 13 and DES 14) and 1:1 betaine:glucose (DES 15). This approach facilitated the safe development of an ISO extraction technique capable of producing a significant quantity of extracts that can be directly incorporated into cosmetic formulations. DES 1 was selected as the most efficient solvent for the extraction of isoquercetin (ISO yield 861 µg/g), found in GB, CC and CJ leaf extracts. Finally, using the central composite design approach, Jin et al. collected data on specific bioactivities with several leaf extract mixtures and identified the best-performing in terms of increasing antioxidant activity and anti-tyrosinase and anti-elastase effects.

The study by Hsieh et al. [89] highlighted the potential of natural DES as alternative solvents to volatile organic solvents (VOS), with the same or even better efficiency. In particular, the authors extracted gingerols from *Zingiber officinale Roscoe* (ginger) powder to obtain extracts that could be incorporated directly into formulations, without the need for work-up steps for product isolation. NaDES were designed and prepared by the ultrasonication assisted method, with three different hydrogen bond acceptors (choline chloride, betaine and L-carnitine) and five polyalcohols (triethylene glycol, ethylene glycol, 1,3-propanediol, glycerol, and 1,3-butanediol) as hydrogen bond donors in opportune molar ratios: choline chloride-triethylene glycol 1:4 (CC-TriG), choline chloride-ethylene glycol 1:2 (CC-EG), choline chloride-1,3-propanediol 1:4 (CC-PG), choline chloride-glycerol 1:2 (CC-gly), choline chloride-1,3-butanediol 1:4 (CC-ButG), betaine-triethylene glycol 1:4 (Bet-TriG), betaine-ethylene glycol 1:2 (Bet-EG), betaine-1,3-propanediol 1:4 (Bet-PG), betaine-glycerol 1:2 (Bet-gly), betaine-1,3-butanediol 1:4 (Bet-ButG), L-carnitine-triethylene glycol 1:4 (Lcat-TriG), L-carnitine-ethylene glycol 1:2 (Lcat-EG), L-carnitine-1,3-propanediol 1:4 (Lcat-PG), L-carnitine-glycerol 1:2 (Lcat-gly), and L-carnitine-1,3-butanediol 1:4 (Lcat-ButG). Ultrasonication-assisted extractions were carried out after diluting all of the NaDES samples with 75% water v/v, reducing viscosities for a more effective extraction. Three NaDES (Bet-ButG, Lcat-Trig and Lcat-ButG) resulted in the most efficient eutectic systems, containing the highest concentration of gingerols. Finally, the authors suggested the best operative conditions for UAE (50 °C for 30 min with a 30/1 solvent/solid ratio *v*/*w*) that could maintain the antioxidant activity of gingerols and prevent phenol degradation.

Rocha et al. [90] emphasized the effectiveness of NaDES-based extracts from botanical matrices as cosmetic ingredients. By the heating and stirring technique, the research team prepared three eutectic solvents composed as follows: lactic acid-glycerol 1:1 and 10% water (NADES 1), lactic acid-glycine 5:1 and 13% water (NADES 2), and lactic acid-sodium citrate 4:1 and 31% water (NADES 3). NADES 1–3 were subjected to an accurate physicochemical characterization (melting point, pH, density, refractive index, surface tension, viscosity, conductivity, and polarity) after a freeze-drying process. Then, cork extraction was conducted for each NaDES in a high-pressure closed system, leading to three NaDES-based samples (Extract 1–3) with antioxidant and antibacterial properties. Once assessed with regard to their antioxidant activity, transdermal permeability, and cytotoxicity, all samples were added to two commercial cosmetic products. The new complexes (Formulation A and Formulation B) showed an enhanced antioxidant activity and no cytotoxicity on keratinocytes (for extract concentrations up to 10 mg/mL). Furthermore, Rocha et al. suggested that Extract 2 (corresponding to NADES 2 solvent) would be the most suitable for inclusion in cosmetic formulations.

Marijan et al. [91] conducted an extraction using NaDES to derive bioactive compounds from flowering aerial parts of *Lotus corniculatus* (*LC*), *Medicago lupulina* (*ML*), and *Knautia arvensis* (*KA*), as well as from leaves of *Plantago major* (*PM*) selected from urban parks. In this work, the authors demonstrated that organic waste from urban areas can contain useful minerals for skin health. The UA extraction was exploited using an NaDES composed by glycerol, betaine and glucose (in a weight ratio of 20:4:1) and then diluted with water in the proportion 8:2 (DES80) or 4:6 (DES40) in order to investigate two different polarities of the solvent [17,30]. Furthermore, for a comparative evaluation, Marijan et al. utilized environmentally friendly extraction solvents made by dissolving hydroxypropyl-β-cyclodextrin (HPβCD) or γ-cyclodextrin (γCD) in aqueous solutions. The -different metals identified by extraction in plants were calcium, very abundant in *PM*, zinc, especially present in *KA*, iron, in *ML* and less in *LC*. The last two extracts (obtained by DES80) were the richest in phenols; in *ML* samples, the authors identified quercetin, kaempferol, luteolin and apigenin, while in *LC* samples, only kaempferol was detected. Differently, in *KA* and *PM* extracts, the highest concentration of phenols was obtained from DES40. In particular, in *PM*, all of the studied phenols except luteolin were identified, while in *KA*, only apigenin and luteolin were detected. The results obtained by Marijan et al. seem to indicate *LC extract* as a good anti-tyrosinase agent and *KA* as a better anti-elastase one. In conclusion, the solvents they used, in addition to contributing to bioactivity, allowed for the recovery of bioactive compounds and metals in organic waste from urban parks, which can be exploited to produce ecological cosmetic formulations with added high value.

Another research group, Alishlah et al. [92], optimized a UA extraction of oxyresveratrol from the root powder of *Morus alba* (mulberry) urea and glycerin eutectic systems. The aim of this study was the formulation of an efficient skin whitening cosmetic lotion containing the NaDES-based extracts. The heating and stirring method was selected for the preparation and evaluation of NaDES with a urea-glycerin molar ratio of 1:1, 1:2, and 1:3. UAE was performed with different extraction times (i.e., 10, 15 and 20 min) using 20 mL NaDES for 1 g mulberry powder; HPLC was used for the identification and quantitation of oxyresveratrol after extraction. The best results in terms of oxyresveratrol recovery (2.42 mg/g dry powder) were obtained in 15 min by NaDES with urea-glycerin 1:3. Therefore, this sample was used at a percentage of 35% *w*/*w* to formulate three oil-in-water emulsions (whitening skin lotions: formula A, formula B and formula C) containing stearic acid (1%), isopropyl myristate (5%), propylene glycol (15%) cetyl alcohol (2% A, 4% B, or 6%C), Tween 80 (3.88% A, 4.13% B, or 4.29% C) and glyceryl monostearate (1.12% A, 0.87% B, or 0.71% C). Based on physical evaluations, formula A was chosen as the best formulation for the development of a cosmetic bleaching product for the skin.

Oktaviyanti et al. [93] developed and optimized a green ultrasound-assisted deep eutectic solvent extraction of *Ixora javanica* flowers to obtain a natural antioxidant and skin lightening agent to be used in the cosmetic field. The researchers evaluated the extraction efficiency of 11 NaDES regarding flavonoids and anthocyanins, and the antioxidant and anti-tyrosinase activity of the obtained extracts. By use of the heating method, choline chloride (Ch) was coupled in opportune molar ratios with different HBDs (polyols and organic acids) to form the following eutectic systems: ChPg (choline-propylene glycol 1:1), ChGl (choline-glycerol 1:2), ChEg (choline-ethylene glycol 1:2), ChPeg (choline-polyethylene glycol 1:2),

ChSb (choline-sorbitol 1:1), ChPd (choline-1,3-propanediol 1:3), ChOa (choline-oxalic acid 1:1), ChLa (choline-lactic acid 1:2), ChGa (choline-glycolic acid 1:2), ChMa (choline-malic acid 1:1) and ChCa (choline-citric acid 1:1). According to the authors, the best extractive NaDES solvent for *I. javanica* flowers was ChPg, which demonstrated the best-performing anti-tyrosinase activity. The design and optimization of the extraction parameters to maximize flavonoids recovery was achieved by the response surface methodology (RSM); the best-performing conditions required an extraction temperature at 57 °C for 5 min with a matrix-solvent ratio of 1:50 g/mL. The authors concluded that NaDES can be used as useful green alternative organic solvents for bioactive compound extractions from natural models to be added in cosmetic formulations.

In addition to natural models, agrifood waste also represents a promising renewable source of bioactive cosmetic ingredients. In this regard, Punzo et al. [94] studied NaDES for the extraction of polyphenols, from freeze-dried red grape pomace, for topical applications. NaDES, prepared by the heating and stirring method, were obtained by three HBDs (urea, citric acid and ethylene glycol), selected on the basis of their proven skin compatibility and mixed in optimal molar ratio with betaine (HBA), as follows: betaine-citric acid 1:1 (BET-CA), betaine-ethylene glycol 1:2 (BET-EG), and betaine-urea 1:2 (BET-U). Among the samples obtained and used directly as topical formulations, BET-CA extracts (the richest in malvidin), showed the best antioxidant and anti-inflammatory activity at concentrations able to permeate the skin. Therefore, this formulation was indicated by Punzo et al. as the most suitable ingredient for anti-aging cosmetic formulations. Moreover, NaDES were proven as excellent extractants and carriers for polyphenols; the researchers assessed and confirmed the in vitro safety of NaDES extracts by means of human 3D keratinocytes. The authors concluded that their findings could support the use of NaDES as promising cosmetic ingredients and carriers in new drug delivery systems for topical applications, since they can affect the permeation of active molecules.

3. NADES in the Pharmaceutical Field

As stated in the reported literature, NaDES are not only considered as green alternatives to conventional organic solvents, but they also promote and enhance the extraction of bioactive compounds from natural models, or agrifood waste, suitable for cosmetic application. More recently, several publications have referred to the exploitation of NaDES in pharmaceutical technology to solubilize and stabilize a wide range of pharmaceutical systems. Moreover, they can be applied in hydrogels and film formation and as carriers to deliver bioactive compounds in many other innovative pharmaceutical forms.

In this regard, Delgado-Rangel et al. [95] used NaDES, without crosslinkers, to create 3D pure and porous materials constituted by chitosan (CTS), to be used against *V. cholerae* biofilm. In particular, the research group optimized an environmentally friendly method that allowed the formation of porous monoliths and films, underlining the versatility of application of NaDES-assisted phase separation processes. The preparation of CTS matrix films was carried out in three steps by evaporation-induced phase separation. The solution, based on 2% CTS and acetic acid, was mixed with a NaDES obtained by the heating method and composed by a mixture of chloride choline-urea in a molar ratio of 1:2 (CCU-DES). After the evaporation of the acidic aqueous solvent from the CTS CCU-DES mixture, the plasticized CTS film structure was obtained. Between the different weight ratios of CTS/CUU analyzed, Delgado-Rangel et al. selected the equal weight ratio, as it allowed obtaining films with the most suitable porous structure. In addition, the research group observed that the thermal stability of CTS was affected by residual NaDES within its porous structure. They concluded that, as regards films formed by chitosan in this specific example, the porosity influenced *V. cholerae* growth.

Differently, the research group of Alkhawaja et al. [96] used NaDES as a carrier of a phosphodiesterase 5 (PDE-5) enzyme inhibitor, namely tadalafil (TDF), with the aim of developing a formulation to be applied on burns and able to prevent the systemic absorption of the drug. By stirring at room temperature, the authors prepared seven

NaDES formulations, based on malonic acid (MA) and choline chloride (CC) in different molar ratios, with and without propylene glycol (PG) to provide different viscosities. B01, B02, and B03 formulations were composed by the simple mixture of MA-CC 1:1, 1:2, and 2:1, respectively. Once prepared, these NaDES were mixed with PG at different ratios to obtain the B04 formulation (B01-PG 1:1), B05 formulation (B01-PG 1:2), B06 formulation (B02-PG 1:1), and B07 formulation (B02-PG 1:2). The characterization of blank NaDES, obtained by evaluation of spreadability and measurement of contact angle, allowed the selection of B01 and B04 as potential topical formulations. Subsequently, Alkhawaja and co-workers improved the aqueous solubility of TDF by mixing it into B01. Moreover, a new formulation (F01) was developed by incorporating lidocaine (LCD) into the NaDES samples containing TFD, to also provide a local anesthetic effect. Due to topical effects, B01 and B04 were chosen to formulate topical preparations with or without lidocaine. The authors concluded that F01 delays the healing process, thereby lowering the probability of scarring that may result from burn wounds. Moreover, the presence of NaDES in the formulations, having antimicrobial activities, reduces the risk of bacterial infections.

Filip et al. [97] coupled NaDES with hydroxypropylcellulose (HPC) to produce self-assembled hydrogels (HPC-NaDES), compatible with the human gingival fibroblast (HGF) cell line, for applications in the pharmaceutical field. In particular, the authors obtained HPC-NaDES 17% and HPC-NaDES 29% aqueous solutions by adding to a 14% HPC solution the NaDES previously prepared. Choline chloride (ChCl) was mixed in different molar ratios with four HBDs and small water amounts to obtain ChCl-urea 1:2 (U), ChCl-glycerol 1:2 (GL), ChCl-lactic acid 1:1 (LA), and ChCl-citric acid 1:1 (CA). HPC-NaDES were then characterized by FT-IR, H^1NMR, DSC, TGA measurements, and rheological tests. According to the authors, the HPC-NaDES physicochemical properties are influenced by different parameters such as hydrogen bond interactions between HBA and HBD, content of NaDES, and the water amount. Stronger hydrogen bonds were observed in HPC-CA and HPC-GL compared to the other hydrogels obtained. All HPC-NaDES hydrogels exhibited a pseudo-plastic behavior. Furthermore, the latter possessed thermo-thickening characteristics since the HPC in aqueous systems has a lower critical temperature than the solution itself. Finally, the disk diffusion methods [98,99] enabled the determination of antibacterial and antifungal activities, showing this order of efficacy: ChCl-CA > ChCl-LA > ChCl-U > ChCl-GL.

A research team that exploited the solubilizing abilities of NaDES (Mustafa et al. [100]) screened various types of eutectic mixtures to solubilize poorly water-soluble drugs and produce liquid formulations for parenteral administration and gastric tube feeding. In particular, the authors conducted tests of NaDES solubilization on some insoluble drugs such as nitrofurantoin, trimethoprim, griseofulvin, methylphenidate, and spironolactone, and on water-unstable ones (trichloroacetaldehyde monohydrate or chloral hydrate). They observed good drug solubility in eutectic systems based on choline chloride or betaine, coupled with different HBDs, such as organic acids, sugars, and polyols. Good results were achieved for methylphenidate, trimethoprim, griseofulvin, spironolactone, and nitrofurantoin. In addition, the stability of NaDES samples containing drugs was tested at 4 °C for up to 4 months. The results suggested that methylphenidate and trimethoprim are better solubilized in acidic NaDES, while pure acetic or lactic acids are more effective for spironolactone and griseofulvin solubilization. Nitrofurantoin could be dissolved only by a mixture of choline chloride–acetic-acid–proline–water (1:1:1:5 molar ratio) at a concentration of 5 mg/mL. Unstable drugs dissolved at the maximum concentration of 250 mg/mL. Therefore, Mustafa et al. suggested that NaDES represent promising solvents to optimize liquid formulations with poorly water-soluble drugs, but further investigations are needed.

Li et al. [101] exploited the features of NaDES with the aim of improving the antibacterial properties of a hydrogel to be used as a wound dressing. Specifically, they prepared a hydrogel of sodium hyaluronate (SH), coated with dopamine (DA), using a NaDES composed by choline chloride and glucose. After combining SH and DA, Li et al. added N-Hydroxysuccinimide (NHS) and 1-ethyl-3-(3-(dimethylamino)propyl) carbodiimide (EDC) as coupling agents to the mixture using the techniques described by Lee et al. [102]. The resulting product, i.e., the SH conjugate with dopamine (DASH), was then purified and lyophilized. Then, DASH and NaDES were combined in a mass ratio of 4:175 to form a DES-DASH hydrogel. Subsequently, a DES-DASH@Ag hydrogel containing a silver nitrate solution was prepared and tested for its antibacterial activity against *S. aureus* and *E. coli*. The results showed a nontoxic behavior towards NIH-3 T3 fibroblast cell lines and the ability to support wound healing in mouse skin within 12 days of surgery. Thus, Li et al. suggested a future use of the DES-DASH@Ag hydrogel as a topical application for wound dressing.

Sokolova et al. [103] exploited the plasticizing effect of chitosan of (CS) with NaDES to create CS/DES films. According to the method described by Samarov et al. [104], they prepared NaDES by mixing malonic acid (MA) with choline chloride (ChCl). The CS/DES films (with a thickness of 20 μm) were obtained by casting at room temperature a mixture of CS and water with a NaDES content ranging from 0 to 82% (*w*/*w*), in Petri disks. Film characterization was performed by means of Fourier transform infrared spectroscopy (FT-IR), scanning electron microscopy (SEM), atomic force microscopy (AFM), water absorption isotherms, mechanical measurements, thermogravimetric analysis (TA) and differential scanning calorimetry (DSC). The analysis of water absorption isotherms, AFM data, and FTIR spectra indicated that during NaDES formation, MA and ChCl strongly interact, as well as CS and DES. The other results obtained by Sokolova et al. indicated a glass transition temperature between +2.0 and −2.3 °C, with maximum elongation at break of 62% shown by a film containing 67% by weight of NaDES. The increase in NaDES content (from 0 to 82%) led to a decrease in elasticity at tension from 800 MPa down to 16 MPa. Additionally, film with 82% NaDES demonstrated an elastic modulus with a bimodal trend. Finally, all of the studied films were found to be soluble in water at room temperature.

A further paper, in which NADES were used to endow plasticizing properties to chitosan films, was presented by Pontillo et al. [105]. The authors investigated the solubility of chitosan in NaDES aqueous solutions of choline chloride: lactic acid (ChCl-LA, molar ratio 1:1.5) and betaine: lactic acid (bet-LA, 1:2), demonstrating that chitosan can be dissolved in 1% NaDES *w*/*v* water solutions. Films prepared by the casting technique were compact, with elasticity properties comparable to films obtained by chitosan dissolved in 1% acetic acid (F/AA). Films containing NaDES or physical mixtures were more elastic and pliable. Films prepared with NaDES solutions (F/bet:LA NaDES and F/ChCl:LA NaDES) were significantly thicker than the F/AA films. The influence of acids on films' mechanical properties was confirmed by comparing different organic acids: the presence of lactic acid instead of acetic acid increases both the resistance of burst and the maximum elongation of chitosan films. Interesting results were obtained observing SEM morphology: F/AA films presented small holes, possibly due to the evaporation of the acetic acid, while the two F/bet:LA NaDES and F/ChCl:LA NaDES seemed to have a smooth surface with small wrinkled structured pores. The results suggested that properties of the films can be easily adjusted to fit the requirements useful for a wide range of applications; in particular, the new mixtures could represent promising alternatives for biomedical wound healing patches that usually lack in terms of elasticity.

Cerdá-Bernard et al. [106] investigated chitosan/alginate hydrogels to stabilize freeze-dried *C. sativus* flower extracts obtained by NaDES-UAE. The study aim was to exploit unused flower waste, reducing its environmental impact while stabilizing high added-value ingredients to screen their potential practical applications. In particular, they proposed an innovative extraction method that afforded the preparation of stable bioactive compounds with excellent antioxidant activity. NaDES obtained by the heating method [107] were based on different molar ratios of betaine-lactic acid (80%)-water 1:2:2.5 (Bet/LA/Water), glucose-lactic acid (80%)-water 1:5:6.2 (Glu/LA/Water), betaine-glycerol-water 1:3:1 (Bet/Gly/Water), L-proline-citric acid-water 2:1:3 (Pro/CA/Water), and L-proline-glycerol 1:2 (Pro/Gly), this last proving to be the best sample. The optimal UAE parameters for the extraction of saffron floral by-products and stigmas considered a process time of 20 min at 180 W and 90% Pro/Gly. Showing excellent antioxidant properties, these extracts were subsequently incorporated into a 0.3% chitosan/2% alginate hydrogel for stability improvement and to study their possible application as food formulations. Hydrogels containing extracts were then evaluated with regard to their water uptake and water retention capacities and total phenolic content (TPC) during in vitro digestion. Hydrogels with saffron stigma NaDES extracts showed an increased TPC after an hour of intestinal digestion, with constant levels up to 2 h. Otherwise, the hydrogel with saffron floral by-product NaDES extract showed an increase in TPC within the first 2 h. Therefore, Cerdá et al. proposed NaDES-UAE as an optimal combination for the recovery of bioactive compounds from saffron flower waste and suggested possible uses of their hydrogels for the cosmetic, food and pharmaceutical areas.

Silva et al. [108] presented a drug delivery system in which curcumin, dissolved in NaDES based on choline chloride (CC) and glycerol (GLY), was encapsulated into beads obtained by ionotropic gelation with chitosan and alginate. Beads were produced using an extrusion-dripping method. The main goal of the study was to develop curcumin-loaded hydrogel beads with an improved solubility and stability during transit along the gastrointestinal tract. In this context, NaDES can offer a green and promising alternative to overcome solubility hurdles and the need for removing organic solvent.

Wang et al. [109] investigated a hydrogel entirely composed by natural ingredients (sodium hyaluronate—SH, dopamine—DA, chitosan—CS, aloe vera—AV and NaDES) to be applied as a green and degradable wound dressing formulation. The hydrogel showed good cytocompatibility on NIH-3T3 fibroblast cells, and antibacterial properties against both Gram-positive (*S. aureus*) and Gram-negative (*E. coli*) bacteria. Sample surface morphologies were characterized by scanning electron microscopy (SEM); hydrogel swelling and in vitro degradation studies were assessed by mass change in phosphate buffered saline (PBS) solutions at 37 °C, and the dynamic rheological performances were evaluated by a strain-controlled rheometer. The results obtained by the NaDES-SH-CS/DA/AV hydrogel showed good cytocompatibility on NIH-3T3 fibroblast cells and good antibacterial properties. Moreover, the formulation promoted skin tissue regeneration with good wound healing effects on mouse skin within 12 days of surgery.

A new approach to NaDES application in the pharmaceutical and cosmetic fields seems to be exploitable by transforming these natural solvents into eutectogels for active ingredient delivery. Zeng et al. [110] proposed this opportunity using xanthan gum, a well-known polysaccharidic gelling agent produced by bacterial fermentation. This low-cost, biocompatible and biodegradable polymeric excipient is widely used in hydrogel preparations for biomedical and technological applications. Recently, it attracted a great deal of attention as a biomaterial for tissue scaffold preparation (extracellular matrix) in tissue engineering studies. The authors prepared highly biodegradable, thermostable eutectogels, by gelation with xanthan gum, of four low-viscosity NaDES containing choline chloride as the HBA and glycerol, xylitol, sorbitol or citric acid as the HBD. Gelation was obtained at low concentrations of xanthan gum (less than 5%). Morphology of the xanthan gum eutectogels was observed by optical and electron microscopy, and the possible gel formation mechanism was investigated by Fourier-transform infrared spectroscopy (FT-IR) and X-ray diffraction (XRD). The rheological properties were also studied, and the results showed excellent thermostability of the eutectogels in a temperature range of from 60 to 80 °C, with unchanged weight, keeping the gel stored at 80 °C for 10 h. By comparison with xanthan gum hydrogels, the resulting eutectogels were more stable in response to temperature increases, providing good rheological characteristics that were maintained over time.

4. Conclusions

In this review, the potential of NaDES as alternative green solvents in the extraction of natural active ingredients and as drug carriers was presented and explored. The most significant papers of the last 5 years regarding cosmetic and pharmaceutical formulations were discussed.

The main difficulties in the application of NaDES in industrial extraction processes are often represented by the high viscosities and by the separation of the solute after extraction. This second drawback is usually overcome by water addition or increasing the temperature, since these substances are thermo and pH switchable.

This review could help to disclose new opportunities and challenges for NaDES implementation not only as green alternative solvents but also as potential useful pathways to deliver bioactive ingredients in innovative formulations.

In conclusion, their application versatility, safety, biodegradability, biocompatibility and natural origin support NaDES as solvents of the future in the food, cosmetic and pharmaceutical fields.

Author Contributions: E.R., D.C. and C.V. wrote parts of the original text. F.S.R.d.C. and G.Z. served as text supervisors and revised the manuscript. All authors have read and agreed to the published version of the manuscript.

Funding: This work received no external funding.

Institutional Review Board Statement: Not applicable.

Informed Consent Statement: Not applicable.

Data Availability Statement: Not applicable.

Conflicts of Interest: The authors declare no conflicts of interest.

References

1. Chemat, F.; Vian, M.A.; Cravotto, G. Green Extraction of Natural Products: Concept and Principles. *Int. J. Mol. Sci.* **2012**, *13*, 8615–8627. [CrossRef]
2. Dai, Y. Natural Deep Eutectic Solvents and Their Application in Natural Product Research and Development. 2013. Available online: https://scholarlypublications.universiteitleiden.nl/handle/1887/21787 (accessed on 7 January 2024).
3. Teles, A.R.R.; Capela, E.V.; Carmo, R.S.; Coutinho, J.A.P.; Silvestre, A.J.D.; Freire, M.G. Solvatochromic Parameters of Deep Eutectic Solvents Formed by Ammonium-Based Salts and Carboxylic Acids. *Fluid. Phase Equilibria* **2017**, *448*, 15–21. [CrossRef]
4. Osch, D.J.G.P.; van Kollau, L.J.B.M.; Bruinhorst, A.; van den Asikainen, S.; Rocha, M.A.A.; Kroon, M.C. Ionic Liquids and Deep Eutectic Solvents for Lignocellulosic Biomass Fractionation. *Phys. Chem. Chem. Phys.* **2017**, *19*, 2636–2665. [CrossRef] [PubMed]
5. Mero, A.; Koutsoumpos, S.; Giannios, P.; Stavrakas, I.; Moutzouris, K.; Mezzetta, A.; Guazzelli, L. Comparison of Physicochemical and Thermal Properties of Choline Chloride and Betaine-Based Deep Eutectic Solvents: The Influence of Hydrogen Bond Acceptor and Hydrogen Bond Donor Nature and Their Molar Ratios. *J. Mol. Liq.* **2023**, *377*, 121563. [CrossRef]
6. Florindo, C.; Oliveira, F.S.; Rebelo, L.P.N.; Fernandes, A.M.; Marrucho, I.M. Insights into the Synthesis and Properties of Deep Eutectic Solvents Based on Cholinium Chloride and Carboxylic Acids. *ACS Sustain. Chem. Eng.* **2014**, *2*, 2416–2425. [CrossRef]
7. Zhang, Q.; Vigier, K.D.O.; Royer, S.; Jérôme, F. Deep Eutectic Solvents: Syntheses, Properties and Applications. *Chem. Soc. Rev.* **2012**, *41*, 7108–7146. [CrossRef] [PubMed]
8. Welton, T. Room-Temperature Ionic Liquids. Solvents for Synthesis and Catalysis. *Chem. Rev.* **1999**, *99*, 2071–2084. [CrossRef] [PubMed]
9. Visser, A.E.; Swatloski, R.P.; Rogers, R.D. pH-Dependent Partitioning in Room Temperature Ionic Liquids Provides a Link to Traditional Solvent Extraction Behavior. *Green Chem.* **2000**, *2*, 1–4. [CrossRef]
10. Zainal-Abidin, M.H.; Hayyan, M.; Hayyan, A.; Jayakumar, N.S. New Horizons in the Extraction of Bioactive Compounds Using Deep Eutectic Solvents: A Review. *Anal. Chim. Acta* **2017**, *979*, 1–23. [CrossRef] [PubMed]
11. El Abedin, S.Z.; Endres, F. Ionic Liquids: The Link to High-Temperature Molten Salts? *Acc. Chem. Res.* **2007**, *40*, 1106–1113. [CrossRef]
12. Abbott, A.P.; Capper, G.; Davies, D.L.; Rasheed, R.K.; Tambyrajah, V. Novel Solvent Properties of Choline Chloride/Urea Mixtures. *Chem. Commun.* **2003**, *1*, 70–71. [CrossRef] [PubMed]
13. Kudłak, B.; Owczarek, K.; Namieśnik, J. Selected Issues Related to the Toxicity of Ionic Liquids and Deep Eutectic Solvents—A Review. *Environ. Sci. Pollut. Res.* **2015**, *22*, 11975–11992. [CrossRef] [PubMed]
14. Ijardar, S.P.; Singh, V.; Gardas, R.L. Revisiting the Physicochemical Properties and Applications of Deep Eutectic Solvents. *Molecules* **2022**, *27*, 1368. [CrossRef] [PubMed]
15. Abbott, A.P.; Boothby, D.; Capper, G.; Davies, D.L.; Rasheed, R.K. Deep Eutectic Solvents Formed between Choline Chloride and Carboxylic Acids: Versatile Alternatives to Ionic Liquids. *J. Am. Chem. Soc.* **2004**, *126*, 9142–9147. [CrossRef] [PubMed]
16. Wagle, D.V.; Deakyne, C.A.; Baker, G.A. Quantum Chemical Insight into the Interactions and Thermodynamics Present in Choline Chloride Based Deep Eutectic Solvents. *J. Phys. Chem. B* **2016**, *120*, 6739–6746. [CrossRef] [PubMed]

17. Dai, Y.; van Spronsen, J.; Witkamp, G.-J.; Verpoorte, R.; Choi, Y.H. Natural Deep Eutectic Solvents as New Potential Media for Green Technology. *Anal. Chim. Acta* **2013**, *766*, 61–68. [CrossRef] [PubMed]
18. Choi, Y.H.; van Spronsen, J.; Dai, Y.; Verberne, M.; Hollmann, F.; Arends, I.W.C.E.; Witkamp, G.-J.; Verpoorte, R. Are Natural Deep Eutectic Solvents the Missing Link in Understanding Cellular Metabolism and Physiology? *Plant Physiol.* **2011**, *156*, 1701–1705. [CrossRef]
19. Liu, Y.; Friesen, J.B.; McAlpine, J.B.; Lankin, D.C.; Chen, S.-N.; Pauli, G.F. Natural Deep Eutectic Solvents: Properties, Applications, and Perspectives. *J. Nat. Prod.* **2018**, *81*, 679–690. [CrossRef]
20. Obluchinskaya, E.; Daurtseva, A.; Pozharitskaya, O.; Flisyuk, E.; Shikov, A. Natural Deep Eutectic Solvents as Alternatives for Extracting Phlorotannins from Brown Algae. *Pharm. Chem. J.* **2019**, *53*, 243–247. [CrossRef]
21. Obluchinskaya, E.D.; Pozharitskaya, O.N.; Shevyrin, V.A.; Kovaleva, E.G.; Flisyuk, E.V.; Shikov, A.N. Optimization of Extraction of Phlorotannins from the Arctic Fucus Vesiculosus Using Natural Deep Eutectic Solvents and Their HPLC Profiling with Tandem High-Resolution Mass Spectrometry. *Mar. Drugs* **2023**, *21*, 263. [CrossRef]
22. Dai, Y.; Verpoorte, R.; Choi, Y.H. Natural Deep Eutectic Solvents Providing Enhanced Stability of Natural Colorants from Safflower (Carthamus Tinctorius). *Food Chem.* **2014**, *159*, 116–121. [CrossRef]
23. Xin, R.; Qi, S.; Zeng, C.; Khan, F.I.; Yang, B.; Wang, Y. A Functional Natural Deep Eutectic Solvent Based on Trehalose: Structural and Physicochemical Properties. *Food Chem.* **2017**, *217*, 560–567. [CrossRef]
24. Faggian, M.; Sut, S.; Perissutti, B.; Baldan, V.; Grabnar, I.; Dall'Acqua, S. Natural Deep Eutectic Solvents (NADES) as a Tool for Bioavailability Improvement: Pharmacokinetics of Rutin Dissolved in Proline/Glycine after Oral Administration in Rats: Possible Application in Nutraceuticals. *Molecules* **2016**, *21*, 1531. [CrossRef]
25. Sut, S.; Faggian, M.; Baldan, V.; Poloniato, G.; Castagliuolo, I.; Grabnar, I.; Perissutti, B.; Brun, P.; Maggi, F.; Voinovich, D.; et al. Natural Deep Eutectic Solvents (NADES) to Enhance Berberine Absorption: An In Vivo Pharmacokinetic Study. *Molecules* **2017**, *22*, 1921. [CrossRef] [PubMed]
26. Mano, F.; Aroso, I.M.; Barreiros, S.; Borges, J.P.; Reis, R.L.; Duarte, A.R.C.; Paiva, A. Production of Poly(Vinyl Alcohol) (PVA) Fibers with Encapsulated Natural Deep Eutectic Solvent (NADES) Using Electrospinning. *ACS Sustain. Chem. Eng.* **2015**, *3*, 2504–2509. [CrossRef]
27. Vieira Sanches, M.; Freitas, R.; Oliva, M.; Mero, A.; De Marchi, L.; Cuccaro, A.; Fumagalli, G.; Mezzetta, A.; Colombo Dugoni, G.; Ferro, M.; et al. Are Natural Deep Eutectic Solvents Always a Sustainable Option? A Bioassay-Based Study. *Environ. Sci. Pollut. Res.* **2023**, *30*, 17268–17279. [CrossRef] [PubMed]
28. Dai, Y.; van Spronsen, J.; Witkamp, G.-J.; Verpoorte, R.; Choi, Y.H. Ionic Liquids and Deep Eutectic Solvents in Natural Products Research: Mixtures of Solids as Extraction Solvents. *J. Nat. Prod.* **2013**, *76*, 2162–2173. [CrossRef] [PubMed]
29. Fernández, M.d.L.Á.; Boiteux, J.; Espino, M.; Gomez, F.J.V.; Silva, M.F. Natural Deep Eutectic Solvents-Mediated Extractions: The Way Forward for Sustainable Analytical Developments. *Anal. Chim. Acta* **2018**, *1038*, 1–10. [CrossRef]
30. Jeong, K.M.; Ko, J.; Zhao, J.; Jin, Y.; Yoo, D.E.; Han, S.Y.; Lee, J. Multi-Functioning Deep Eutectic Solvents as Extraction and Storage Media for Bioactive Natural Products That Are Readily Applicable to Cosmetic Products. *J. Clean. Prod.* **2017**, *151*, 87–95. [CrossRef]
31. Liu, W.; Zhang, K.; Qin, Y.; Yu, J. A Simple and Green Ultrasonic-Assisted Liquid–Liquid Microextraction Technique Based on Deep Eutectic Solvents for the HPLC Analysis of Sesamol in Sesame Oils. *Anal. Methods* **2017**, *9*, 4184–4189. [CrossRef]
32. Milano, F.; Giotta, L.; Guascito, M.R.; Agostiano, A.; Sblendorio, S.; Valli, L.; Perna, F.M.; Cicco, L.; Trotta, M.; Capriati, V. Functional Enzymes in Nonaqueous Environment: The Case of Photosynthetic Reaction Centers in Deep Eutectic Solvents. *ACS Sustain. Chem. Eng.* **2017**, *5*, 7768–7776. [CrossRef]
33. Zahrina, I.; Nasikin, M.; Krisanti, E.; Mulia, K. Deacidification of Palm Oil Using Betaine Monohydrate-Based Natural Deep Eutectic Solvents. *Food Chem.* **2018**, *240*, 490–495. [CrossRef] [PubMed]
34. Jeong, K.M.; Zhao, J.; Jin, Y.; Heo, S.R.; Han, S.Y.; Yoo, D.E.; Lee, J. Highly Efficient Extraction of Anthocyanins from Grape Skin Using Deep Eutectic Solvents as Green and Tunable Media. *Arch. Pharm. Res.* **2015**, *38*, 2143–2152. [CrossRef]
35. Guo, N.; Kou, P.; Jiang, Y.-W.; Wang, L.-T.; Niu, L.-J.; Liu, Z.-M.; Fu, Y.-J. Natural Deep Eutectic Solvents Couple with Integrative Extraction Technique as an Effective Approach for Mulberry Anthocyanin Extraction. *Food Chem.* **2019**, *296*, 78–85. [CrossRef]
36. Panić, M.; Gunjević, V.; Cravotto, G.; Radojčić Redovniković, I. Enabling Technologies for the Extraction of Grape-Pomace Anthocyanins Using Natural Deep Eutectic Solvents in up-to-Half-Litre Batches Extraction of Grape-Pomace Anthocyanins Using NADES. *Food Chem.* **2019**, *300*, 125185. [CrossRef]
37. Popovic, B.M.; Micic, N.; Potkonjak, A.; Blagojevic, B.; Pavlovic, K.; Milanov, D.; Juric, T. Novel Extraction of Polyphenols from Sour Cherry Pomace Using Natural Deep Eutectic Solvents—Ultrafast Microwave-Assisted NADES Preparation and Extraction. *Food Chem.* **2022**, *366*, 130562. [CrossRef]
38. Grillo, G.; Gunjević, V.; Radošević, K.; Redovniković, I.R.; Cravotto, G. Deep Eutectic Solvents and Nonconventional Technologies for Blueberry-Peel Extraction: Kinetics, Anthocyanin Stability, and Antiproliferative Activity. *Antioxidants* **2020**, *9*, 1069. [CrossRef] [PubMed]
39. Cai, C.; Li, F.-F.; Liu, L.; Tan, Z. Deep Eutectic Solvents Used as the Green Media for the Efficient Extraction of Caffeine from Chinese Dark Tea. *Sep. Purif. Technol.* **2019**, *227*, 115723. [CrossRef]

40. Alioui, O.; Sobhi, W.; Tiecco, M.; Alnashef, I.M.; Attoui, A.; Boudechicha, A.; Kumar Yadav, K.; Fallatah, A.M.; Elboughdiri, N.; Jeon, B.-H.; et al. Theoretical and Experimental Evidence for the Use of Natural Deep Eutectic Solvents to Increase the Solubility and Extractability of Curcumin. *J. Mol. Liq.* **2022**, *359*, 119149. [CrossRef]
41. Zurob, E.; Cabezas, R.; Villarroel, E.; Rosas, N.; Merlet, G.; Quijada-Maldonado, E.; Romero, J.; Plaza, A. Design of Natural Deep Eutectic Solvents for the Ultrasound-Assisted Extraction of Hydroxytyrosol from Olive Leaves Supported by COSMO-RS. *Sep. Purif. Technol.* **2020**, *248*, 117054. [CrossRef]
42. Bajkacz, S.; Adamek, J. Evaluation of New Natural Deep Eutectic Solvents for the Extraction of Isoflavones from Soy Products. *Talanta* **2017**, *168*, 329–335. [CrossRef]
43. Chen, S.; Xiao, L.; Li, S.; Meng, T.; Wang, L.; Zhang, W. The Effect of Sonication-Synergistic Natural Deep Eutectic Solvents on Extraction Yield, Structural and Physicochemical Properties of Pectins Extracted from Mango Peels. *Ultrason. Sonochem.* **2022**, *86*, 106045. [CrossRef]
44. Lin, L. Phenolic Acid Extraction from Orange Peel with Natural Deep Eutectic Solvents. *J. Phys. Conf. Ser.* **2022**, *2321*, 012020. [CrossRef]
45. Zannou, O.; Pashazadeh, H.; Ghellam, M.; Ali Redha, A.; Koca, I. Enhanced Ultrasonically Assisted Extraction of Bitter Melon (Momordica Charantia) Leaf Phenolic Compounds Using Choline Chloride-Acetic Acid-Based Natural Deep Eutectic Solvent: An Optimization Approach and In Vitro Digestion. *Biomass Convers. Biorefinery* **2022**, *3*. [CrossRef]
46. Ünlü, A.E. Green and Non-conventional Extraction of Bioactive Compounds from Olive Leaves: Screening of Novel Natural Deep Eutectic Solvents and Investigation of Process Parameters. *Waste Biomass Valorization* **2021**, *12*, 5329–5346. [CrossRef]
47. Chanioti, S.; Tzia, C. Extraction of Phenolic Compounds from Olive Pomace by Using Natural Deep Eutectic Solvents and Innovative Extraction Techniques. *Innov. Food Sci. Emerg. Technol.* **2018**, *48*, 228–239. [CrossRef]
48. Fanali, C.; Gallo, V.; Della Posta, S.; Dugo, L.; Mazzeo, L.; Cocchi, M.; Piemonte, V.; De Gara, L. Choline Chloride–Lactic Acid-Based NADES As an Extraction Medium in a Response Surface Methodology-Optimized Method for the Extraction of Phenolic Compounds from Hazelnut Skin. *Molecules* **2021**, *26*, 2652. [CrossRef] [PubMed]
49. Manuela, P.; Drakula, S.; Cravotto, G.; Verpoorte, R.; Hruškar, M.; Radojčić Redovniković, I.; Radošević, K. Biological Activity and Sensory Evaluation of Cocoa By-Products NADES Extracts Used in Food Fortification. *Innov. Food Sci. Emerg. Technol.* **2020**, *66*, 102514. [CrossRef]
50. Lanjekar, K.J.; Gokhale, S.; Rathod, V.K. Utilization of Waste Mango Peels for Extraction of Polyphenolic Antioxidants by Ultrasound-Assisted Natural Deep Eutectic Solvent. *Bioresour. Technol. Rep.* **2022**, *18*, 101074. [CrossRef]
51. Vieira, C.; Rebocho, S.; Craveiro, R.; Paiva, A.; Duarte, A.R.C. Selective Extraction and Stabilization of Bioactive Compounds from Rosemary Leaves Using a Biphasic NADES. *Front. Chem.* **2022**, *10*, 954835. [CrossRef]
52. Hong, J.; Deng, M.; Zhao, L. Natural Deep Eutectic Solvent Combined with Ultrasonic Enhancement: A Green Extraction Strategy for Solanesol in Tobacco Leaves. *Ind. Crops Prod.* **2022**, *187*, 115355. [CrossRef]
53. Xu, Z.; Cai, Y.; Ma, Q.; Zhao, Z.; Yang, D.; Xu, X. Optimization of Extraction of Bioactive Compounds from Baphicacanthus Cusia Leaves by Hydrophobic Deep Eutectic Solvents. *Molecules* **2021**, *26*, 1729. [CrossRef]
54. Silva, N.H.C.S.; Morais, E.S.; Freire, C.S.R.; Freire, M.G.; Silvestre, A.J.D. Extraction of High Value Triterpenic Acids from Eucalyptus Globulus Biomass Using Hydrophobic Deep Eutectic Solvents. *Molecules* **2020**, *25*, 210. [CrossRef]
55. Paiva, A.; Craveiro, R.; Aroso, I.; Martins, M.; Reis, R.L.; Duarte, A.R.C. Natural Deep Eutectic Solvents—Solvents for the 21st Century. *ACS Sustain. Chem. Eng.* **2014**, *2*, 1063–1071. [CrossRef]
56. Radošević, K.; Ćurko, N.; Gaurina Srček, V.; Cvjetko Bubalo, M.; Tomašević, M.; Kovačević Ganić, K.; Radojčić Redovniković, I. Natural Deep Eutectic Solvents as Beneficial Extractants for Enhancement of Plant Extracts Bioactivity. *LWT* **2016**, *73*, 45–51. [CrossRef]
57. Wen, Q.; Chen, J.-X.; Tang, Y.-L.; Wang, J.; Yang, Z. Assessing the Toxicity and Biodegradability of Deep Eutectic Solvents. *Chemosphere* **2015**, *132*, 63–69. [CrossRef] [PubMed]
58. Pena-Pereira, F.; Kloskowski, A.; Namieśnik, J. Perspectives on the Replacement of Harmful Organic Solvents in Analytical Methodologies: A Framework toward the Implementation of a Generation of Eco-Friendly Alternatives. *Green Chem.* **2015**, *17*, 3687–3705. [CrossRef]
59. Anastas, P.; Eghbali, N. Green Chemistry: Principles and Practice. *Chem. Soc. Rev.* **2009**, *39*, 301–312. [CrossRef] [PubMed]
60. Dai, Y.; Witkamp, G.-J.; Verpoorte, R.; Choi, Y.H. Tailoring Properties of Natural Deep Eutectic Solvents with Water to Facilitate Their Applications. *Food Chem.* **2015**, *187*, 14–19. [CrossRef] [PubMed]
61. Martins, M.; Aroso, I.M.; Reis, R.L.; Duarte, A.R.C.; Craveiro, R.; Paiva, A. Enhanced Performance of Supercritical Fluid Foaming of Natural-Based Polymers by Deep Eutectic Solvents. *AIChE J.* **2014**, *60*, 3701–3706. [CrossRef]
62. Wei, Z.; Qi, X.; Li, T.; Luo, M.; Wang, W.; Zu, Y.; Fu, Y. Application of Natural Deep Eutectic Solvents for Extraction and Determination of Phenolics in Cajanus Cajan Leaves by Ultra Performance Liquid Chromatography. *Sep. Purif. Technol.* **2015**, *149*, 237–244. [CrossRef]
63. Wei, Z.-F.; Wang, X.-Q.; Peng, X.; Wang, W.; Zhao, C.-J.; Zu, Y.-G.; Fu, Y.-J. Fast and Green Extraction and Separation of Main Bioactive Flavonoids from Radix Scutellariae. *Ind. Crops Prod.* **2015**, *63*, 175–181. [CrossRef]
64. Gutiérrez, M.; Ferrer, M.L.; Yuste, L.; Rojo, F.; Monte, F. Bacteria Incorporation in Deep-Eutectic Solvents through Freeze-Drying. *Angew. Chem.* **2010**, *49*, 2158–2162. [CrossRef]

65. Florindo, C.; Romero, L.; Rintoul, I.; Branco, L.C.; Marrucho, I.M. From Phase Change Materials to Green Solvents: Hydrophobic Low Viscous Fatty Acid–Based Deep Eutectic Solvents. *ACS Sustain. Chem. Eng.* **2018**, *6*, 3888–3895. [CrossRef]
66. Bajkacz, S.; Adamek, J. Development of a Method Based on Natural Deep Eutectic Solvents for Extraction of Flavonoids from Food Samples. *Food Anal. Methods* **2018**, *11*, 1330–1344. [CrossRef]
67. Gomez, F.J.V.; Espino, M.; Fernández, M.A.; Silva, M.F. A Greener Approach to Prepare Natural Deep Eutectic Solvents. *ChemistrySelect* **2018**, *3*, 6122–6125. [CrossRef]
68. Chemat, F.; Cravotto, G. *Microwave-Assisted Extraction for Bioactive Compounds*; Food Engineering Series; Editions Springer: Berlin/Heidelberg, Germany, 2013.
69. Zhao, B.-Y.; Xu, P.; Yang, F.-X.; Wu, H.; Zong, M.-H.; Lou, W.-Y. Biocompatible Deep Eutectic Solvents Based on Choline Chloride: Characterization and Application to the Extraction of Rutin from Sophora Japonica. *ACS Sustain. Chem. Eng.* **2015**, *3*, 2746–2755. [CrossRef]
70. Gutiérrez, M.; Ferrer, M.L.; Mateo, C.R.; Monte, F. Freeze-Drying of Aqueous Solutions of Deep Eutectic Solvents: A Suitable Approach to Deep Eutectic Suspensions of Self-Assembled Structures. *Langmuir ACS J. Surf. Colloids* **2009**, *25*, 5509–5515. [CrossRef] [PubMed]
71. Francisco, M.; Van Den Bruinhorst, A.; Kroon, M.C. Low-Transition-Temperature Mixtures (LTTMs): A New Generation of Designer Solvents. *Angew. Chem. Int. Ed.* **2013**, *52*, 3074–3085. [CrossRef]
72. Mišan, A.; Nađpal, J.; Stupar, A.; Pojić, M.; Mandić, A.; Verpoorte, R.; Choi, Y.H. The Perspectives of Natural Deep Eutectic Solvents in Agri-Food Sector. *Crit. Rev. Food Sci. Nutr.* **2020**, *60*, 2564–2592. [CrossRef] [PubMed]
73. Craveiro, R.; Aroso, I.; Flammia, V.; Carvalho, T.; Viciosa, M.T.; Dionísio, M.; Barreiros, S.; Reis, R.L.; Duarte, A.R.C.; Paiva, A. Properties and Thermal Behavior of Natural Deep Eutectic Solvents. *J. Mol. Liq.* **2016**, *215*, 534–540. [CrossRef]
74. Mjalli, F.S.; Al-Azzawi, M. Aliphatic Amino Acids as Possible Hydrogen Bond Donors for Preparing Eutectic Solvents. *J. Mol. Liq.* **2021**, *330*, 115637. [CrossRef]
75. Benoit, C.; Virginie, C.; Boris, V. Chapter Twelve—The Use of NADES to Support Innovation in the Cosmetic Industry. In *Advances in Botanical Research*; Verpoorte, R., Witkamp, G.-J., Choi, Y.H., Eds.; Eutectic Solvents and Stress in Plants; Academic Press: Cambridge, MA, USA, 2021; Volume 97, pp. 309–332.
76. Rente, D.; Cvjetko Bubalo, M.; Panić, M.; Paiva, A.; Caprin, B.; Radojcic Redovnikovic, I.; Duarte, A. Review of Deep Eutectic Systems from Laboratory to Industry, Taking the Application in the Cosmetics Industry as an Example. *J. Clean. Prod.* **2022**, *380*, 135147. [CrossRef]
77. Vasyliev, G.; Lyudmyla, K.; Hladun, K.; Skiba, M.; Vorobyova, V. Valorization of Tomato Pomace: Extraction of Value-Added Components by Deep Eutectic Solvents and Their Application in the Formulation of Cosmetic Emulsions. *Biomass Conv. Bioref.* **2022**, *12*, 95–111. [CrossRef]
78. Petkov, H.; Trusheva, B.; Krustanova, S.; Grozdanova, T.; Popova, M.; Alipieva, K.; Bankova, V. Green Extraction of Antioxidants from Natural Sources with Natural Deep Eutectic Solvents. *Proc. Bulg. Acad. Sci.* **2022**, *75*, 1129–1137. [CrossRef]
79. El Maaiden, E.; El Kahia, H.; Nasser, B.; Moustaid, K.; Qarah, N.; Boukcim, H.; Hirich, A.; Kouisni, L.; El Kharrassi, Y. Deep Eutectic Solvent-Ultrasound Assisted Extraction as a Green Approach for Enhanced Extraction of Naringenin from Searsia Tripartita and Retained Their Bioactivities. *Front. Nutr.* **2023**, *10*, 1193509. [CrossRef]
80. Hernández-Corroto, E.; Plaza, M.; Marina, M.L.; García, M.C. Sustainable Extraction of Proteins and Bioactive Substances from Pomegranate Peel (*Punica granatum* L.) Using Pressurized Liquids and Deep Eutectic Solvents. *Innov. Food Sci. Emerg. Technol.* **2020**, *60*, 102314. [CrossRef]
81. Jamaleddine, A.; Urrutigoïty, M.; Bouajila, J.; Merah, O.; Evon, P.; de Caro, P. Ecodesigned Formulations with Tomato Pomace Extracts. *Cosmetics* **2023**, *10*, 7. [CrossRef]
82. Wils, L.; Leman-Loubière, C.; Bellin, N.; Clément-Larosière, B.; Pinault, M.; Chevalier, S.; Enguehard-Gueiffier, C.; Bodet, C.; Boudesocque-Delaye, L. Natural Deep Eutectic Solvent Formulations for Spirulina: Preparation, Intensification, and Skin Impact. *Algal Res.* **2021**, *56*, 102317. [CrossRef]
83. Silva, Y.P.A.; Ferreira, T.A.P.C.; Jiao, G.; Brooks, M.S. Sustainable Approach for Lycopene Extraction from Tomato Processing By-Product Using Hydrophobic Eutectic Solvents. *J. Food Sci. Technol.* **2019**, *56*, 1649–1654. [CrossRef]
84. Fernández, M.d.L.Á.; Espino, M.; Gomez, F.J.V.; Silva, M.F. Novel Approaches Mediated by Tailor-Made Green Solvents for the Extraction of Phenolic Compounds from Agro-Food Industrial by-Products. *Food Chem.* **2018**, *239*, 671–678. [CrossRef] [PubMed]
85. Wu, L.; Li, L.; Chen, S.; Wang, L.; Lin, X. Deep Eutectic Solvent-Based Ultrasonic-Assisted Extraction of Phenolic Compounds from *Moringa oleifera* L. Leaves: Optimization, Comparison and Antioxidant Activity. *Sep. Purif. Technol.* **2020**, *247*, 117014. [CrossRef]
86. Jin, Y.; Jung, D.; Li, K.; Park, K.; Ko, J.; Yang, M.; Lee, J. Application of Deep Eutectic Solvents to Prepare Mixture Extracts of Three Long-Lived Trees with Maximized Skin-Related Bioactivities. *Appl. Sci.* **2019**, *9*, 2581. [CrossRef]
87. Union, P.O. of the EU. 2006/257/EC: Commission Decision of 9 February 2006 Amending Decision 96/335/EC Establishing an Inventory and a Common Nomenclature of Ingredients Employed in Cosmetic Products (Text with EEA Relevance), CELEX1. Available online: https://op.europa.eu/en/publication-detail/-/publication/db30de80-11f8-4358-b1d6-e38d6cf96625 (accessed on 14 August 2023).
88. Yoo, D.E.; Jeong, K.M.; Han, S.Y.; Kim, E.M.; Jin, Y.; Lee, J. Deep Eutectic Solvent-Based Valorization of Spent Coffee Grounds. *Food Chem.* **2018**, *255*, 357–364. [CrossRef] [PubMed]

89. Hsieh, Y.-H.; Li, Y.; Pan, Z.; Chen, Z.; Lu, J.; Yuan, J.; Zhu, Z.; Zhang, J. Ultrasonication-Assisted Synthesis of Alcohol-Based Deep Eutectic Solvents for Extraction of Active Compounds from Ginger. *Ultrason. Sonochem.* **2020**, *63*, 104915. [CrossRef] [PubMed]
90. Rocha, D.; Freitas, D.S.; Magalhães, J.; Fernandes, M.; Silva, S.; Noro, J.; Ribeiro, A.; Cavaco-Paulo, A.; Martins, M.; Silva, C. NADES-Based Cork Extractives as Green Ingredients for Cosmetics and Textiles. *Processes* **2023**, *11*, 309. [CrossRef]
91. Marijan, M.; Jablan, J.; Jakupović, L.; Jug, M.; Marguí, E.; Dalipi, R.; Sangiorgi, E.; Zovko Končić, M. Plants from Urban Parks as Valuable Cosmetic Ingredients: Green Extraction, Chemical Composition and Activity. *Agronomy* **2022**, *12*, 204. [CrossRef]
92. Alishlah, T.; Mun'im, A.; Jufri, M. Optimization of Urea-Glycerin Based NADES-UAE for Oxyresveratrol Extraction from Morus Alba Roots for Preparation of Skin Whitening Lotion. *JYP* **2019**, *11*, 155–160. [CrossRef]
93. Oktaviyanti, N.D.; Kartini; Munim, A. Application and Optimization of Ultrasound-Assisted Deep Eutectic Solvent for the Extraction of New Skin-Lightening Cosmetic Materials from Ixora Javanica Flower. *Heliyon* **2019**, *5*, e02950. [CrossRef]
94. Punzo, A.; Porru, E.; Silla, A.; Simoni, P.; Galletti, P.; Roda, A.; Tagliavini, E.; Samorì, C.; Caliceti, C. Grape Pomace for Topical Application: Green NaDES Sustainable Extraction, Skin Permeation Studies, Antioxidant and Anti-Inflammatory Activities Characterization in 3D Human Keratinocytes. *Biomolecules* **2021**, *11*, 1181. [CrossRef]
95. Delgado-Rangel, L.H.; Huerta-Saquero, A.; Eufracio-García, N.; Meza-Villezcas, A.; Mota-Morales, J.D.; González-Campos, J.B. Deep Eutectic Solvent-Assisted Phase Separation in Chitosan Solutions for the Production of 3D Monoliths and Films with Tailored Porosities. *Int. J. Biol. Macromol.* **2020**, *164*, 4084–4094. [CrossRef]
96. Alkhawaja, B.; Al-Akayleh, F.; Al-Khateeb, A.; Nasereddin, J.; Ghanim, B.Y.; Bolhuis, A.; Jaber, N.; Al-Remawi, M.; Qinna, N.A. Deep Eutectic Liquids as a Topical Vehicle for Tadalafil: Characterisation and Potential Wound Healing and Antimicrobial Activity. *Molecules* **2023**, *28*, 2402. [CrossRef] [PubMed]
97. Filip, D.; Macocinschi, D.; Balan-Porcarasu, M.; Varganici, C.-D.; Dumitriu, R.-P.; Peptanariu, D.; Tuchilus, C.G.; Zaltariov, M.-F. Biocompatible Self-Assembled Hydrogen-Bonded Gels Based on Natural Deep Eutectic Solvents and Hydroxypropyl Cellulose with Strong Antimicrobial Activity. *Gels* **2022**, *8*, 666. [CrossRef]
98. Ma, W. Performance Standards for Antimicrobial Susceptibility Testing Sixteenth Informational Supplement. *M 100-S 16*. 2006. Available online: https://clsi.org/standards/products/microbiology/documents/m100/ (accessed on 7 January 2024).
99. Rex, J.H.; Clinical & Laboratory Standards Institute. *Method for Antifungal Disk Diffusion Susceptibility Testing of Yeasts: Approved Guideline*, 2nd ed.; Clinical and Laboratory Standards Institute: Wayne, PA, USA, 2009.
100. Mustafa, N.R.; Spelbos, V.S.; Witkamp, G.-J.; Verpoorte, R.; Choi, Y.H. Solubility and Stability of Some Pharmaceuticals in Natural Deep Eutectic Solvents-Based Formulations. *Molecules* **2021**, *26*, 2645. [CrossRef]
101. Li, W.; Zhao, X.; Huang, T.; Ren, Y.; Gong, W.; Guo, Y.; Wang, J.; Tu, Q. Preparation of Sodium Hyaluronate/Dopamine/AgNPs Hydrogel Based on the Natural Eutetic Solvent as an Antibaterial Wound Dressing. *Int. J. Biol. Macromol.* **2021**, *191*, 60–70. [CrossRef]
102. Lee, S.-W.; Ryu, J.H.; Do, M.J.; Namkoong, E.; Lee, H.; Park, K. NiCHE Platform: Nature-Inspired Catechol-Conjugated Hyaluronic Acid Environment Platform for Salivary Gland Tissue Engineering. *ACS Appl. Mater. Interfaces* **2020**, *12*, 4285–4294. [CrossRef]
103. Sokolova, M.P.; Smirnov, M.A.; Samarov, A.A.; Bobrova, N.V.; Vorobiov, V.K.; Popova, E.N.; Filippova, E.; Geydt, P.; Lahderanta, E.; Toikka, A.M. Plasticizing of Chitosan Films with Deep Eutectic Mixture of Malonic Acid and Choline Chloride. *Carbohydr. Polym.* **2018**, *197*, 548–557. [CrossRef] [PubMed]
104. Samarov, A.A.; Smirnov, M.A.; Sokolova, M.P.; Popova, E.N.; Toikka, A.M. Choline Chloride Based Deep Eutectic Solvents as Extraction Media for Separation of N-Hexane–Ethanol Mixture. *Fluid. Phase Equilibria* **2017**, *448*, 123–127. [CrossRef]
105. Nefeli Pontillo, A.R.; Koutsoukos, S.; Welton, T.; Detsi, A. Investigation of the Influence of Natural Deep Eutectic Solvents (NaDES) in the Properties of Chitosan-Stabilised Films. *Mater. Adv.* **2021**, *2*, 3954–3964. [CrossRef]
106. Cerdá-Bernad, D.; Pitterou, I.; Tzani, A.; Detsi, A.; Frutos, M.J. Novel Chitosan/Alginate Hydrogels as Carriers of Phenolic-Enriched Extracts from Saffron Floral by-Products Using Natural Deep Eutectic Solvents as Green Extraction Media. *Curr. Res. Food Sci.* **2023**, *6*, 100469. [CrossRef]
107. Tzani, A.; Kalafateli, S.; Tatsis, G.; Bairaktari, M.; Kostopoulou, I.; Pontillo, A.R.N.; Detsi, A. Natural Deep Eutectic Solvents (NaDESs) as Alternative Green Extraction Media for Ginger (Zingiber Officinale Roscoe). *Sustain. Chem.* **2021**, *2*, 576–598. [CrossRef]
108. Silva, J.M.; Silva, E.; Reis, R.L. Therapeutic Deep Eutectic Solvents Assisted the Encapsulation of Curcumin in Alginate-Chitosan Hydrogel Beads. *Sustain. Chem. Pharm.* **2021**, *24*, 100553. [CrossRef]
109. Wang, Y.; Zhang, Y.; Lin, Z.; Huang, T.; Li, W.; Gong, W.; Guo, Y.; Su, J.; Wang, J.; Tu, Q. A Green Method of Preparing a Natural and Degradable Wound Dressing Containing Aloe Vera as an Active Ingredient. *Compos. Part B Eng.* **2021**, *222*, 109047. [CrossRef]
110. Zeng, C.; Zhao, H.; Wan, Z.; Xiao, Q.; Xia, H.; Guo, S. Highly Biodegradable, Thermostable Eutectogels Prepared by Gelation of Natural Deep Eutectic Solvents Using Xanthan Gum: Preparation and Characterization. *RSC Adv.* **2020**, *10*, 28376–28382. [CrossRef] [PubMed]

gels

MDPI

Review

Aloe vera-Based Hydrogels for Wound Healing: Properties and Therapeutic Effects

Mariana Chelu †, Adina Magdalena Musuc *, Monica Popa † and Jose Calderon Moreno *

"Ilie Murgulescu" Institute of Physical Chemistry, 202 Splaiul Independentei, 060021 Bucharest, Romania; mchelu@icf.ro (M.C.); pmonica@icf.ro (M.P.)

* Correspondence: amusuc@icf.ro (A.M.M.); calderon@icf.ro (J.C.M.)

† These authors contributed equally to this work.

Abstract: *Aloe vera*-based hydrogels have emerged as promising platforms for the delivery of therapeutic agents in wound dressings due to their biocompatibility and unique wound-healing properties. The present study provides a comprehensive overview of recent advances in the application of *Aloe vera*-based hydrogels for wound healing. The synthesis methods, structural characteristics, and properties of *Aloe vera*-based hydrogels are discussed. Mechanisms of therapeutic agents released from *Aloe vera*-based hydrogels, including diffusion, swelling, and degradation, are also analyzed. In addition, the therapeutic effects of *Aloe vera*-based hydrogels on wound healing, as well as the reduction of inflammation, antimicrobial activity, and tissue regeneration, are highlighted. The incorporation of various therapeutic agents, such as antimicrobial and anti-inflammatory ones, into *Aloe vera*-based hydrogels is reviewed in detail. Furthermore, challenges and future prospects of *Aloe vera*-based hydrogels for wound dressing applications are considered. This review provides valuable information on the current status of *Aloe vera*-based hydrogels for the delivery of therapeutic agents in wound dressings and highlights their potential to improve wound healing outcomes.

Keywords: *Aloe vera*; hydrogels; wound healing; wound dressings; therapeutic agents

Citation: Chelu, M.; Musuc, A.M.; Popa, M.; Calderon Moreno, J. *Aloe vera*-Based Hydrogels for Wound Healing: Properties and Therapeutic Effects. *Gels* **2023**, *9*, 539. https://doi.org/10.3390/gels9070539

Academic Editor: Kunpeng Cui

Received: 15 June 2023
Revised: 26 June 2023
Accepted: 30 June 2023
Published: 3 July 2023

1. Introduction

Medicinal plants have been used since ancient times. It has even been estimated that nearly 80% of the world's population relies on traditional herbal medicine for primary health care [1]. Herbal therapies have recently shown an upward trend for a variety of ailments in parallel with the development of modern medicine. Many new drugs and treatments derived from medicinal plants are being developed and prescribed today. According to the World Health Organization (WHO), almost 25% of modern medicines are derived from plants that were used in traditional medicine. Additionally, many drugs are synthetic analogues obtained from model compounds isolated from plants [2]. This review summarizes the preparation, structural features, and properties of *Aloe vera*-based hydrogels and recent advances in *Aloe vera*-based hydrogels for wound dressing applications.

Aloe vera (*AV*) belongs to the Liliaceae family, of which the best-known species is *Aloe Barbadensis Miller*, and has been used for thousands of years in traditional medicine [3]. Being one of the most famous medicinal plants in the world, it is considered a miracle gift of nature due to its many therapeutic benefits [4].

References to the medicinal use of the *AV* plant date back 4000 years, but the first inscriptions mentioning the plant were found on a collection of Sumerian clay tablets from 2100 BC [5,6]. Additionally, in the Egyptian Ebers Papyrus of 1552 BC, the plant was mentioned as a laxative [5]. The first populations to identify and appreciate the healing properties of *Aloe* plants were the Egyptians, Romans, Greeks, Arabs, and Indians [7]. There were many legends, which said that the *Aloe* plant was used by the Egyptian Queen Nefertiti (1353 BC), considered "the most beautiful woman who ever lived", and by Queen

Cleopatra VII (69–30 BC) in their usual beauty treatments, but also as medicine. According to legend, in 333 BC, Aristotle advised Alexander the Great to capture the island of Socotra in the Indian Ocean for its famous *AV* plantations, which were needed to treat his wounded soldiers [5].

Starting in the 1950s, *Aloe* leaf gel began to be industrialized and commercialized. The global *AV* extracts market size is projected to grow from USD 2.65 billion in 2023 to USD 4.55 billion by 2030 at a compound annual growth rate of 8.0% during the forecast period [8]. The market demand for *AV* products is now widespread globally and has been steadily increasing, driven by consumer awareness of its various health benefits associated with medicinal and cosmetic properties and the growing preference for natural and organic herbal products, including (i) health and wellness products such as dietary supplements, herbal remedies, and functional beverages, for their potential health benefits such as aiding digestion and supporting the immune system; (ii) skin care products and cosmetics, e.g., lotions, creams, gels, and face masks, due to its soothing and moisturizing properties; (iii) pharmaceuticals: *AV* extracts are used in the production of ointments, creams, and oral medications for burns, wounds, psoriasis, and gastrointestinal disorders; (iv) agriculture and farming: *AV* is used in soil improvement and as a natural fertilizer. Gel-based pharmaceutical and skin care products account for approximately 80% of the market size.

AV is a shrubby plant with fleshy green leaves, conical and filled with a clear, viscous gel. It grows perennially in many areas of the globe [9,10]. *AV* gel has been used for curative and therapeutic purposes, and numerous bioactive components have been discovered in the inner gel. It was believed that the special biological activities of *AV* gel are due to the synergistic effect of the multitude of biochemical components present in its composition. It exhibits numerous biological benefits such as astringent, anti-diabetic, anti-ulcer, antibacterial, anti-inflammatory, antimicrobial, antioxidant, hemostatic, and anti-carcinogenic properties and also effectiveness in treating gastrointestinal disorders [11,12].

AV is a plant often cultivated in people's homes around the world as a natural compound intended for widespread use by both adults and children and recognized in clinical practice as a tool for wound healing [13–17]. *AV* gel has been particularly associated with the treatment of skin injuries such as cuts, burns, frostbite, radiation, and electrical injuries [18–21].

Depending on the evolution of the recovery process, wounds can be classified into two broad categories: acute and chronic wounds [22,23]. Acute wounds are injuries with complete healing within up to 12 weeks [24,25]. In contrast, chronic wounds take more than three months to heal. This may be due to repeated tissue damage or associated physiological conditions such as poor primary treatment, infections, diabetes, malignancy, severe injury, or a compromised immune system [26–28].

Wound care is necessary to prevent or mitigate possible infection, the most common complication for compromised skin. Dressings are mainly applied to prevent microorganisms from reaching the wound, to keep the wounded area hydrated, and to absorb exudates [29,30]. Traditionally, sterile gauze dressings have been widely applied to wounded areas [31–33]. However, they are not always effective because they do not provide hydration, and sometimes their removal becomes painful because they stick to the wounds. Additionally, to prevent the development of infections, different creams and ointments with antimicrobial action are used, which must be removed and reapplied constantly [34–36]. Modern dressings are adapted to different types of injuries and patient typologies to avoid infection and promote scarless healing. They are designed to provide hydration and interact with wounds by releasing bioactive molecules to accelerate the wound-healing process [37,38].

With the adaptation of synthesis methods and the evolution towards ecological chemistry, it is absolutely necessary to use non-toxic solvents for the production of dressings. Thus, dressings such as dermal patches, foams, hydrogels, hydrocolloids, nanoparticles, nanofibers, films, membranes, and three-dimensional (3D) printed scaffolds can be obtained with various bio-based adaptive features [39–44].

Hydrogels are a class of materials often applied in the soft tissue engineering of skin, blood vessels, and muscles [45,46]. With a three-dimensional porous structure, hydrogels are formed by physically or chemically crosslinked bonds of hydrophilic polymers [47–49]. They are also insoluble and have an exceptional capacity to absorb wound exudates and allow oxygen diffusion to accelerate healing [25,50–52]. They can retain several times more water compared to their dry weight and maintain good hydration in the injured area [53,54]. Due to these unique physical properties, hydrogels are the most suitable dressings to cover skin wounds [55–57]. Hydrogel design and development can provide a platform for the encapsulation of cells, antibacterial agents, or bioactive factors. As dressings, hydrogels must be biocompatible, have suitable physical and mechanical properties, and ensure cell proliferation in wounds [58–60].

Throughout history, humans have used native *AV* gel, which has been shown to have exceptional properties in the wound-healing process and in promoting tissue regeneration. The huge potential of *AV* gel is due to the advantages of the biocompatible, bioavailable, and biodegradable matrix, as well as the ability to heal wounds easily and effectively without leaving scars [52,61,62]. Native *AV* gel not only releases bioactive components but also moisturizes the wound to increase flexibility, acts as a barrier against foreign microbes, and helps reduce pain at nerve endings [21].

2. Phytochemical Constituents of *Aloe vera*

Numerous studies have demonstrated the exceptional healing potential of *AV* and identified the many bioactive compounds responsible for wound healing. The structure of the *Aloe* leaf is configured in the form of three layers. The inner layer consists of a transparent gel containing 99% water and 1% solid matter that compresses over 75 different compounds (such as glucomannans, amino acids, lipids, sterols, and vitamins), the middle layer is a bitter latex in the form of yellow juice rich in glycosides and anthraquinones, and the outer layer is a thick cortex that produces carbohydrates and proteins (Figure 1) [63–67].

Figure 1. Raw morphology of the *Aloe vera* plant.

Depending on the species, the influence of climatic conditions, and the diversity of the ecosystem to which they belong, the phytochemical constituents can be different in *AV* plants. Harvested from the inside of the leaves of the *AV* plant, the gel is a gelatinous substance that contains a complex variety of several bioactive compounds, and the analysis of the dry matter of the dry *AV* gel showed that it mainly contains polysaccharides (approx.

55%), sugars (approx. 17%), minerals (approx. 16%), proteins (approx. 7%), lipids (4%), and phenolic compounds (approx. 1%) [4,68–72]. One of the most important compounds of the gel is acemannan, which is used in many pharmacological and biological applications in medical and industrial fields, such as dentistry [73], metabolic disorders [74], cardiovascular diseases [75], and tumor diseases [76]. It has also been used for wound treatment [77] and drug delivery [78,79]. Other constituents, such as amino acids, are building blocks for body and muscle proteins; sugars control cholesterol levels, proper digestion, liver function, and help strengthen bones. Anthraquinones have an antiviral effect, enzymes catalyze the biochemical reactions, inorganic compounds have a role in the proper functioning of several enzymes in various metabolic pathways, vitamins have a strong antioxidant action in neutralizing free radicals, proteins have an antitumor effect, and hormones and sterols promote wound healing.

It is believed that the power to adjust the various biological and therapeutic implications of *AV* gel is due to the synergistic effect of all the active phytochemical components. This unique composition enabled the gel to harmoniously integrate into human tissues, promoting natural healing and regeneration processes. Applied topically to a wound, *AV* gel acts gently but as a potent antimicrobial and anti-inflammatory agent, inhibiting bacterial growth and reducing inflammation [21,80–84]. Table 1 summarizes the main biocomponents of *AV*. Additionally, the active compounds of the gel stimulate the production of new cells and collagen, which is an essential protein in the process of tissue regeneration (Figure 2) [68,85]. Thus, wounds treated with *AV* gel heal faster and without leaving unsightly scars.

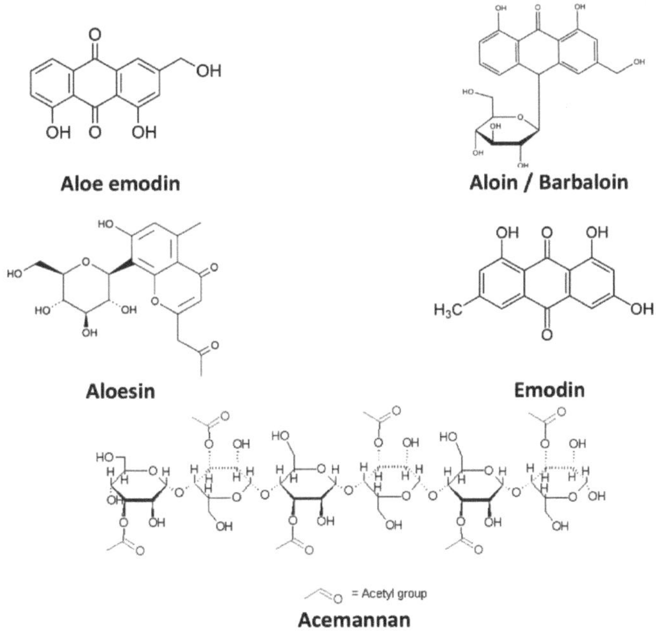

Figure 2. Chemical structure of the main biocomponents isolated from *Aloe vera* [68].

Table 1. Compounds found in *Aloe vera* [64].

Type	Compounds
Anthraquinones/anthrones	Aloe-emodin, aloetic-acid, anthranol, aloin A and B (collectively known as barbaloin) isobarbaloin, emodin, ester of cinnamic acid
Carbohydrates	Pure mannan, acetylated mannan, acetylated glucomannan, glucogalactomannan, galactan, pectic substance, arabinogalactan, galactoglucoarabinomannan, galactogalacturan, xylan, cellulose, acemannan
Enzymes	Alkaline phosphatase, amylase, carboxypeptidase, carboxylase, catalase, cyclooxidase, phosphoenolpyruvate, cyclooxygenase, superoxide dismutase, lipase, oxidase
Inorganic compounds	Calcium, chlorine, phosphorous, chromium, copper, magnesium, iron, manganese, potassium, sodium, zinc
Non-essential and essential amino acids	Alanine, arginine, aspartic acid, glutamic acid, glycine, histidine, hydroxyproline, isoleucine, leucine, lysine, methionine, proline, threonine, tyrosine, valine, phenylalanine
Proteins	Lectins, lectin-like substance
Saccharides	Mannose, glucose, L-rhamnose, aldopentose,
Vitamins	B1, B2, B6, C, β-carotene, choline, folic acid, α-tocopherol
Miscellaneous	Arachidonic acid, γ-linolenic acid, potassium sorbate, steroids (campesterol, cholesterol, β-sitosterol), triglycerides, triterpenoid, gibberellin, lignins, salicylic acid, uric acid

3. Preparation of *Aloe vera* Hydrogels

AV gel can serve as a natural and biocompatible matrix for hydrogel. It can be obtained by extracting the gel from mature *AV* leaves that are healthy and free from any damage or discoloration, removal of the yellow latex layer, which can be irritant, and processing the clear gel in the inner leaf to remove any impurities by washing with distilled water or ethanol. After purification and excess water draining (a concentration of 1–10% (w/v) is typically used for hydrogel formulations), the gel can be mixed with a cross-linking agent, such as a suitable polymer, considering factors such as gelation time, biocompatibility, and stability of the cross-linked hydrogel, to form a hydrogel. Finally, the gel is washed with distilled water to remove any unreacted cross-linking agent or by-products and stored refrigerated in a moisture-sealed container to maintain its moisture content (Figure 3). In Figure 4, the procedure for the *AV* hydrogel network preparation for its use in regenerative medicine is represented [86,87].

The specific procedure for preparing *AV*-based hydrogels can vary depending on the desired application and the chosen cross-linking method. It is essential to follow good laboratory practices and refer to relevant literature or established protocols to ensure the reproducibility and quality of the hydrogel preparation. It is worth mentioning that the incorporation of therapeutic agents, such as antimicrobial and anti-inflammatory agents, into *AV*-based hydrogels can enhance their potential for wound healing and other biomedical applications. By combining *AV* medicinal properties and wound healing effects with the controlled release capabilities of hydrogels, it is possible to develop advanced biomaterials with improved therapeutic outcomes. Therapeutic agents can be added to the *AV* gel solution before or during the cross-linking process. This can be achieved by dissolving the agents in a suitable solvent and then mixing them with the gel solution. The concentration of the agents can be varied to control the release rate and dosage. Additionally, the incorporation of therapeutic agents can be attained by the selection of appropriate therapeutic agents with desired antimicrobial and anti-inflammatory effects based on the specific application. Examples of antimicrobial agents include silver nanoparticles [88], antibiotics [89], or natural antimicrobial compounds [90], while anti-inflammatory agents may include corticosteroids [91] or non-steroidal anti-inflammatory drugs (NSAIDs) [92].

Figure 3. Scheme of the preparation of *Aloe vera* gel.

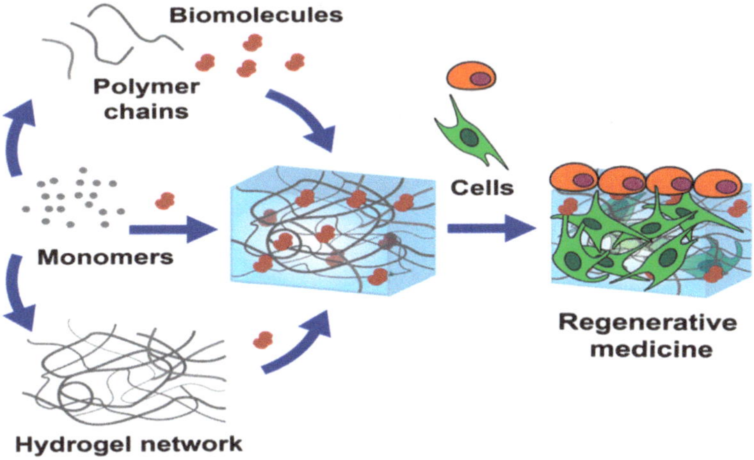

Figure 4. Graphical representation of *Aloe vera* hydrogel network preparation [86].

The cross-linking and gelation of *AV* hydrogels are crucial steps in the preparation process to convert the AV gel solution into a solid hydrogel matrix. Cross-linking is the process of creating covalent or physical bonds between polymer chains, resulting in a three-dimensional network that gives the hydrogel its structural stability and enhanced mechanical properties. Gelation refers to the transformation of the liquid gel solution into a solid gel form. The gelation process involves mixing the *AV* gel solution with an appropriate concentration of cross-linking agent and allowing it to react for a specific period. The *AV* gel solution containing therapeutic agents can be cross-linked using a suitable method, such as chemical cross-linking or physical cross-linking. Cross-linking agents are substances that promote the formation of covalent bonds between polymer chains, resulting in a three-dimensional network structure. This network improves the gel's strength, elasticity, and resistance to dissolution in aqueous environments, making it suitable for wound healing applications. Various cross-linking mechanisms and agents can be utilized for *AV*-based hydrogels. Chemical cross-linking may involve the addition of a cross-linking agent that reacts with hydroxyl groups in *AV* to form covalent bonds, leading to gelation; while physical cross-linking can be achieved through temperature, pH-incorporating temperature, or pH-responsive polymers, the hydrogel forms as the polymer chains undergo a conformational change by simply heating the gel solution to a specific temperature or adjusting the pH. Certain polymers, such as alginate, can undergo ion-induced gelation in the presence of divalent cations such as calcium ions. Calcium chloride ($CaCl_2$) is commonly used to initiate gelation in *Aloe vera*-alginate composite hydrogels. The gelation occurs as the calcium ions form ionic cross-links with the alginate chains [93]. A few commonly employed cross-linking methods are: (i) temperature-induced gelation: *AV* polymers can undergo gelation when the temperature is raised above a critical point, forming a physical cross-linked network; (ii) ionic gelation: addition of multivalent cations, such as calcium ions (Ca^{2+}), can induce gelation by creating ionic interactions between the *AV* polysaccharides; (iii) natural agents such as glutaraldehyde, genipin, and tannic acid can be used to chemically cross-link *AV* hydrogels, these agents react with the functional groups present in the polymer chains, forming stable covalent bonds; (iv) carbodiimides such as 1-ethyl-3-(3-dimethylaminopropyl) carbodiimide (EDC), can facilitate the formation of amide bonds between carboxylic acid groups of *AV* polymers and amine groups from other molecules, resulting in cross-linking; (v) radiation-induced cross-linking: hydrogels can be cross-linked using ionizing radiation, such as gamma rays or electron beams, these high-energy radiations cause the formation of free radicals within the polymer chains, leading to cross-linking. These cross-linking mechanisms and agents help improve the mechanical integrity, swelling behavior, and biocompatibility of *AV*-based hydrogels used in wound dressings. They promote the stability of the hydrogel structure, prevent rapid dissolution in contact with wound exudate, and ensure the sustained release of beneficial components from *AV* for wound healing purposes. It is important to note that the specific choice of cross-linking agent and method may depend on factors such as desired properties, safety considerations, and compatibility with the wound healing environment. During cross-linking and gelation, it is important to control parameters such as temperature, pH, and reaction time to achieve the desired gel properties. The gelation time can be influenced by factors such as the concentration of cross-linking agents, AV gel concentration, and the specific method used. It is crucial to optimize these parameters to obtain hydrogels with desirable properties, such as mechanical strength, swelling behavior, and drug release characteristics. After gelation, it is common to wash the hydrogel to remove any unreacted cross-linking agents or by-products. The resulting *AV* hydrogel can be characterized and evaluated for its physical, chemical, and biological properties, such as gelation time, swelling behavior, mechanical strength, and drug release profile. In vitro and in vivo studies can be conducted to assess the antimicrobial and anti-inflammatory efficacy of the hydrogel, as well as its biocompatibility, to ensure its suitability for various applications, including wound healing, drug delivery, and tissue engineering. Figure 5 presents a schematic illustration of the synthesis and characterization of three composite hydrogels with different concentrations of *AV*, 5%, 10%, and 20% (w/v), and the assessment of their properties [94]. The natural polymer-based hydrogels with high *AV* content, from 38% to 71% by weight in dry gel, demonstrated improved pharmacotechnical properties, including swelling ratio, spreadability, elasticity, and tensile strength. The hydrogel with *AV* content of 10% (w/v) in solution and 55% by weight in dry gel exhibited the highest strength, elasticity, and absorption capacity and also a slightly higher spreadability, indicating it for application in wound care [94].

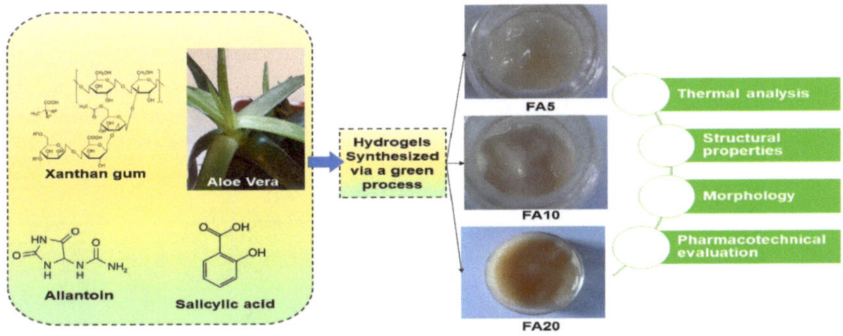

Figure 5. Synthesis, optical imaging of three composite hydrogels with different concentrations of *Aloe vera* (5%, 10%, and 20% *w/v*), and evaluation of their properties [94].

4. Biological and Pharmacological Effects of *Aloe vera*

AV gel has multiple functions. It can be used in the food field due to its proven biological properties such as antioxidant, antiviral, antibacterial, antifungal, and antiochratoxigenic activity against *Aspergillus carbonarius*, *Aspergillus niger*, *Penicillium digitatum*, *Penicillium expansum*, and *Botrytis cinerea* [63,95–98]. It is widely used to produce gel-containing healthy drinks and juices, including sports drinks [99]. It can be a functional food in the activation of lipolysis and the prevention of metabolic changes related to obesity since the phytosterols of *Aloe* gel are effective in reducing visceral fat due to the interaction with cholesterol and also has an effect on glucose metabolism, reducing blood sugar in the experimental mouse model [100]. It acts in intestinal disorders (combats constipation) due to its laxative, anti-dysenteric, anti-hemorrhoidal, and cicatrizing properties [101–103]. Moreover, even *AV* flowers are consumed more often today, knowing that diets rich in antioxidants reduce the risks of cardiovascular diseases and cancers [104].

Additionally, *AV* gel can be used in the medical field due to its demonstrated pharmacological effects on several components of the metabolic syndrome, such as effects against dyslipidemia, hyperglycemia, hypertension, and obesity [105]. Numerous studies have highlighted the beneficial anti-inflammatory, anti-diabetic, immunomodulatory, and anticancer (neoplastic disease) capacity [106–108].

At the same time, it has been studied for its active capabilities, such as hepatoprotective, anti-ulcer, anti-arthritic, and anti-rheumatic properties [109–111]. Many investigations have shown that the dental uses of *AV* are multiple, with a positive impact on the oral area [112–114]. In the case of broken, avulsed teeth, the extract (50%) of *AV* determined the increase in the cell viability of the stem cells in the dental pulp. This result is due to polysaccharides and especially acemannan, which have a positive effect on the growth factor, the expressions of specific osteogenic genes, and DNA synthesis [115,116].

AV has a crucial contribution in reducing pain, combating inflammation, moisturizing the wound, improving the quantitative and qualitative composition of collagen, and improving the migration of neighboring epithelial cells of the wound [117]. *AV* has valuable pharmaceutical properties both through the contained gel and the whole leaf extract, which include the possibility of co-administration of bioavailable vitamins to humans. In a study on human subjects, *Aloe* was found to increase the absorption of both vitamins C and E through a slower absorption mechanism, and the vitamins last longer in plasma with *Aloe*. *Aloe* is said to be the only supplement known to improve the absorption of both vitamins and should be considered a true supplement [118]. Figure 6 presents a graphical representation of the interrelationship between the properties and composition of *AV*.

Figure 6. Graphical representation of the correlation between properties and composition of *Aloe vera*.

The versatile nature of *AV* gel has significant potential in the field of pharmaceutical applications, particularly in improving the absorption capabilities of poorly absorbed orally administered drugs. Different formulations can encapsulate poorly absorbed drugs, while *AV* gel acts as a stabilizing and enhancing agent [119–121]. Due to its outstanding efficacy and compatibility with different drug carriers, the use of *AV* can be further expanded in potential applications and provides a flexible platform for optimizing oral drug delivery.

The release of therapeutic agents from *AV*-based hydrogels can occur through several mechanisms, including diffusion, swelling, and degradation of the hydrogel matrix. These mechanisms play a crucial role in controlling the release rate and duration of the therapeutic agents. Here is an overview of these mechanisms:

Diffusion-controlled release: Diffusion is the most common mechanism for the release of therapeutic agents from hydrogels. The hydrogel matrix acts as a barrier, and therapeutic agents diffuse through the gel network. The release rate is governed by the concentration gradient between the hydrogel and the surrounding medium. The diffusion coefficient of the therapeutic agent in the hydrogel matrix, as well as the pore size and structure of the hydrogel, influence the release kinetics. Factors such as the molecular weight and solubility of the therapeutic agent also affect diffusion-controlled release [122].

Swelling-controlled release: *AV*-based hydrogels have the ability to absorb water and swell, affecting the release of therapeutic agents. When the hydrogel comes into contact with an aqueous medium, it absorbs water and swells, leading to an expansion of the gel network. The swelling of the hydrogel creates channels or pores, facilitating the release of therapeutic agents. The release rate depends on the degree of swelling, which can be influenced by factors such as hydrogel composition, cross-linking density, and environmental conditions (e.g., pH and temperature) [123].

Degradation-controlled release: Some *AV*-based hydrogels can undergo controlled degradation over time. The hydrogel matrix degrades through processes such as hydrolysis, enzymatic degradation, or biodegradation, leading to the release of therapeutic agents. The degradation rate is influenced by factors such as the composition of the hydrogel, cross-linking density, the presence of enzymes or catalysts, and the physicochemical environment. As the hydrogel degrades, the therapeutic agents are gradually released into the surrounding medium [124,125].

These release mechanisms can occur individually or in combination, depending on the specific formulation and properties of the *AV*-based hydrogel, as well as the characteristics of the therapeutic agents. The choice of cross-linking agents, gel composition, and hydrogel architecture can be tailored to optimize the release profile, achieving sustained or controlled release over a desired period. The release of kinetics can also be influenced by external factors such as temperature, pH, and mechanical forces. Additionally, the interactions between the therapeutic agents and the hydrogel matrix, such as electrostatic or chemical interactions, can also impact the release behavior. Therefore, it is essential to carefully design and characterize *AV*-based hydrogels to achieve the desired release profile for specific therapeutic applications.

AV-based formulations have both inhibitory and stimulatory properties that can influence inflammatory processes and wound healing. Its inhibitory system refers to its capacity to reduce inflammation and exhibit anti-inflammatory activity. On the other hand, its stimulatory system refers to its power to promote wound healing. Together, these dual systems allow *AV* to modulate the complex interplay between wound healing and inflammation beneficially. Both the native gel and hydrogels based on *AV* showed beneficial effects and proved effective in different applications, in oral and topical therapies. They accelerate the rate of wound closure and skin healing and alleviate mucocutaneous problems, including gingivitis. As a natural medicine, it is used in oral mouthwashes, toothpaste, submucosal fibrosis, vaginal atrophy in menopausal women, and mucosal lesions induced by chemotherapy and radiotherapy or in veterinary practice. Here, we highlight some main beneficial effects of *AV* hydrogels in wound healing.

4.1. Reduction of Inflammation

Psoriasis is an immune disease, provoked by an unclear cause, which is characterized by inflammation caused by the dysfunction of the immune system and is manifested by an itchy rash, most commonly on the knees, elbows, trunk, and scalp. This disease can cause inflammation in the body and can also affect other organs or tissues in the body. Worldwide, approximately 125 million people suffer from this disease. Plaque psoriasis is associated with several comorbidities, including inflammatory arthritis, cardiometabolic disease, and depression. The American Academy of Dermatology—National Psoriasis Foundation guidelines recommend biologics as alternatives for the first-line treatment of moderate to severe plaque psoriasis due to their therapeutic efficacy and acceptable safety profiles [126]. *AV* has often been used for topical applications in the treatment of psoriasis. A study on rats, in which hydrogels based on *AV* mucilage were developed and prepared with 80% w/w of gel for topical applications, demonstrated good efficiency in controlling hyperkeratinization, showing a 61% reduction of the stratum corneum on the tested animals. The results confirmed the keratolytic action of *AV* hydrogel, which can be used to treat psoriasis. The effect of *AV* leaf extract has been attributed to polysaccharides, rich in glucomannan and acemannan, pectic compounds, cellulose, and hemicelluloses, which determine most of the plant's therapeutic properties [127]. The antipsoriatic properties of *AV* have been combined with the healing activity of Natural Rubber Latex to produce new economic occlusive dressings recommended for the treatment of psoriasis symptoms. In total, 58.8% of loaded *AV*, present on the surface and inside the dressing, was released after 4 days. An in vitro study on human dermal fibroblasts and sheep blood, respectively, confirmed the biocompatibility and hemocompatibility of the new dressings, the preservation of approximately 70% of the free antioxidant properties of *AV*, and the total content of phenolic compounds 2.31 times higher in these dressings compared to natural rubber latex without *AV* [128].

4.2. Prevention of Bacterial Infection

Chitosan and *AV* films encapsulating thymol were prepared to be used in preventing the possibility of bacterial infection and showed a high thymol encapsulation efficiency of 95.3% with good dispersibility. Test results against various pathogenic microbes such as

Bacillus, *Staphylococcus*, *Escherichia*, *Pseudomonas*, *Klebsiella*, and *Candida* showed that the films were effective against bacterial colonization in a thymol concentration-dependent manner. The addition of *AV* increased the water absorption of the films, which is one of the primary factors of healthy wound healing and helped by improving the antioxidant activity and in vitro release efficiency of thymol [129]. New polymer composite films based on polyvinyl alcohol and *AV* have been prepared for wound healing and prevention of surgical wound infections. Films tested for antibacterial and antifungal activity against *E. coli*, *P. aeruginosa*, *Aspergillus flavus*, and *Aspergillus tubingensis* showed antimicrobial activity against all strains; the lowest concentration of *AV* (5%) showed the highest activity against all strains. Sutures of wounds covered with films based on polyvinyl alcohol and *AV* showed that the new composites have antibacterial effects and the potential to be used in the prevention of infections at the surgical site and can be used for wound healing purposes [130]. Films based on alginate, *AV* gel, honey, and cellulose nanocrystals can be used for applications as antibacterial dressings. The morphological, swelling, mechanical, and biological properties of the films prepared and tested against the Gram-negative organisms *Salmonella typhi*, *Klebsiella pneumoniae*, *Escherichia coli*, and the Gram-positive organism *Staphylococcus aureus* were estimated. The films showed superior biocompatibility, good mechanical properties, and excellent antibacterial capabilities [131]. Blended nanofiber membranes for new types of antibacterial wound dressings were made based on polycaprolactone/chitosan/*Aloe vera* (PCL/CS/AV) nanofiber (NFM) by electrospinning. The characterizations and tests carried out showed that the addition of *AV* increased the hydrophilicity and the pore size of the membranes and led to the improvement of the antibacterial performance against *Streptococcus aureus* and *E. coli* and the biocompatibility in 5 days. The membranes produced were proposed as suitable for short-term dressing or acute wounds (1–4 days) [132]. Nanofiber membranes were developed based on natural, biocompatible, and biodegradable composites from *AV* extract, pullulan, chitosan, and citric acid, through Forcespinning® technology. The morpho-structural characterization and thermogravimetric analysis of the membranes indicated their good properties, as well as good water absorption capacities and synergistic antibacterial activity against *Escherichia coli*, which promoted cell attachment and growth. Due to their porous structure and large surface area, the membranes can be recommended as potential dressing applications due to their ability to absorb excessive blood and exudates, their thermal stability, and the protection they offer against infection [133]. Novel sodium alginate/poly(vinyl alcohol) (SA/PVA) hydrogel dressing films enriched with *AV* were produced by a simple method. The influence of different amounts (5, 10, 15, 20, and 25%, v/v) of *AV* solution on the chemical structure and properties of sodium alginate/poly(vinyl alcohol) hydrogel films was studied. The structural, morphological, mechanical, and thermal characterization confirmed that rigid and thermally stable three-dimensional structures were obtained. The results regarding the release profile of the polysaccharides from the hydrogel matrix showed that the active substance was released in a prolonged, gradual manner, even for a week. It was shown that the presence of *AV* within the cross-linked polymer network improved the active substance delivery properties of the hydrogel films. At the same time, the cytotoxicity of the materials was studied, and the results indicated good adhesion properties and a lack of toxicity. In vitro experiments on normal human dermal fibroblasts showed very good cell attachment to *AV* hydrogel discs, which promoted cell spreading and proliferation. As such, SA/PVA/*AV* sustained-release *AV* films have been proposed for applications such as interactive wound dressings [134]. Recent studies have concluded that *AV* gel is an effective antibacterial agent to prevent wound infection caused by various bacteria: *P. Aeruginosa* [135], *Campylobacter rectus*, *Provetella intermedia* [136], and *Escherichia coli* (*E. coli*) [137].

4.3. Skin Regeneration

The skin is part of the body's integumentary system and consists of the epidermis and dermis, with a subcutaneous fatty layer, the hypodermis [138,139]. It protects us against external factors and prevents bacteria and germs from entering the body and blood and causing infections [140,141]. At the same time, the skin is vulnerable and can be affected by acute or chronic wounds [142]. Wound healing is a complex physiological process, which is achieved through four explicit phases: hemostasis, inflammation, proliferation, and remodeling and involves the epidermis-containing keratinocyte, melanocyte, and Langerhans cells, dermis, including fibroblast, neutrophil, mast cell, and dermal dendritic cells, and the hypodermis, which contains mesenchymal stem cells (Figure 7) [22,143–145].

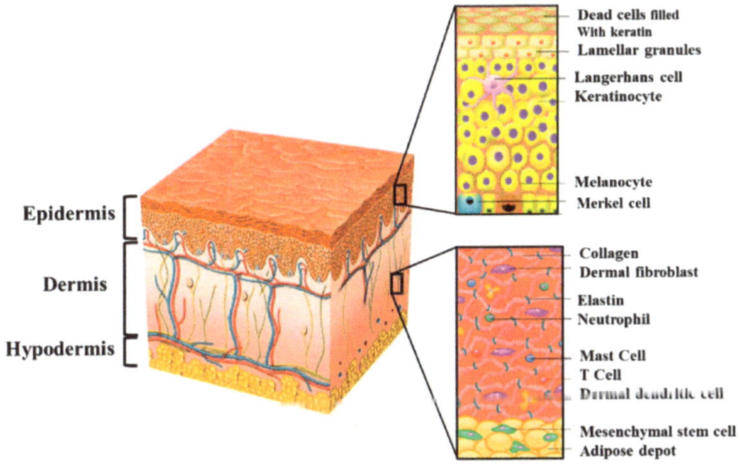

Figure 7. Structure of human skin: the epidermis (which contains keratinocytes, melanocytes, and Langerhans cells) and dermis (which includes fibroblasts, neutrophils, mast cells, and dermal dendritic cells), as well as subcutaneous hypodermis (which contains mesenchymal stem cells) [145].

The wound-healing process consists of four highly integrated and overlapping phases: (i) hemostasis, (ii) inflammation, (iii) proliferation, and (iv) tissue remodeling or resolution [146]. Figure 8 shows the main stages of the normal wound-healing process [145]. Each stage is characterized by key molecular and cellular events and is coordinated by a series of secreted factors that are recognized and released by wound response cells. Hemostasis is the first stage. It involves coagulation, which changes the blood from a liquid to a gel. The inflammation phase begins at the time of injury and lasts up to four days. As inflammatory cells undergo apoptosis, wound healing progresses to the proliferative phase. This phase begins approximately three days after the injury and overlaps with the inflammatory phase, while the tissue remodeling phase, characterized by the formation of granulation tissue, angiogenesis (formation of blood vessels), wound contraction, and the process of epithelialization, can continue for six months to one year after the injury, which leads to the formation of scar tissue. Many variables can disrupt one or more phases of this process, thereby producing inadequate or incorrect healing of skin wounds. The main elements that affect wound healing are oxygenation, infection, age, stress, diabetes, obesity, drugs, alcoholism, smoking, repeated trauma, diet, and poor blood circulation [147–149]. Infection is the most common complication for injured skin; therefore, prevention or mitigation of infection is of utmost importance.

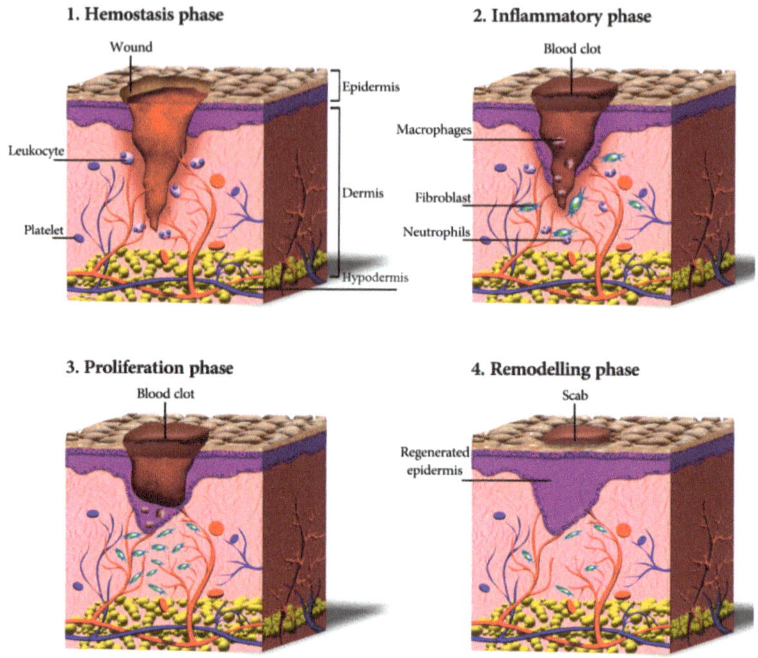

Figure 8. The phases of the wound-healing process [145].

Using an ecological preparation method, a natural, degradable, and environmentally friendly hydrogel dressing was developed using *AV* as an active ingredient. The hydrogel dressing was prepared using only natural ingredients, composed of sodium hyaluronate (SH), dopamine (DA), chitosan (CS), and *AV*, and using a natural deep eutectic green solvent (DES) as the green solvent. The newly synthesized hydrogel showed good cytocompatibility tested on NIH-3T3 fibroblast cells and antibacterial properties against both Gram-positive (*S. aureus*) and Gram-negative (*E. coli*). Additionally, in a study on mice, the hydrogel promoted the regeneration of skin tissue and healed the skin wound after surgery within 12 days. The authors concluded that the newly prepared hydrogel, which is natural, degradable, and ecological and uses *AV* as an active ingredient, shows great potential in wound healing applications [150]. A study on the emergency treatment of vaginal tissue by local application of *AV* and alginate hydrogel for the release of mesenchymal stem cells derived from the maternal endometrium with the aim of promoting maternal injury relief and early healing was carried out in a simulated injury model at birth. It was observed that in the absence of therapy, fibrotic healing can occur in many cases. Local injection of hydrogel-containing mesenchymal cells significantly improves smooth muscle and elastin content, as well as decreases tissue stiffness after 6 weeks. The findings of the study highlighted that immediate treatment of severe vaginal birth trauma with therapeutic mesenchymal stem cells delivered in *AV* and alginate hydrogel might become a potential new treatment strategy for faster healing of birth injuries and prevention of pelvic organ prolapse (Figure 9) [151].

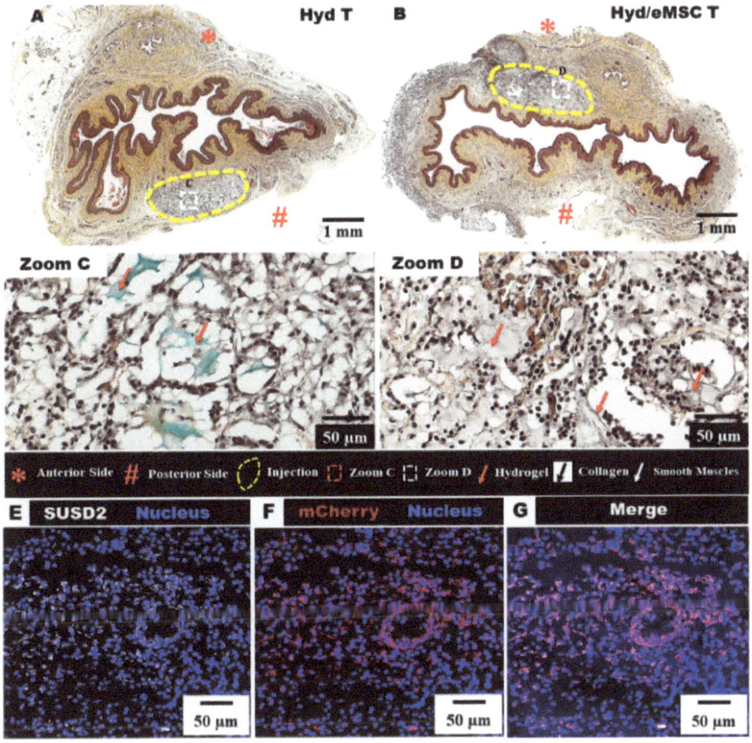

Figure 9. The retention of injected treatments with hydrogel and SUSD2 + mCherry + eMSC (**A**) Hyd T; (**B**) Hyd/eMSC T (yellow dotted lines); (**C**,**D**) red arrows—zoom area of hydrogel and black arrows—zoom area of collagen; (**E**) SUSD2, (**F**) mCherry, and (**G**) merge image of SUSD2 + mCherry in rat vaginal sections after 1 week. Reprinted with permission from ref. [151] Copyright 2023, Elsevier.

Another study explored the potential for acute and chronic wound healing using piperine as a new bioactive compound. New systems of bioactive hydrogels based on carbopol 934 containing piperine mixed with *AV* gels of different gel strengths were prepared and characterized (Figure 10). The developed formulation system was investigated in an excisional wound healing model in the rat model. The results of the in vivo study and histopathological examination showed that the piperine-containing bioactive hydrogel system compared with the piperine-free bioactive hydrogel system, leads to early and intrinsic wound healing (Figure 11). Thus, the findings of the study emphasized that the new piperine-containing bioactive hydrogel is a promising therapeutic approach for the application of wound healing [152].

Studying the influence of a commercial hydrogel formulation based on *AV* with 1,2-propanediol (propanediol) and triethanolamine (TEA) on skin wound healing was investigated in female Wistar rats. Additionally, the study aimed to show that the presence of specific additives, propanediol and triethanolamine, does not exert any negative effect on wound healing.

The results showed that the prepared hydrogel had a positive effect on inflammation, angiogenesis, and wound contraction and reduced the total healing time by 29%, with the total closure of the wound being achieved in 15 days (Figure 12). The paper highlighted the influence of the bioactive components of *AV*, related to rhamnogalacturonan and pectin-like acemannan, which improved the healing process of skin wounds [153].

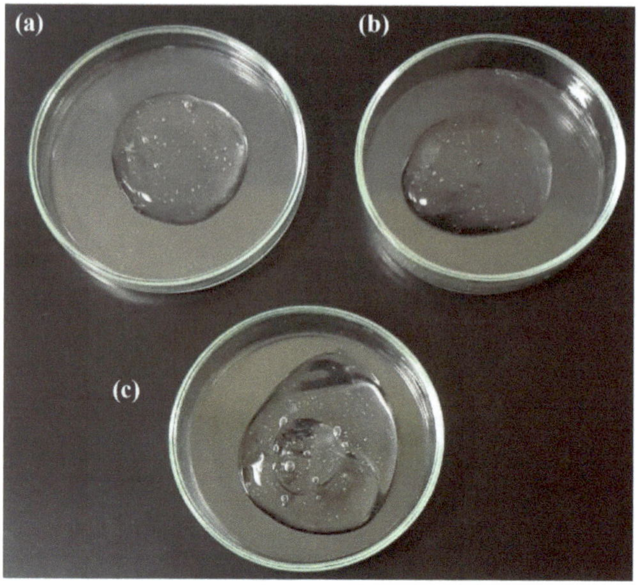

Figure 10. Images of different types of prepared hydrogel systems. (**a**) Hydrogel based only on Carbopol 934. (**b**) Placebo bioactive hydrogel (except piperine). (**c**) The bioactive hydrogel contains Carbopol 934, *Aloe vera*, and piperine [152].

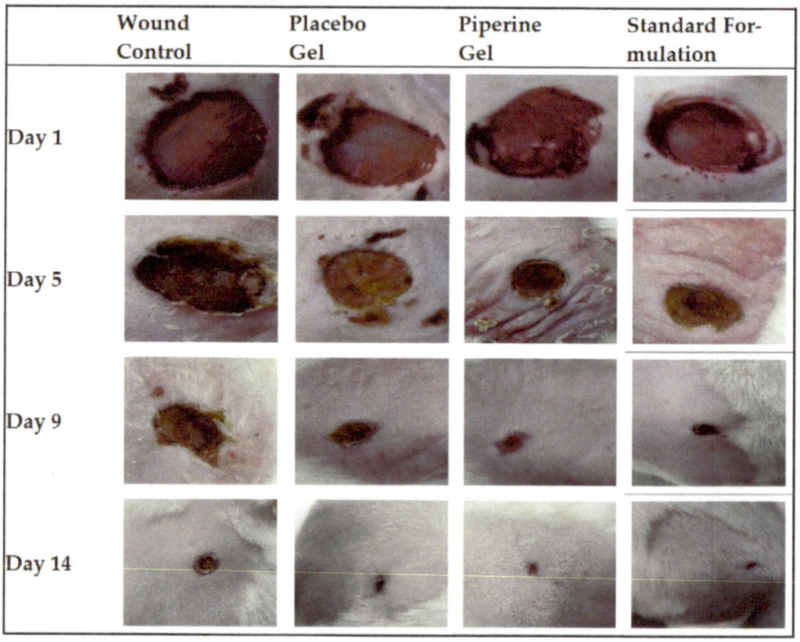

Figure 11. Phases of the wound-healing process in the untreated (control), placebo gel, piperine gel, and marketed standard formulation groups of rats. Photo for day 1, day 5, day 9, and day 14 of treatment [152].

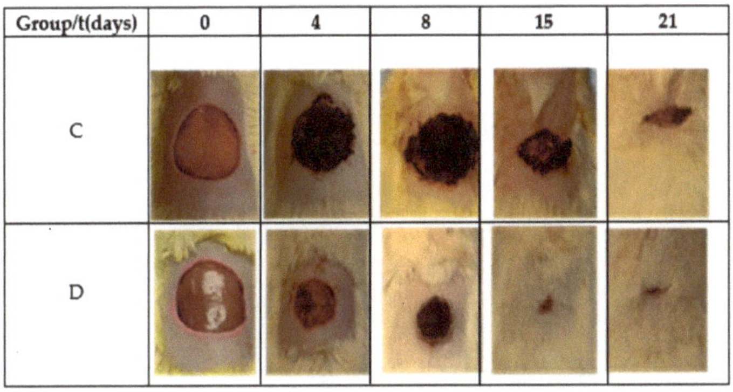

Figure 12. Images during the wound-healing process in female Wistar rats [153].

A novelty in the area of efficient ecological materials is the new system of biocompatible hydrogels based on *AV* that was prepared by a completely green synthesis method for wound healing applications (Figure 13).

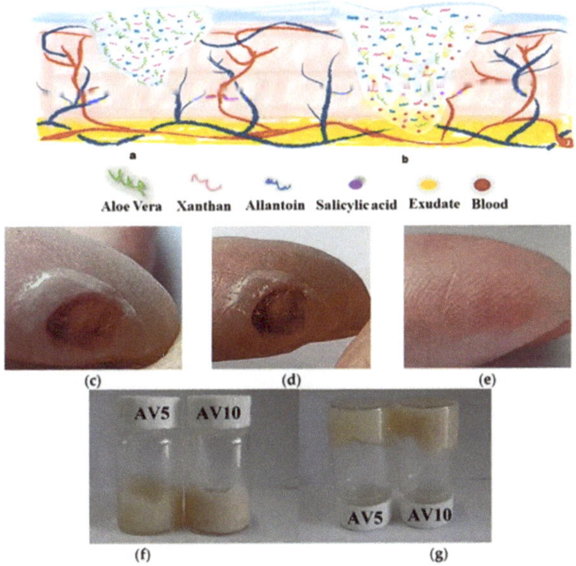

Figure 13. *Aloe vera*-based hydrogel for wound healing dressing: (**a**) dry hydrogel (**b**) wet hydrogel. The healing process (**c**) initial time; (**d**) after 5 min; (**e**) after 20 days; (**f**,**g**) inverted vial method [154].

Hydrogels with different concentrations of *AV* (5 and 10%, respectively) also contain other natural components such as salicylic acid, allantoin, and xanthan gum. The hydrogels' rheological properties, morphology, cell viability, biocompatibility, and cytotoxicity, were studied. The preliminary examinations showed that the hydrogels are very well supported on a wound, without stinging even more; they quickly penetrated the tissue and ensured good hydration of the area. Testing the antibacterial activity of the hydrogels was evaluated both on Gram-positive strains, *Staphylococcus aureus*, and on Gram-negative strains, *Pseudomonas aeruginosa*. The results showed that they have good antibacterial properties (Figure 14i). Moreover, the in vitro scratch test demonstrated the suitable ability of these "green" hydrogels to accelerate cell proliferation and migration and induce closure of a wounded area, making them suitable for wound healing applications (Figure 14ii) [154].

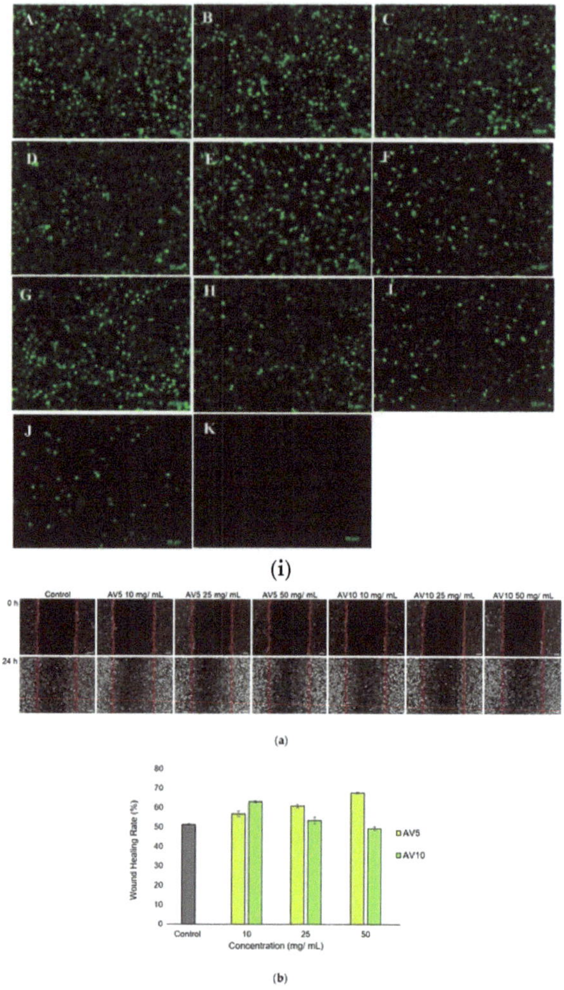

Figure 14. (i) Live/dead fluorescent images of L929, control, (**A**)—untreated and treated with AV5 (**B–F**) and AV10 (**G–K**) hydrogels at different concentrations for 48 h. (**B**,**G**)—10 mg/mL; (**C**,**H**)—25 mg/mL; (**D**,**I**)—50 mg/mL; (**E**,**J**)—75 mg/mL; (**F**,**K**)—100 mg/mL. (ii) Light microscope images (**a**) after in vitro generation of a wound for 24 h. (**b**) ImageJ analysis of wound closure percentage [154].

4.4. Healing Burns

A clinical study was conducted on 30 patients with similar types of second-degree burns in two places on different parts of the body. This research was conducted to evaluate the effectiveness of *AV* cream for partial thickness burns and to compare its results with those of silver sulfadiazine. Each patient had one burn treated randomly with topical silver sulfadiazine ointment and one treated with *Aloe* cream. The mean time to re-epithelialization and healing of partial-thickness burns was significantly shorter for the *Aloe* group at 15.9 ± 2 days versus 18.73 ± 2.65 days for the SSD group ($p < 0.0001$). Both sites were negative for microbial contamination on days 3, 7, and 13. Study results showed that *AV* cream promoted better wound healing with smaller lesions and had shorter healing times than silver sulfadiazine [155]. A similar international study was accomplished on 50 patients with second-degree burns and evaluated the effectiveness of *AV* gel compared with 1% silver sulfadiazine cream as a special dressing for the treatment of superficial and partial thickness burn wounds. The study used 98% unrefined gel from the inner leaf of the plant. Thermal burn patients bandaged with *AV* gel showed improvements compared to those bandaged with silver sulfadiazine cream in terms of early wound epithelialization, early pain relief, and cost-effectiveness of treatment management [156]. Another double-blind, randomized clinical trial in 11 patients treated once daily for 14 days compared the efficacy of herbal *AV* cream with 1% silver sulfadiazine in reducing the pain of second-degree burns. The herbal cream was prepared from *AV* gel and essential oils of *Lavandula stoechas* and *Pelargonium roseum*. In total, 56 patients were treated with herbal cream, and another 55 were treated with silver sulfadiazine 1%. Study results demonstrated that pain intensity at 14 days was significantly reduced in both groups compared to baseline ($p < 0.001$). However, a greater reduction in pain from baseline to the 7- and 14-day mark was observed in the herbal cream group ($p = 0.014$ and $p = 0.05$). One case of infection was reported in the herbal cream group; however, it cleared up with continued treatment. The findings of this clinical trial showed that the herbal cream was superior to silver sulfadiazine in relieving pain for superficial second-degree burns [157]. In an additional clinical case study, the therapeutic impact of *AV* gel on chronic skin burns in a 17-year-old patient with a rejected skin graft is presented. This is a before–after comparative study design in a case of fire burn in which initiation of *AV* gel treatment is accompanied by the promotion of wound repair. Before being treated with gel, the patient who had suffered burns on 30–40% of her body surface for 10 days had a healthy skin graft operation on her previous chest, which was rejected after 5 days. Following chronic unhealed skin lesions, the patient was treated with *AV* gel for 21 days continuously. The skin healing process began with the formation of granulation tissue and epithelization of the wounds. During the treatment, no sign of skin infection and no topical side effects of *AV* gel, such as allergic reactions and itching, were observed. This study on the impact of *AV* gel in the healing of burns can be considered a cheap and quick effect of substitution therapy instead of surgery [158].

4.5. Protection against Chemoradiation Secondary Effects in Cancer Treatment

A multicenter, randomized, double-blind, controlled trial was performed on 120 patients with head and neck cancer treated with concurrent chemoradiation. Patients received either *AV* gel or placebo gel and were assessed for adverse levels of skin toxicity with the Radiation-Induced Skin Reaction Rating Scale (RISRAS). At the 5th and 6th week of treatment, grades moderate to severe erythematous skin at values of 13.6% and 24.1% versus 27.8 and 42.6% were observed for members of the *AV* gel group and the placebo group, respectively ($p = 0.05$ for the 5th week and $p = 0.038$ for the 6th week). At week 7, in the placebo group, moderate to severe cases of wet scaling were observed in eight patients (19.0%) ($p = 0.001$), as well as a burning sensation with RISRAS scores of 3–4, representing only 11.9% of patients ($p = 0.016$). The study authors concluded that there was no prophylactic efficacy for radiation-induced dermatitis in the *AV* gel group compared with the placebo group but that topical applications of *AV* gel along with a routine skin care program from starting radiation would reduce the severity of any burning sensations, along with the incidence of erythematous, moist scaling of the skin in head and neck cancer patients receiving concurrent chemoradiation [159].

4.6. Summary of Clinical Effects of AV on Prevention and Healing of Skin Wounds

An earlier systematic study [121] concluded that *AV* helps to retain skin moisture and integrity and prevents skin ulcers due to its content of mucopolysaccharides, amino acids, zinc, and water. Furthermore, *AV* was found to be 'much more effective and less costly compared to the currently available alternative treatments' in terms of quality and speed of wound healing. Considering the tendency to promote traditional medicine as well as the rare side effects of AV, the use of this medicinal plant for the healing of skin wounds is recommended.

AV gel has been demonstrated to be active in wound healing through several reported mechanisms [31], including increased epithelial cell viability, proliferation, and migration, moisture retention [160], increased quantity and cross-linking of collagen [161], and hindering inflammation through the decrease of proinflammatory cytokines [162–166]. The various active components of *AV* include acemannan, aloesin, aloe-emodin, aloin, emodin, and glucomannan [68]. Acemannan is known to stimulate epidermal keratinocytes and the production of fibrotic cytokines [167,168]. Glucomannan, a water-soluble mucopolysaccharide, stimulates fibroblast growth factor production and the activity and proliferation of these cells, leading to the increased amount of collagen on the wound site with enhanced transversal connections [21,64,169]. Emodin emodinolin, anthraquinone derivatives found in *AV*, act as competitive inhibitors of thromboxane synthetase and have significant anti-inflammatory properties [21]. The anti-inflammatory properties of *AV* are related to the inhibition of proinflammatory cytokines [162,164,165], hindering ROS production [162,164], and blocking the signalling of JAK1-STAT1/3 [68]. The anti-inflammatory effects and increased collagen production and cross-linking promote the rearrangement of epithelial tissues [12], reducing the wounded area and accelerating the healing process [170]. Various studies have confirmed that topical *AV* creams heal first- and second-degree burns in less than half the time than standard treatment with silver sulfadiazine [21,171–173]. *AV* has an anti-erythema activity similar to that of the positive control group (i.e., hydrocortisone gel) after 6 days of treatment [174]. *AV* gel has also demonstrated potent angiogenic activity, an essential process in wound healing, attributed to angiogenic compounds such as beta-sitosterol [175,176]. Table 2 summarizes various beneficial effects of *AV* compounds for wound healing reported in clinical studies.

Table 2. Beneficial effects for wound healing of *AV* gels.

Enhanced Reported	References
Cell viability	Sholehvar et al. [115], Liu et al. [177]
Epitelial cell proliferation	Moriyama et al. [167], Hashemi et al. [170], Shanmugan et al. [178], Teplicki et al. [179]
Epitelial cell midration	Teplicki et al. [179], Negahdari et al. [180], Wahedi et al. [181], Muller et al. [182]
Moisture retention	Dal'Belo et al. [160], Hamman et al. [183]
Keratinocyte proliferation	Moriyama et al. [167]
Collagen quantity	Hekmatpou et al. [21], Rahman et al. [64], Nabipour et al. [121], Abdel Hamid et al. [169], Hashemi et al. [170], Shanmugan et al. [178]
Collagen cross-linking	Hekmatpou et al. [21], Rahman et al. [64], Abdel Hamid et al. [169], Shanmugan et al. [178]
GSH activity	Liu et al. [177]
SOD activity	Liu et al. [177]
Antioxidant enzyme activity	Anilakumar et al. [184], Hassanpour et al. [185]
Accelerated wound healing	Moriyama et al. [167], Maenthaisong et al. [171], Somboonwong et al. [173], Shanmugan et al. [178], Negahdari et al. [180], Wahedi et al. [178–181], Hormozi et al. [186], Ali et al. [187]
Growth factors production	Hashemi et al. [170], Wahedi et al. [181]
Wound closure	Curto et al. [188]
Lysosomal stabilization	Paul et al. [165], DeOliveira et al. [189]
Stimulate fibrotic cytokines	Wahedi et al. [181], Zeng et al. [190]
Angiogenesis	Moon et al. [175], Choi et al. [176]
Block the signaling of JAK1-STAT1/3	Sánchez et al. [68]

Table 2. *Cont.*

Enhanced Reported	References
Thromboxane reduction	Zeng et al. [21], Hekmatpou et al. [189]
Hindering IL-6	Ma et al. [162], Jiang et al. [164]
Hindering IL-8	Leng et al. [163], Na et al. [191]
Hindering IL-12	Ahluwalia et al. [163], Leng et al. [166]
TNF alpha levels reduced	Leng et al. [163], Jiang et al. [164], Paul et al. [165], Ahluwalia et al. [166]
Erythema reduction	Fox et al. [174], Reuter et al. [192]
Pain reduction	Hekmatpou et al. [21], Rompicherla et al. [119]
T cell proliferation suppressed	Li et al. [193]
Lipid peroxidation reduced	Liu et al. [177]
Proinflammatory cytokines reduced	Ma et al. [162], Leng et al. [163], Jiang et al. [164], Paul et al. [165], Ahluwalia et al. [166]
Type IV collagen degradation	Curto et al. [188]
ROS production hindered	Ma et al. [162], Jiang et al. [164]
Inflammation reduction	Hekmatpou et al. [21], Paul et al. [165]

Recent studies on *AV* gels with added therapeutic agents have reported the positive interaction between graphene oxide/reduced graphene oxide (GO/rGO) and *AV* hydrogels to be a strongly promising strategy for the advancement of therapeutic approaches for wound healing (Figure 15) [178].

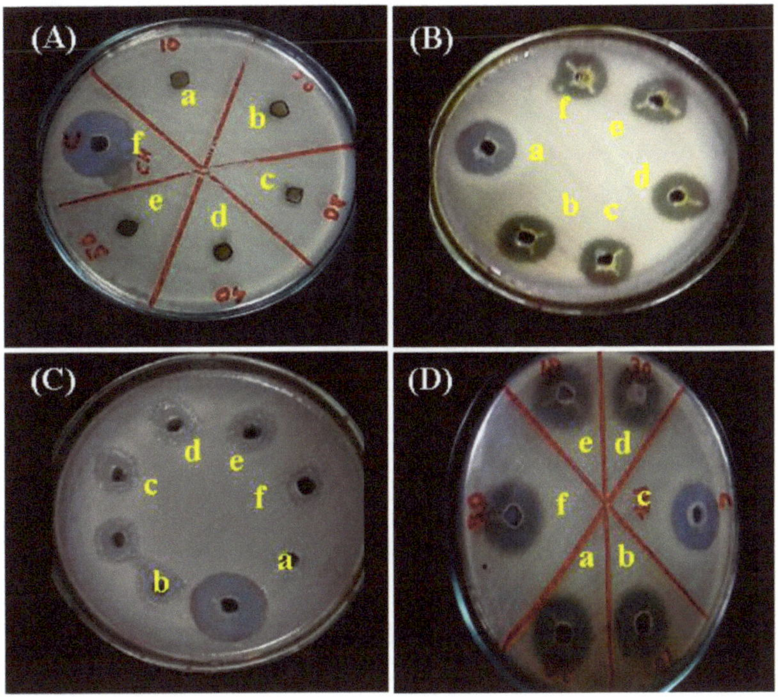

Figure 15. Agar assay (**a**) hydrogel; (**b**) hydrogel + GO; (**c**) hydrogel + rGO; (**d**) *Aloe vera* gel; (**e**) *Aloe vera* gel + GO; (**f**) *Aloe vera* gel + rGO against (**A**) *Pseudomonas aeruginosa*, (**B**) *Bacillus subtilis*, (**C**) *Staphylococcus aureus*, and (**D**) *E. coli*. [178].

Jales et al. further confirmed the great potential of *AV* mucilaginous hydrogel with a high keratolytic effect that can be used in psoriasis treatment [127]. Puliero et al. investigated the use of *AV* extracts for ocular therapeutic or preventive purposes. They demonstrated that the best lenses allowing for the high and stable release of *AV* extract to the corneal surface are those composed of ionic hydrogels [194]. Capsaicin, a powerful anti-inflammatory and analgesic agent, poorly water-soluble, was successfully incorporated into *AV* gel for topical drug delivery and to reduce skin irritation caused by capsaicin [119]. The *AV* gels softness, biocompatibility, and fast spreading or penetrating capacity are particularly useful features to encapsulate and deliver various nanoparticles with antimicrobial properties (e.g., ZnO or TiO_2) [195], drugs, cell culture, both for wound healing, and bio-sensing applications [196]. The combination of *AV* and *Rheum palmatum* root can promote the migration of human primary fibroblasts (Figure 16) [182].

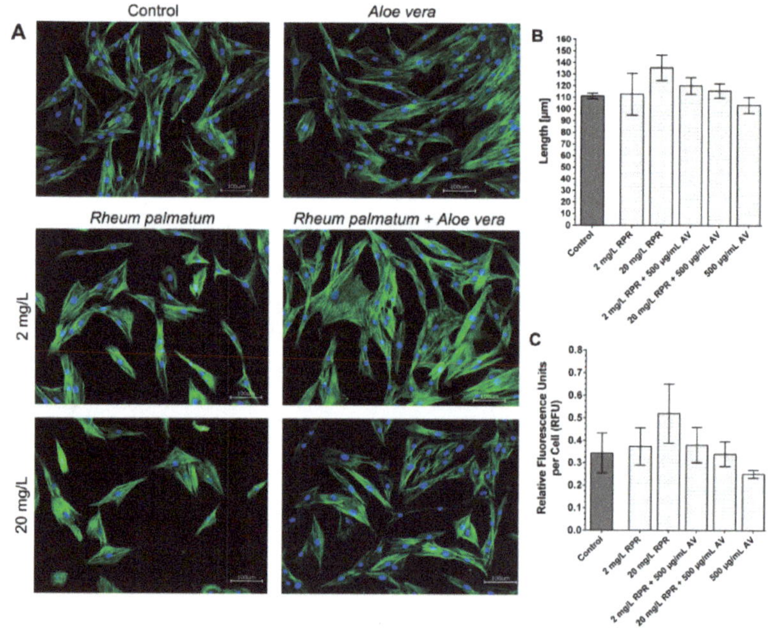

Figure 16. Actin staining of fibroblasts with phalloidin (**A**); the average length of the fibroblasts (**B**); fluorescent actin staining (**C**) [182].

None of the dressings available on the market today are fully capable of reproducing all the characteristics of native skin. An asymmetric bilayer membrane with a top dense polycaprolactone layer that provides mechanical support and a bottom porous layer of chitosan and *AV*, aiming to improve the healing process, was designed to mimic both layers of the skin [197]. The results obtained revealed the potential of these asymmetric membranes to be applied as wound dressings in the future.

5. Side Effects

No serious adverse reactions were demonstrated following the topical application of *AV* inner gel products. *AV* used in dietary supplements appears to be safe [198]. The inner gel was evaluated by the Cosmetic Ingredient Review Expert Panel as noncytotoxic [199]. However, due to the cytotoxicity, mutagenicity, and carcinogenicity of anthraquinones, it is crucial to monitor the content of these phenolic compounds in *AV* whole leaf extract and latex [200,201]. Topical and oral use of *AV* whole leaf extract in humans can cause adverse clinical effects: skin irritation, hives, cramping, and diarrhea to those who are allergic to plants in the lily family, for example, onion and tulips [202–204].

6. Conclusions and Future Prospective Studies

It is important to apply modern delivery techniques to develop affordable products based on efficacious traditional natural medicines for wound healing and to improve their therapeutic effect.

Further research is needed to ensure that these formulations reach the pharmaceutical market. Chemotherapy treatments for cancer are associated with the presence of ulcers in the oral mucosa that causes pain, bleeding, and difficulty swallowing or speaking. There is no effective standard treatment, and few studies have been published on the therapeutic effects of natural products such as *AV* to improve the local retention period.

Future treatments may arise from medicinal plants, which have fewer side effects and improved bioavailability for the wound-healing process. In addition, in the future, a great challenge is represented by the development of an intelligent treatment that presents anti-inflammatory, antimicrobial, and antioxidant cumulative properties for the treatment of all types of wounds. Furthermore, the commercialization and use in preclinical research and clinical practice of natural products used in wound healing must be increased significantly to discover the potential of these products, considered natural bioactive molecules, in the treatment and regeneration of skin tissue. Future research should be considered to find new natural bioactive compounds related to their usage in the wound-healing process and their ability to act as substitutes for existing antibiotics.

By incorporating therapeutic agents into *AV*-based hydrogels, it is possible to develop multifunctional biomaterials that provide sustained release of agents, promote wound healing, reduce inflammation, and prevent or treat microbial infections. However, it is important to note that the specific formulation and efficacy of such hydrogels may vary depending on the therapeutic agents chosen, their concentration, crosslinking method, and other factors. Extensive research and testing (rheological analysis, drug release profiles, permeability, and stability studies) are required to optimize the formulation and ensure its safety and effectiveness for clinical use and to promote human well-being worldwide.

Author Contributions: Conceptualization, M.C., A.M.M. and J.C.M.; methodology, M.P., M.C. and J.C.M.; data curation, M.C. and A.M.M.; resources, J.C.M.; writing—original draft preparation, M.C., A.M.M., M.P. and J.C.M.; writing—review and editing, A.M.M. and J.C.M.; visualization, M.P.; supervision, A.M.M. and J.C.M. All authors have read and agreed to the published version of the manuscript.

Funding: This research received no external funding.

Institutional Review Board Statement: Not applicable.

Informed Consent Statement: Not applicable.

Data Availability Statement: Not applicable.

Conflicts of Interest: The authors declare no conflict of interest.

References

1. Ekor, M. The growing use of herbal medicines: Issues relating to adverse reactions and challenges in monitoring safety. *Front. Pharmacol.* **2014**, *10*, 177. [CrossRef] [PubMed]
2. Khan, M.S.A.; Ahmad, I. *Chapter 1-Herbal Medicine: Current Trends and Future Prospects*; New Look to Phytomedicine; Khan, M.S.A., Ahmad, I., Chattopadhyay, D., Eds.; Academic Press: Cambridge, MA, USA, 2019; pp. 3–13, ISBN 9780128146194. [CrossRef]
3. Surjushe, A.; Vasani, R.; Saple, D.G. *Aloe vera*: A short review. *Indian J. Dermatol.* **2008**, *53*, 163–166. [CrossRef] [PubMed]
4. Kumar, R.; Singh, A.K.; Gupta, A.; Bishayee, A.; Pandey, A.K. Therapeutic potential of *Aloe vera*—A miracle gift of nature. *Phytomedicine* **2019**, *60*, 152996. [CrossRef] [PubMed]
5. Manvitha, K.; Bidya, B. *Aloe vera*: A wonder plant its history, cultivation and medicinal uses. *J. Pharm. Phytochem.* **2014**, *2*, 85–88.
6. Mehta, D.I. "History OF *Aloe vera*" – (A Magical Plant). *IOSR J. Humanit. Soc. Sci.* **2017**, *22*, 21–24.
7. Gao, Y.; Kuok, K.I.; Jin, Y.; Wang, R. Biomedical applications of *Aloe vera*. *Crit. Rev. Food Sci. Nutr.* **2019**, *59*, S244–S256. [CrossRef]
8. Available online: https://www.fortunebusinessinsights.com/aloe-vera-extract-market-103893 (accessed on 20 May 2023).
9. Deep, A.; Kumar, D.; Bansal, N.; Narasimhan, B.; Marwaha, R.K.; Sharma, P.C. Understanding mechanistic aspects and therapeutic potential of natural substances as anticancer agents. *Phytomed. Plus* **2023**, *3*, 100418. [CrossRef]

10. Noori, A.S.; Mageed, N.F.; Abdalameer, N.K.; Mohammed, M.K.; Mazhir, S.N.; Ali, A.H.; Jaber, N.A.; Mohammed, S.H. The histological effect of activated *Aloe vera* extract by microwave plasma on wound healing. *Chem. Phys. Lett.* **2022**, *807*, 140112. [CrossRef]
11. Kim, S.-T.; Pressman, P.; Clemens, R.; Moore, A.; Hamilton, R.; Hayes, A.W. The absence of genotoxicity of *Aloe vera* beverages: A review of the literature. *Food Chem. Toxicol.* **2023**, *174*, 113628. [CrossRef]
12. Sánchez-Machado, D.I.; López-Cervantes, J.; Sendón, R.; Sanches-Silva, A. *Aloe vera*: Ancient knowledge with new frontiers. *Trends Food Sci. Technol.* **2017**, *61*, 94–102. [CrossRef]
13. Altinkaynak, C.; Haciosmanoglu, E.; Ekremoglu, M.; Hacioglu, M.; Özdemir, N. Anti-microbial, anti-oxidant and wound healing capabilities of *Aloe vera*-incorporated hybrid nanoflowers. *J. Biosci. Bioeng.* **2023**, *135*, 321–330. [CrossRef] [PubMed]
14. Hattingh, A.; Laux, J.-P.; Willers, C.; Hamman, J.; Steyn, D.; Hamman, H. In vitro wound healing effects of combinations of *Aloe vera* gel with different extracts of Bulbine frutescens. *S. Afr. J. Bot.* **2023**, *158*, 254–264. [CrossRef]
15. Movaffagh, J.; Khatib, M.; Bazzaz, B.S.F.; Taherzadeh, Z.; Hashemi, M.; Moghaddam, A.S.; Tabatabaee, S.A.; Azizzadeh, M.; Jirofti, N. Evaluation of wound-healing efficiency of a functional Chitosan/*Aloe vera* hydrogel on the improvement of re-epithelialization in full thickness wound model of rat. *J. Tissue Viabil.* **2022**, *31*, 649–656. [CrossRef] [PubMed]
16. Abdel-Mohsen, A.M.; Frankova, J.; Abdel-Rahman, R.M.; Salem, A.A.; Sahffie, N.M.; Kubena, I.; Jancar, J. Chitosan-glucan complex hollow fibers reinforced collagen wound dressing embedded with *Aloe vera*. II. Multifunctional properties to promote cutaneous wound healing. *Int. J. Pharm.* **2020**, *582*, 119349. [CrossRef] [PubMed]
17. Ghorbani, M.; Nezhad-Mokhtari, P.; Ramazani, S. *Aloe vera*-loaded nanofibrous scaffold based on Zein/Polycaprolactone/Collagen for wound healing. *Int. J. Biol. Macromol.* **2020**, *153*, 921–930. [CrossRef]
18. Farid, A.; Haridyy, H.; Ashraf, S.; Ahmed, S.; Safwat, G. *Aloe vera* gel as a stimulant for mesenchymal stem cells differentiation and a natural therapy for radiation induced liver damage. *J. Radiat. Res. Appl. Sci.* **2022**, *15*, 270–278. [CrossRef]
19. Razia, S.; Park, H.; Shin, E.; Shim, K.-S.; Cho, E.; Kang, M.C.; Kim, S.Y. Synergistic effect of *Aloe vera* flower and Aloe gel on cutaneous wound healing targeting MFAP4 and its associated signaling pathway: In-vitro study. *J. Ethnopharmacol.* **2022**, *290*, 115096. [CrossRef]
20. Sharifi, E.; Chehelgerdi, M.; Fatahian-Kelishadrokhi, A.; Yazdani-Nafchi, F.; Ashrafi-Dehkordi, K. Comparison of therapeutic effects of encapsulated Mesenchymal stem cells in *Aloe vera* gel and Chitosan-based gel in healing of grade-II burn injuries. *Regen. Ther.* **2021**, *21*, 30–37. [CrossRef]
21. Hekmatpou, D.; Mehrabi, F.; Rahzani, K.; Aminiyan, A. The Effect of *Aloe vera* Clinical Trials on Prevention and Healing of Skin Wound: A Systematic Review. *Iran J. Med. Sci.* **2019**, *44*, 1–9. [CrossRef]
22. Tejiram, S.; Kavalukas, S.L.; Shupp, J.W.; Barbul, A. 1-Wound healing. In *Ågren, Wound Healing Biomaterials*; Magnus, S., Ed.; Woodhead Publishing: Sawston, UK, 2016; pp. 3–39. [CrossRef]
23. Clark, R.A.F.; Musillo, M.; Stransky, T. Chapter 70—Wound repair: Basic biology to tissue engineering. In *Principles of Tissue Engineering*, 5th ed.; Lanza, R., Langer, R., Vacanti, J.P., Atala, A., Eds.; Academic Press: Cambridge, MA, USA, 2020; pp. 1309–1329. [CrossRef]
24. Frykberg, R.G.; Banks, J.; Deptuła, M.; Karpowicz, P.; Wardowska, A.; Sass, P.; Sosnowski, P.; Mieczkowska, A.; Filipowicz, N.; Dzierżyńska, M.; et al. Challenges in the Treatment of Chronic Wounds. *Adv. Wound Care* **2015**, *4*, 560–582. [CrossRef]
25. Nguyen, H.M.; Le, T.T.N.; Nguyen, A.T.; Le, H.N.T.; Pham, T.T. Biomedical materials for wound dressing: Recent advances and applications. *RSC Adv.* **2023**, *13*, 5509–5528. [CrossRef]
26. Yaşayan, G.; Nejati, O.; Ceylan, A.F.; Karasu, C.; Ugur, P.K.; Bal-Öztürk, A.; Zarepour, A.; Zarrabi, A.; Mostafavi, E. Tackling chronic wound healing using nanomaterials: Advancements, challenges, and future perspectives. *Appl. Mater. Today* **2023**, *32*, 101829. [CrossRef]
27. Verdes, M.; Mace, K.; Margetts, L.; Cartmell, S. Status and challenges of electrical stimulation use in chronic wound healing. *Curr. Opin. Biotechnol.* **2022**, *75*, 102710. [CrossRef] [PubMed]
28. Bormann, D.; Gugerell, A.; Ankersmit, H.J.; Mildner, M. Therapeutic Application of Cell Secretomes in Cutaneous Wound Healing. *J. Investig. Dermatol.* **2023**, *143*, 893–912. [CrossRef] [PubMed]
29. Middelkoop, E.; Sheridan, R.L. Skin Substitutes and 'the next level'. In *Total Burn Care*; Elsevier: Amsterdam, The Netherlands, 2018; pp. 167–173.e2. [CrossRef]
30. Shi, C.; Wang, C.; Liu, H.; Li, Q.; Li, R.; Zhang, Y.; Liu, Y.; Shao, Y.; Wang, J. Selection of Appropriate Wound Dressing for Various Wounds. *Front. Bioeng. Biotechnol.* **2020**, *8*, 182. [CrossRef]
31. Alven, S.; Khwaza, V.; Oyedeji, O.O.; Aderibigbe, B.A. Polymer-Based Scaffolds Loaded with *Aloe vera* Extract for the Treatment of Wounds. *Pharmaceutics* **2021**, *13*, 961. [CrossRef] [PubMed]
32. Hasan, S.; Hasan, M.A.; Hassan, M.U.; Amin, M.; Javed, T.; Fatima, L. Biopolymers in diabetic wound care management: A potential substitute to traditional dressings. *Eur. Polym. J.* **2023**, *189*, 111979. [CrossRef]
33. Corrales-Orovio, R.; Carvajal, F.; Holmes, C.; Miranda, M.; González-Itier, S.; Cárdenas, C.; Vera, C.; Schenck, T.L.; Egaña, J.T. Development of a photosynthetic hydrogel as potential wound dressing for the local delivery of oxygen and bioactive molecules. *Acta Biomater.* **2023**, *155*, 154–166. [CrossRef]
34. Smith, R.; Russo, J.; Fiegel, J.; Brogden, N. Antibiotic Delivery Strategies to Treat Skin Infections When Innate Antimicrobial Defense Fails. *Antibiotics* **2020**, *9*, 56. [CrossRef]

35. Popova, T.P.; Ignatov, I.; Petrova, T.E.; Kaleva, M.D.; Huether, F.; Karadzhov, S.D. Antimicrobial Activity In Vitro of Cream from Plant Extracts and Nanosilver, and Clinical Research In Vivo on Veterinary Clinical Cases. *Cosmetics* **2022**, *9*, 122. [CrossRef]
36. Moretta, A.; Scieuzo, C.; Petrone, A.M.; Salvia, R.; Manniello, M.D.; Franco, A.; Lucchetti, D.; Vassallo, A.; Vogel, H.; Sgambato, A.; et al. Antimicrobial Peptides: A New Hope in Biomedical and Pharmaceutical Fields. *Front. Cell. Infect. Microbiol.* **2021**, *11*, 668632. [CrossRef]
37. Ghasemi, A.H.; Farazin, A.; Mohammadimehr, M.; Naeimi, H. Fabrication and characterization of biopolymers with antibacterial nanoparticles and Calendula officinalis flower extract as an active ingredient for modern hydrogel wound dressings. *Mater. Today Commun.* **2022**, *31*, 103513. [CrossRef]
38. Huda, R.M.; Rashdan, H.R.; El-Naggar, M.E. Chapter 2-Traditional and modern wound dressings. In *Developments in Applied Microbiology and Biotechnology, Antimicrobial Dressings*; Khan, R., Gowri, S., Eds.; Academic Press: Cambridge, MA, USA, 2023; pp. 21–42. [CrossRef]
39. Ferroni, L.; Gardin, C.; D'Amora, U.; Calzà, L.; Ronca, A.; Tremoli, E.; Ambrosio, L.; Zavan, B. Exosomes of mesenchymal stem cells delivered from methacrylated hyaluronic acid patch improve the regenerative properties of endothelial and dermal cells. *Biomater. Adv.* **2022**, *139*, 213000. [CrossRef] [PubMed]
40. Feketshane, Z.; Adeyemi, S.A.; Ubanako, P.; Ndinteh, D.T.; Ray, S.S.; Choonara, Y.E.; Aderibigbe, B.A. Dissolvable sodium alginate-based antibacterial wound dressing patches: Design, characterization, and in vitro biological studies. *Int. J. Biol. Macromol.* **2023**, *232*, 123460. [CrossRef] [PubMed]
41. Fligge, M.; Letofsky-Papst, I.; Bäumers, M.; Zimmer, A.; Breitkreutz, J. Personalized dermal patches–Inkjet printing of prednisolone nanosuspensions for individualized treatment of skin diseases. *Int. J. Pharm.* **2023**, *630*, 122382. [CrossRef] [PubMed]
42. Raepsaet, C.; Alves, P.; Cullen, B.; Gefen, A.; Lázaro-Martínez, J.L.; Lev-Tov, H.; Najafi, B.; Santamaria, N.; Sharpe, A.; Swanson, T.; et al. The development of a core outcome set for clinical effectiveness studies of bordered foam dressings in the treatment of complex wounds. *J. Tissue Viability* **2023**. [CrossRef]
43. Genesi, B.P.; de Melo Barbosa, R.; Severino, P.; Rodas, A.C.D.; Yoshida, C.M.P.; Mathor, M.B.; Lopes, P.S.; Viseras, C.; Souto, E.B.; da Silva, C.F. *Aloe vera* and copaiba oleoresin-loaded chitosan films for wound dressings: Microbial permeation, cytotoxicity, and in vivo proof of concept. *Int. J. Pharm.* **2023**, *634*, 122648. [CrossRef]
44. Rajati, H.; Alvandi, H.; Rahmatabadi, S.S.; Hosseinzadeh, L.; Arkan, E. A nanofiber-hydrogel composite from green synthesized AgNPs embedded to PEBAX/PVA hydrogel and PA/Pistacia atlantica gum nanofiber for wound dressing. *Int. J. Biol. Macromol.* **2023**, *226*, 1426–1443. [CrossRef]
45. Revete, A.; Aparicio, A.; Cisterna, B.A.; Revete, J.; Luis, L.; Ibarra, E.; González, E.A.S.; Molino, J.; Reginensi, D. Advancements in the Use of Hydrogels for Regenerative Medicine: Properties and Biomedical Applications. *Int. J. Biomater.* **2022**, *2022*, 3606765. [CrossRef]
46. Caló, E.; Khutoryanskiy, V.V. Biomedical applications of hydrogels: A review of patents and commercial products. *Eur. Polym. J.* **2015**, *65*, 252–267. [CrossRef]
47. Taaca, K.L.M.; Prieto, E.I.; Vasquez, M.R., Jr. Current Trends in Biomedical Hydrogels: From Traditional Crosslinking to Plasma-Assisted Synthesis. *Polymers* **2022**, *14*, 2560. [CrossRef] [PubMed]
48. Bustamante-Torres, M.; Romero-Fierro, D.; Arcentales-Vera, B.; Palomino, K.; Magaña, H.; Bucio, E. Hydrogels Classification According to the Physical or Chemical Interactions and as Stimuli-Sensitive Materials. *Gels* **2021**, *7*, 182. [CrossRef] [PubMed]
49. Palencia, M.; Lerma, T.A.; Garcés, V.; Mora, M.A.; Martínez, J.M.; Palencia, S.L. Chapter 9-Eco-friendly hydrogels. In *Advances in Green and Sustainable Chemistry, Eco-Friendly Functional Polymers*; Palencia, M., Lerma, T.A., Garcés, V., Mora, M.A., Martínez, J.M., Palencia, S.L., Eds.; Elsevier: Amsterdam, The Netherlands, 2021; pp. 141–153. [CrossRef]
50. Firlar, I.; Altunbek, M.; McCarthy, C.; Ramalingam, M.; Camci-Unal, G. Functional Hydrogels for Treatment of Chronic Wounds. *Gels* **2022**, *8*, 127. [CrossRef] [PubMed]
51. Ruffo, M.; Parisi, O.I.; Dattilo, M.; Patitucci, F.; Malivindi, R.; Pezzi, V.; Tzanov, T.; Puoci, F. Synthesis and evaluation of wound healing properties of hydro-diab hydrogel loaded with green-synthetized AGNPS: In vitro and in ex vivo studies. *Drug Deliv. Transl. Res.* **2022**, *12*, 1881–1894. [CrossRef]
52. Solanki, D.; Vinchhi, P.; Patel, M.M. Design Considerations, Formulation Approaches, and Strategic Advances of Hydrogel Dressings for Chronic Wound Management. *ACS Omega* **2023**, *8*, 8172–8189. [CrossRef]
53. Ahmed, E.M. Hydrogel: Preparation, characterization, and applications: A review. *J. Adv. Res.* **2015**, *6*, 105–121. [CrossRef]
54. Louf, J.-F.; Lu, N.B.; O'connell, M.G.; Cho, H.J.; Datta, S.S. Under pressure: Hydrogel swelling in a granular medium. *Sci. Adv.* **2021**, *7*, 2711. [CrossRef]
55. Jia, B.; Li, G.; Cao, E.; Luo, J.; Zhao, X.; Huang, H. Recent progress of antibacterial hydrogels in wound dressings. *Mater. Today Bio* **2023**, *19*, 100582. [CrossRef]
56. Fernandes, A.; Rodrigues, P.M.; Pintado, M.; Tavaria, F.K. A systematic review of natural products for skin applications: Targeting inflammation, wound healing, and photo-aging. *Phytomedicine* **2023**, *115*, 154824. [CrossRef]
57. He, J.J.; McCarthy, C.; Camci-Unal, G. Development of Hydrogel-Based Sprayable Wound Dressings for Second-and Third-Degree Burns. *Adv. NanoBiomed Res.* **2021**, *1*, 2100004. [CrossRef]
58. Li, S.; Dong, S.; Xu, W.; Tu, S.; Yan, L.; Zhao, C.; Ding, J.; Chen, X. Antibacterial Hydrogels. *Adv. Sci.* **2018**, *5*, 1700527. [CrossRef] [PubMed]

59. Pérez-Luna, V.H.; González-Reynoso, O. Encapsulation of Biological Agents in Hydrogels for Therapeutic Applications. *Gels* **2018**, *4*, 61. [CrossRef] [PubMed]
60. Alkekhia, D.; LaRose, C.; Shukla, A. β-Lactamase-Responsive hydrogel drug delivery platform for bacteria-triggered cargo release. *ACS Appl. Mater. Interfaces* **2022**, *14*, 27538–27550. [CrossRef] [PubMed]
61. Baghersad, S.; Hivechi, A.; Bahrami, S.H.; Milan, P.B.; Siegel, R.A.; Amoupour, M. Optimal *Aloe vera* encapsulated PCL/Gel nanofiber design for skin substitute application and the evaluation of its in vivo implantation. *J. Drug Deliv. Sci. Technol.* **2022**, *74*, 103536. [CrossRef]
62. Balaji, A.; Vellayappan, M.V.; John, A.A.; Subramanian, A.P.; Jaganathan, S.K.; SelvaKumar, M.; Faudzi, A.A.B.M.; Supriyanto, E.; Yusof, M. Biomaterials based nano-applications of *Aloe vera* and its perspective: A review. *RSC Adv.* **2015**, *5*, 86199–86213. [CrossRef]
63. Yadeta, A.T. Food applications of Aloe species: A review. *J. Plant Sci. Phytopathol.* **2022**, *6*, 24–32. [CrossRef]
64. Rahman, S.; Carter, P.; Bhattarai, N. *Aloe vera* for Tissue Engineering Applications. *J. Funct. Biomater.* **2017**, *8*, 6. [CrossRef] [PubMed]
65. Minjares-Fuentes, R.; Femenia, A. Chapter 3.4-*Aloe vera*. In *Nonvitamin and Nonmineral Nutritional Supplements*; Nabavi, S.M., Silva, A.S., Eds.; Academic Press: Cambridge, MA, USA, 2019; pp. 145–152. [CrossRef]
66. Enachi, E.; Boev, M.; Bahrim, G. *Aloe vera* plant-an important source of bioactive compounds with functional value. *Innov. Rom. Food Biotechnol.* **2020**, *19*, 1–20. Available online: https://www.gup.ugal.ro/ugaljournals/index.php/IFRB/article/view/4309 (accessed on 22 May 2023).
67. Khan, R.U.; Naz, S.; De Marzo, D.; Dimuccio, M.M.; Bozzo, G.; Tufarelli, V.; Losacco, C.; Ragni, M. *Aloe vera*: A Sustainable Green Alternative to Exclude Antibiotics in Modern Poultry Production. *Antibiotics* **2023**, *12*, 44. [CrossRef]
68. Sánchez, M.; González-Burgos, E.; Iglesias, I.; Gómez-Serranillos, M.P. Pharmacological Update Properties of *Aloe vera* and its Major Active Constituents. *Molecules* **2020**, *25*, 1324. [CrossRef]
69. Massoud, D.; Alrashdi, B.M.; Fouda, M.M.A.; El-Kott, A.; Soliman, S.A.; Abd-Elhafeez, H.H. *Aloe vera* and wound healing: A brief review. *Braz. J. Pharm. Sci.* **2022**, *58*, e20837. [CrossRef]
70. Pressman, P.; Clemens, R.; Hayes, A.W. *Aloe vera* at the frontier of glycobiology and integrative medicine: Health implications of an ancient plant. *SAGE Open Med.* **2019**, *7*, 2050312119875921. [CrossRef] [PubMed]
71. Babu, S.N.; Noor, A. Bioactive constituents of the genus Aloe and their potential therapeutic and pharmacological applications: A review. *J. Appl. Pharm. Sci.* **2020**, *10*, 133–145.
72. Suriati, L. Nano Coating of Aloe-Gel Incorporation Additives to Maintain the Quality of Freshly Cut Fruits. *Front. Sustain. Food Syst.* **2022**, *6*, 914254. [CrossRef]
73. Godoy, D.J.D.; Chokboribal, J.; Pauwels, R.; Banlunara, W.; Sangvanich, P.; Jaroenporn, S.; Thunyakitpisal, P. Acemannan increased bone surface, bone volume, and bone density in a calvarial defect model in skeletally-mature rats. *J. Dent. Sci.* **2018**, *13*, 334–341. [CrossRef]
74. Wang, X.-F.; Chen, X.; Tang, Y.; Wu, J.-M.; Qin, D.-L.; Yu, L.; Yu, C.-L.; Zhou, X.-G.; Wu, A.-G. The Therapeutic Potential of Plant Polysaccharides in Metabolic Diseases. *Pharmaceuticals* **2022**, *15*, 1329. [CrossRef] [PubMed]
75. Sabbaghzadegan, S.; Golsorkhi, H.; Soltani, M.H.; Kamalinejad, M.; Bahrami, M.; Kabir, A.; Dadmehr, M. Potential protective effects of *Aloe vera* gel on cardiovascular diseases: A mini-review. *Phytother. Res.* **2021**, *35*, 6101–6113. [CrossRef]
76. Lissoni, P.; Rovelli, F.; Brivio, F.; Zago, R.; Colciago, M.; Messina, G.; Mora, A.; Porro, G. A randomized study of chemotherapy versus biochemotherapy with chemotherapy plus Aloe arborescens in patients with metastatic cancer. *Vivo* **2009**, *23*, 171–175.
77. Thant, A.A.; Ruangpornvisuti, V.; Sangvanich, P.; Banlunara, W.; Limcharoen, B.; Thunyakitpisal, P. Characterization of a bioscaffold containing polysaccharide acemannan and native collagen for pulp tissue regeneration. *Int. J. Biol. Macromol.* **2023**, *225*, 286–297. [CrossRef]
78. Madrid, R.R.M.; Mathews, P.D.; Pimenta, B.V.; Mertins, O. Biopolymer–Lipid Hybrid Cubosome for Delivery of Acemannan. *Mater. Proc.* **2023**, *14*, 56. [CrossRef]
79. Suciati, T.; Rachmawati, P.; Soraya, E.; Mahardhika, A.B.; Hartarti, R.; Anggadiredja, K. A novel acemannan-chitosan modified lipid nanoparticles as intracellular delivery vehicles of antibiotic. *J. Appl. Pharm. Sci.* **2018**, *8*, 1–11.
80. Cataldi, V.; Di Bartolomeo, S.; Di Campli, E.; Nostro, A.; Cellini, L.; Di Giulio, M. In vitro activity of *Aloe vera* inner gel against microorganisms grown in planktonic and sessile phases. *Int. J. Immunopathol. Pharmacol.* **2015**, *28*, 595–602. [CrossRef]
81. Leitgeb, M.; Kupnik, K.; Knez, Ž.; Primožič, M. Enzymatic and Antimicrobial Activity of Biologically Active Samples from *Aloe arborescens* and *Aloe barbadensis*. *Biology* **2021**, *10*, 765. [CrossRef]
82. Yahya, R.; Al-Rajhi, A.M.H.; Alzaid, S.Z.; Al Abboud, M.A.; Almuhayawi, M.S.; Al Jaouni, S.K.; Selim, S.; Ismail, K.S.; Abdelghany, T.M. Molecular Docking and Efficacy of *Aloe vera* Gel Based on Chitosan Nanoparticles against *Helicobacter pylori* and Its Antioxidant and Anti-Inflammatory Activities. *Polymers* **2022**, *14*, 2994. [CrossRef] [PubMed]
83. Yazdani, N.; Hossini, S.E.; Edalatmanesh, M.A. Anti-inflammatory Effect of *Aloe vera* Extract on Inflammatory Cytokines of Rats Fed with a High-Fat Diet (HFD). *Jundishapur J. Nat. Pharm. Prod.* **2022**, *17*, e114323. [CrossRef]
84. Fani, M.; Kohanteb, J. Inhibitory activity of *Aloe vera* gel on some clinically isolated cariogenic and periodontopathic bacteria. *J. Oral Sci.* **2012**, *54*, 15–21. [CrossRef] [PubMed]
85. Haghani, F.; Arabnezhad, M.-R.; Mohammadi, S.; Ghaffarian-Bahraman, A. *Aloe vera* and Streptozotocin-Induced Diabetes Mellitus. *Rev. Bras. Farm.* **2022**, *32*, 174–187. [CrossRef] [PubMed]

86. Chimisso, V.; Aleman Garcia, M.A.; Yorulmaz Avsar, S.; Dinu, I.A.; Palivan, C.G. Design of Bio-Conjugated Hydrogels for Regenerative Medicine Applications: From Polymer Scaffold to Biomolecule Choice. *Molecules* **2020**, *25*, 4090. [CrossRef]
87. Pereira, R.; Mendes, A.; Bártolo, P. Alginate/*Aloe vera* Hydrogel Films for Biomedical Applications. *Procedia CIRP* **2013**, *5*, 210–215. [CrossRef]
88. Liu, Y.; Fan, J.; Lv, M.; She, K.; Sun, J.; Lu, Q.; Han, C.; Ding, S.; Zhao, S.; Wang, G.; et al. Photocrosslinking silver nanoparticles–*Aloe vera*–silk fibroin composite hydrogel for treatment of full-thickness cutaneous wounds. *Regen. Biomater.* **2021**, *8*, rbab048. [CrossRef]
89. Yang, K.; Han, Q.; Chen, B.; Zheng, Y.; Zhang, K.; Li, Q.; Wang, J. Antimicrobial hydrogels: Promising materials for medical application. *Int. J. Nanomed.* **2018**, *13*, 2217–2263. [CrossRef]
90. Azahra, S.; Parisa, N.; Fatmawati, F.; Amalia, E.; Larasati, V. Antibacterial Efficacy of *Aloe vera* Sap Against Staphylococcus aureus and Escherichia coli. *Biosci. Med. J. Biomed. Transl. Res.* **2019**, *3*, 29–37. [CrossRef]
91. Huang, C.; Dong, L.; Zhao, B.; Lu, Y.; Huang, S.; Yuan, Z.; Luo, G.; Xu, Y.; Qian, W. Anti-inflammatory hydrogel dressings and skin wound healing. *Clin. Transl. Med.* **2022**, *12*, e1094. [CrossRef] [PubMed]
92. Kim, M.W.; Kang, J.-H.; Shin, E.; Shim, K.-S.; Kim, M.J.; Lee, C.-K.; Yoon, Y.S.; Oh, S.H. Processed *Aloe vera* gel attenuates non-steroidal anti-inflammatory drug (NSAID)-induced small intestinal injury by enhancing mucin expression. *Food Funct.* **2019**, *10*, 6088–6097. [CrossRef] [PubMed]
93. Bialik-Wąs, K.; Raftopoulos, K.N.; Pielichowski, K. Alginate Hydrogels with *Aloe vera*: The Effects of Reaction Temperature on Morphology and Thermal Properties. *Materials* **2022**, *15*, 748. [CrossRef] [PubMed]
94. Chelu, M.; Popa, M.; Ozon, E.A.; Cusu, J.P.; Anastasescu, M.; Surdu, V.A.; Moreno, J.C.; Musuc, A.M. High-Content *Aloe vera* Based Hydrogels: Physicochemical and Pharmaceutical Properties. *Polymers* **2023**, *15*, 1312. [CrossRef]
95. Kahramanoğlu, I.; Chen, C.; Chen, J.; Wan, C. Chemical Constituents, Antimicrobial Activity, and Food Preservative Characteristics of *Aloe vera* Gel. *Agronomy* **2019**, *9*, 831. [CrossRef]
96. Radha, M.H.; Laxmipriya, N.P. Evaluation of biological properties and clinical effectiveness of *Aloe vera*: A systematic review. *J. Tradit. Complement. Med.* **2014**, *5*, 21–26. [CrossRef]
97. Pop, R.M.; Puia, I.C.; Puia, A.; Chedea, V.S.; Levai, A.M.; Bocsan, I.C.; Buzoianu, A.D. Pot *Aloe vera* gel–a natural source of antioxidants. *Not. Bot. Horti Agrobot. Cluj Napoca* **2022**, *50*, 12732. [CrossRef]
98. Flores-López, M.L.; Romaní, A.; Cerqueira, M.A.; Rodríguez-García, R.; Jasso de Rodríguez, D.; Vicente, A.A. Compositional features and bioactive properties of whole fraction from *Aloe vera* processing. *Ind. Crops Prod.* **2016**, *91*, 179–185. [CrossRef]
99. Ahlawat, K.S.; Khatkar, B.S. Processing, food applications and safety of *Aloe vera* products: A review. *J. Food Sci. Technol.* **2011**, *48*, 525–533. [CrossRef] [PubMed]
100. Rahoui, W.; Merzouk, H.; El Haci, I.A.; Bettioui, R.; Azzi, R.; Benali, M. Beneficial effects of *Aloe vera* gel on lipid profile, lipase activities and oxidant/antioxidant status in obese rats. *J. Funct. Foods* **2018**, *48*, 525–532. [CrossRef]
101. Cavasana, A.L.; dos Santos, C.H.M.; Dourado, D.M.; Guimarães, F.D.S.; Barros, F.H.R.; de Campos, G.C.O.; Leme, G.A.L.; da Silva, L.D.M.; Wahl, L.M.; Gutterres, N.B.D.A.; et al. Effectiveness of the *Aloe vera* extract in the treatment of fistula-in-ano. *J. Coloproctology* **2020**, *40*, 67–72. [CrossRef]
102. Khedmat, H.; Karbasi, A.; Amini, M.; Aghaei, A.; Taheri, S. *Aloe vera* in treatment of refractory irritable bowel syndrome: Trial on Iranian patients. *J. Res. Med. Sci.* **2013**, *18*, 732.
103. Le Phan, T.H.; Park, S.Y.; Jung, H.J.; Kim, M.W.; Cho, E.; Shim, K.-S.; Shin, E.; Yoon, J.-H.; Maeng, H.-J.; Kang, J.-H.; et al. The Role of Processed *Aloe vera* Gel in Intestinal Tight Junction: An In Vivo and In Vitro Study. *Int. J. Mol. Sci.* **2021**, *22*, 6515. [CrossRef]
104. López-Cervantes, J.; Sánchez-Machado, D.I.; Cruz-Flores, P.; Mariscal-Domínguez, M.F.; de la Mora-López, G.S.; Campas-Baypoli, O.N. Antioxidant capacity, proximate composition, and lipid constituents of *Aloe vera* flowers. *J. Appl. Res. Med. Aromat. Plants* **2018**, *10*, 93–98. [CrossRef]
105. Shakib, Z.; Shahraki, N.; Razavi, B.M.; Hosseinzadeh, H. *Aloe vera* as an herbal medicine in the treatment of metabolic syndrome: A review. *Phytother. Res.* **2019**, *33*, 2649–2660. [CrossRef]
106. Parlati, L.; Voican, C.S.; Perlemuter, K.; Perlemuter, G. *Aloe vera* -induced acute liver injury: A case report and literature review. *Clin. Res. Hepatol. Gastroenterol.* **2017**, *41*, e39–e42. [CrossRef]
107. Guha, P.; Subhashis, P.; Das, A.; Halder, B.; Bhattacharjee, S.; Chaudhuri, T.K. Analyses of Human and Rat Clinical Parameters in Rheumatoid Arthritis Raise the Possibility of Use of Crude *Aloe vera* Gel in Disease Amelioration. *Immunome Res.* **2014**, *10*, 81. [CrossRef]
108. Majumder, R.; Das, C.K.; Mandal, M. Lead bioactive compounds of *Aloe vera* as potential anticancer agent. *Pharmacol. Res.* **2019**, *148*, 104416. [CrossRef]
109. El Salam, H.A.; Megahed, H.; El Aziz, S.A.; Sobhy, M.; El-Mouaty, H.A. Comparative Study of the Possible Hepatoprotective Effect of Each of N-acetylcysteine, Coenzyme Q10 and *Aloe vera* Gel in Acute Acetaminophen Induced Hepatotoxicity in Albino Rats. (Histological and Biochemical Study). *Ain Shams J. Forensic Med. Clin. Toxicol.* **2014**, *23*, 115–138. [CrossRef]
110. Şehitoğlu, M.H.; Karaboga, I.; Kiraz, A.; Kiraz, H.A. The hepatoprotective effect of *Aloe vera* on ischemia-reperfusion injury in rats. *North. Clin. Istanb.* **2018**, *6*, 203–209. [CrossRef] [PubMed]
111. Salehi, B.; Albayrak, S.; Antolak, H.; Kręgiel, D.; Pawlikowska, E.; Sharifi-Rad, M.; Uprety, Y.; Fokou, P.V.T.; Yousef, Z.; Zakaria, Z.A.; et al. Aloe Genus Plants: From Farm to Food Applications and Phytopharmacotherapy. *Int. J. Mol. Sci.* **2018**, *19*, 2843. [CrossRef] [PubMed]

112. Sujatha, G.; Kumar, G.S.; Muruganandan, J.; Prasad, T.S. *Aloe vera* in Dentistry. *J. Clin. Diagn. Res.* **2014**, *8*, ZI01–ZI02. [CrossRef]
113. Mangaiyarkarasi, S.P.; Manigandan, T.; Elumalai, M.; Cholan, P.K.; Kaur, R.P. Benefits of *Aloe vera* in dentistry. *J Pharm Bioallied Sci.* **2015**, *7* (Suppl. S1), S255–S259. [CrossRef]
114. Nair, G.R.; Naidu, G.S.; Jain, S.; Nagi, R.; Makkad, R.S.; Jha, A. Clinical Effectiveness of *Aloe vera* in the Management of Oral Mucosal Diseases- A Systematic Review. *J. Clin. Diagn. Res.* **2016**, *10*, ZE01–ZE07. [CrossRef]
115. Sholehvar, F.; Mehrabani, D.; Yaghmaei, P.; Vahdati, A. The effect of Aloe vera*gel* on viability of dental pulp stem cells. *Dent. Traumatol.* **2016**, *32*, 390–396. [CrossRef]
116. Songsiripradubboon, S.; Kladkaew, S.; Trairatvorakul, C.; Sangvanich, P.; Soontornvipart, K.; Banlunara, W.; Thunyakitpisal, P. Stimulation of Dentin Regeneration by Using Acemannan in Teeth with Lipopolysaccharide-induced Pulp Inflammation. *J. Endod.* **2017**, *43*, 1097–1103. [CrossRef]
117. Aparicio Salcedo, S.V.; Carranza Aldana, B.S.; Chávez Salas, S.A.; Quispe Tinco, L.S.; Palomino Zevallos, C.A.; Peralta Medina, A.N.; Robles Esquerre, J.M. Pharmacological efficacy of *Aloe vera* in wound healing: A narrative review: Eficacia farmacológica del *Aloe vera* en la cicatrización de heridas: Una revisión narrativa. *Rev. Fac. Med. Hum.* **2023**, *23*, 110–120. [CrossRef]
118. Vinson, J.; Al Kharrat, H.; Andreoli, L. Effect of *Aloe vera* preparations on the human bioavailability of vitamins C and E. *Phytomedicine* **2005**, *12*, 760–765. [CrossRef]
119. Rompicherla, N.C.; Joshi, P.; Shetty, A.; Sudhakar, K.; Amin, H.I.M.; Mishra, Y.; Mishra, V.; Albutti, A.; Alhumeed, N. Design, Formulation, and Evaluation of *Aloe vera* Gel-Based Capsaicin Transemulgel for Osteoarthritis. *Pharmaceutics* **2022**, *14*, 1812. [CrossRef] [PubMed]
120. Valizadeh, A.; Darvishi, M.H.; Amani, A.; Zarchi, A.A.K. Design and development of novel formulation of *Aloe vera* nanoemulsion gel contained erythromycin for topical antibacterial therapy: In vitro and in vivo assessment. *J. Drug Deliv. Sci. Technol.* **2022**, *74*, 103519. [CrossRef]
121. Nabipour, H.; Rohani, S. Zirconium metal organic framework/*Aloe vera* carrier loaded with naproxen as a versatile platform for drug delivery. *Chem. Pap.* **2023**, *77*, 3461–3470. [CrossRef]
122. Lin, C.-C.; Metters, A.T. Hydrogels in controlled release formulations: Network design and mathematical modeling. *Adv. Drug Deliv. Rev.* **2006**, *58*, 1379–1408. [CrossRef]
123. Bettini, R.; Colombo, P.; Massimo, G.; Catellani, P.L.; Vitali, T. Swelling and drug release in hydrogel matrices: Polymer viscosity and matrix porosity effects. *Eur. J. Pharm. Sci.* **1994**, *2*, 213–219. [CrossRef]
124. Dadashzadeh, A.; Imani, R.; Moghassemi, S.; Omidfar, K.; Abolfathi, N. Study of hybrid alginate/gelatin hydrogel-incorporated niosomal *Aloe vera* capable of sustained release of *Aloe vera* as potential skin wound dressing. *Polym. Bull.* **2020**, *77*, 387–403. [CrossRef]
125. Meng, R.; Zhu, H.; Deng, P.; Li, M.; Ji, Q.; He, H.; Jin, L.; Wang, B. Research progress on albumin-based hydrogels: Properties, preparation methods, types and its application for antitumor-drug delivery and tissue engineering. *Front. Bioeng. Biotechnol.* **2023**, *11*, 1137145. [CrossRef]
126. Armstrong, A.W.; Read, C. Pathophysiology, clinical presentation, and treatment of psoriasis: A review. *JAMA* **2020**, *323*, 1945–1960. [CrossRef]
127. Jales, S.T.L.; Barbosa, R.D.M.; de Albuquerque, A.C.; Duarte, L.H.V.; da Silva, G.R.; Meirelles, L.M.A.; da Silva, T.M.S.; Alves, A.F.; Viseras, C.; Raffin, F.N.; et al. Development and Characterization of *Aloe vera* Mucilaginous-Based Hydrogels for Psoriasis Treatment. *J. Compos. Sci.* **2022**, *6*, 231. [CrossRef]
128. Herculano, R.D.; dos Santos, T.O.; de Barros, N.R.; Brasil, G.S.P.; Scontri, M.; Carvalho, B.G.; Mecwan, M.; Farhadi, N.; Kawakita, S.; Perego, C.H.; et al. *Aloe vera*-loaded natural rubber latex dressing as a potential complementary treatment for psoriasis. *Int. J. Biol. Macromol.* **2023**, *242*, 124779. [CrossRef]
129. Sharma, K.; Munjal, M.; Sharma, R.K.; Sharma, M. Thymol encapsulated chitosan-*Aloe vera* films for antimicrobial infection. *Int. J. Biol. Macromol.* **2023**, *30*, 123897. [CrossRef]
130. Ghafoor, B.; Ali, M.N.; Ansari, U.; Bhatti, M.F.; Mir, M.; Akhtar, H.; Darakhshan, F. New Biofunctional Loading of Natural Antimicrobial Agent in Biodegradable Polymeric Films for Biomedical Applications. *Int. J. Biomater.* **2016**, *2016*, 6964938. [CrossRef] [PubMed]
131. Thomas, T.; Nath, M.S.; Mathew, N.; Reshmy, R.; Philip, E.; Latha, M.S. Alginate film modified with aloevera gel and cellulose nanocrystals for wound dressing application: Preparation, characterization and in vitro evaluation. *J. Drug Deliv. Sci. Technol.* **2020**, *59*, 101894. [CrossRef]
132. Yin, J.; Xu, L. Batch preparation of electrospun polycaprolactone/chitosan/*Aloe vera* blended nanofiber membranes for novel wound dressing. *Int. J. Biol. Macromol.* **2020**, *160*, 352–363. [CrossRef]
133. Barbosa, R.; Villarreal, A.; Rodriguez, C.; De Leon, H.; Gilkerson, R.; Lozano, K. *Aloe vera* extract-based composite nanofibers for wound dressing applications. *Mater. Sci. Eng. C* **2021**, *124*, 112061. [CrossRef]
134. Bialik-Wąs, K.; Pluta, K.; Malina, D.; Barczewski, M.; Malarz, K.; Mrozek-Wilczkiewicz, A. Advanced SA/PVA-based hydrogel matrices with prolonged release of *Aloe vera* as promising wound dressings. *Mater. Sci. Eng. C* **2021**, *120*, 111667. [CrossRef]
135. Goudarzi, M.; Fazeli, M.; Azad, M.; Seyedjavadi, S.S.; Mousavi, R. *Aloe vera* Gel: Effective Therapeutic Agent against Multidrug-Resistant *Pseudomonas aeruginosa* Isolates Recovered from Burn Wound Infections. *Chemother. Res. Pract.* **2015**, *2015*, 639806. [CrossRef]

136. Susanto, C.; Purba, M.R.; Mahrani, R.; Efendi, R. Efficacy of *Aloe vera* Hydrogel in Inhibition The Growth of Campylobacter rectus and Provetella intermedia Bacteria. *Biosci. Med. J. Biomed. Transl. Res.* **2021**, *5*, 784–790. [CrossRef]
137. Singh, S.; Anjum, S.; Joy, J.; Gupta, B. Polysaccharide–*Aloe vera* Bioactive Hydrogels as Wound Care System. In *Cellulose-Based Superabsorbent Hydrogels. Polymers and Polymeric Composites: A Reference Series*; Mondal, M., Ed.; Springer: Cham, Switzerland, 2018. [CrossRef]
138. Abdo, J.M.; Sopko, N.A.; Milner, S.M. The applied anatomy of human skin: A model for regeneration. *Wound Med.* **2020**, *28*, 100179. [CrossRef]
139. Hofmann, E.; Schwarz, A.; Fink, J.; Kamolz, L.-P.; Kotzbeck, P. Modelling the Complexity of Human Skin In Vitro. *Biomedicines* **2023**, *11*, 794. [CrossRef]
140. Swaney, M.H.; Kalan, L.R. Living in Your Skin: Microbes, Molecules, and Mechanisms. *Infect. Immun.* **2021**, *89*, e00695-20. [CrossRef] [PubMed]
141. Byrd, A.L.; Belkaid, Y.; Segre, J.A. The human skin microbiome. *Nat. Rev. Microbiol.* **2018**, *16*, 143–155. [CrossRef]
142. Sen, C.K. Human Wound and its Burden: Updated 2020 Compendium of Estimates. *Adv. Wound Care* **2021**, *10*, 281–292. [CrossRef] [PubMed]
143. Grubbs, H.; Manna, B. Wound Physiology. In *StatPearls*; StatPearls Publishing: Treasure Island, FL, USA, 2023. Available online: https://www.ncbi.nlm.nih.gov/books/NBK518964/ (accessed on 22 May 2023).
144. Harper, D.; Young, A.; McNaught, C.-E. The physiology of wound healing. *Surgery* **2014**, *32*, 445–450. [CrossRef]
145. Tavakoli, S.; Klar, A.S. Advanced Hydrogels as Wound Dressings. *Biomolecules* **2020**, *10*, 1169. [CrossRef]
146. Gosain, A.; DiPietro, L.A. Aging and wound healing. *World J. Surg.* **2004**, *28*, 321–326. [CrossRef] [PubMed]
147. Merabet, N.; Lucassen, P.J.; Crielaard, L.; Stronks, K.; Quax, R.; Sloot, P.M.; la Fleur, S.E.; Nicolaou, M. How exposure to chronic stress contributes to the development of type 2 diabetes: A complexity science approach. *Front. Neuroendocr.* **2022**, *65*, 100972. [CrossRef]
148. Eming, S.A.; Murray, P.J.; Pearce, E.J. Metabolic orchestration of the wound healing response. *Cell Metab.* **2021**, *33*, 1726–1743. [CrossRef]
149. El-Ashram, S.; El-Samad, L.M.; Basha, A.A.; El Wakil, A. Naturally-derived targeted therapy for wound healing: Beyond classical strategies. *Pharmacol. Res.* **2021**, *170*, 105749. [CrossRef]
150. Wang, Y.; Zhang, Y.; Lin, Z.; Huang, T.; Li, W.; Gong, W.; Guo, Y.; Su, J.; Wang, J.; Tu, Q. A green method of preparing a natural and degradable wound dressing containing *Aloe vera* as an active ingredient. *Compos. Part B Eng.* **2021**, *222*, 109047. [CrossRef]
151. Paul, K.; Darzi, S.; Del Borgo, M.P.; Cousins, F.L.; Werkmeister, J.A.; Gargett, C.E.; Mukherjee, S. Vaginal delivery of tissue engineered endometrial mesenchymal stem/stromal cells in an *Aloe vera*-alginate hydrogel alleviates maternal simulated birth injury. *Appl. Mater. Today* **2021**, *22*, 100890. [CrossRef]
152. Alsareii, S.A.; Ahmad, J.; Umar, A.; Ahmad, M.Z.; Shaikh, I.A. Enhanced In Vivo Wound Healing Efficacy of a Novel Piperine-Containing Bioactive Hydrogel in Excision Wound Rat Model. *Molecules* **2023**, *28*, 545. [CrossRef] [PubMed]
153. Meza-Valle, K.Z.; Saucedo-Acuña, R.A.; Tovar-Carrillo, K.L.; Cuevas-González, J.C.; Zaragoza-Contreras, E.A.; Melgoza-Lozano, J. Characterization and Topical Study of *Aloe vera* Hydrogel on Wound-Healing Process. *Polymers* **2021**, *13*, 3958. [CrossRef]
154. Chelu, M.; Musuc, A.M.; Aricov, L.; Ozon, E.A.; Iosageanu, A.; Stefan, L.M.; Prelipcean, A.-M.; Popa, M.; Moreno, J.C. Antibacterial *Aloe vera* Based Biocompatible Hydrogel for Use in Dermatological Applications. *Int. J. Mol. Sci.* **2023**, *24*, 3893. [CrossRef]
155. Khorasani, G.; Hosseinimehr, S.J.; Azadbakht, M.; Zamani, A.; Mahdavi, M.R. Aloe versus silver sulfadiazine creams for second-degree burns: A randomized controlled study. *Surg. Today* **2009**, *39*, 587–591. [CrossRef] [PubMed]
156. Shahzad, M.N.; Ahmed, N. Effectiveness of *Aloe vera* gel compared with 1% silver sulphadiazine cream as burn wound dressing in second degree burns. *J. Pak. Med. Assoc.* **2013**, *63*, 225–230.
157. Panahi, Y.; Beiraghdar, F.; Akbari, H.; Bekhradi, H.; Taghizadeh, M.; Sahebkar, A. A herbal cream consisting of *Aloe vera*, Lavandula stoechas, and Pelargonium roseum as an alternative for silver sulfadiazine in burn management. *Asian Biomed.* **2012**, *6*, 273–278. [CrossRef]
158. Avijgan, M.; Alinaghian, M.; Esfahani, M.H. *Aloe vera* Gel as a Traditional and Complementary Method for Chronic Skin Burn: A Case Report. *Adv. Infect. Dis.* **2017**, *7*, 19–25. [CrossRef]
159. Tungkasamit, T.; Chakrabandhu, S.; Samakgarn, V.; Kunawongkrit, N.; Jirawatwarakul, N.; Chumachote, A.; Chitapanarux, I. Reduction in severity of radiation-induced dermatitis in head and neck cancer patients treated with topical *Aloe vera* gel: A randomized multicenter double-blind placebo-controlled trial. *Eur. J. Oncol. Nurs.* **2022**, *59*, 102164. [CrossRef]
160. Dal'Belo, S.E.; Gaspar, L.R.; Maia Campos, P.M. Moisturizing effect of cosmetic formulations containing *Aloe vera* extract in different concentrations assessed by skin bioengineering techniques. *Ski. Res. Technol.* **2006**, *12*, 241–246. [CrossRef]
161. Zhang, L.; Tizard, I.R. Activation of a mouse macrophage cell line by acemannan: The major carbohydrate fraction from *Aloe vera* gel. *Immunopharmacology* **1996**, *35*, 119–128. [CrossRef] [PubMed]
162. Ma, Y.; Tang, T.; Sheng, L.; Wang, Z.; Tao, H.; Zhang, Q.; Qi, Z. Aloin suppresses lipopolysaccharide-induced inflammation by inhibiting JAK1-STAT1/3 activation and ROS production in RAW264. 7 cells. *Int. J. Mol. Med.* **2018**, *42*, 1925–1934.
163. Leng, H.; Pu, L.; Xu, L.; Shi, X.; Ji, J.; Chen, K. Effects of Aloe polysaccharide, a polysaccharide extracted from *Aloe vera*, on TNF-α-induced HaCaT cell proliferation and the underlying mechanism in psoriasis. *Mol. Med. Rep.* **2018**, *18*, 3537–3543. [CrossRef]

164. Jiang, K.; Guo, S.; Yang, C.; Yang, J.; Chen, Y.; Shaukat, A.; Deng, G. Barbaloin protects against lipopolysaccharide (LPS)-induced acute lung injury by inhibiting the ROS-mediated PI3K/AKT/NF-κB pathway. *Int. Immunopharm.* **2018**, *64*, 140–150. [CrossRef] [PubMed]
165. Paul, S.; Modak, D.; Chattaraj, S.; Nandi, D.; Sarkar, A.; Roy, J.; Chaudhuri, T.K.; Bhattacharjee, S. *Aloe vera* gel homogenate shows anti-inflammatory activity through lysosomal membrane stabilization and downregulation of TNF-α and Cox-2 gene expressions in inflammatory arthritic animals. *Futur. J. Pharm. Sci.* **2021**, *7*, 12. [CrossRef]
166. Ahluwalia, B.; Magnusson, M.K.; Isaksson, S.; Larsson, F.; Öhman, L. Effects of Aloe barbadensis Mill. extract (AVH200®) on human blood T cell activity in vitro. *J. Ethnopharm.* **2016**, *179*, 301–309. [CrossRef]
167. Moriyama, M.; Moriyama, H.; Uda, J.; Kubo, H.; Nakajima, Y.; Goto, A.; Akaki, J.; Yoshida, I.; Matsuoka, N.; Hayakawa, T. Beneficial effects of the genus aloe on wound healing, cell proliferation, and differentiation of epidermal keratinocytes. *PLoS ONE* **2016**, *11*, e0164799. [CrossRef]
168. Liu, L.Y.; Chen, X.D.; Wu, B.; Jiang, Q. Influence of Aloe polysaccharide on proliferation and hyaluronic acid and hydroxyproline secretion of human fibroblasts in vitro. *J. Chin. Integr. Med.* **2010**, *8*, 256–262. [CrossRef]
169. Hamid, A.A.A.; Soliman, M.F. Effect of topical *Aloe vera* on the process of healing of full-thickness skin burn: A histological and immunohistochemical study. *J. Histol. Histopathol.* **2015**, *2*, 3. [CrossRef]
170. Hashemi, S.A.; Madani, S.A.; Abediankenari, S. The Review on properties of *Aloe vera* in healing of cutaneous wounds. *BioMed Res. Int.* **2015**, *2015*, 714216. [CrossRef]
171. Maenthaisong, R.; Chaiyakunapruk, N.; Niruntraporn, S.; Kongkaew, C. The efficacy of *Aloe vera* used for burn wound healing: A systematic review. *Burns* **2007**, *33*, 713–718. [CrossRef]
172. Moghbel, A.; Ghalambor, A.; Allipanah, S. Wound healing and toxicity evaluation of *Aloe vera* cream on outpatients with second degree burns. *Iran. J. Pharm. Sci. Summer* **2007**, *3*, 157–160.
173. Somboonwong, J.; Thanamittramanee, S.; Jariyapongskul, A.; Patumraj, S. Therapeutic effects of *Aloe vera* on cutaneous microcirculation and wound healing in second degree burn model in rats. *J. Med. Assoc. Thail.* **2000**, *83*, 417–425.
174. Fox, L.T.; du Plessis, J.; Gerber, M.; van Zyl, S.; Boneschans, B.; Hamman, J.H. In Vivo skin hydration and anti-erythema effects of *Aloe vera*, Aloe ferox and Aloe marlothii gel materials after single and multiple applications. *Pharmacogn. Mag.* **2014**, *10* (Suppl. S2), S392–S403. [CrossRef]
175. Moon, E.-J.; Lee, Y.M.; Lee, O.-H.; Lee, M.-J.; Lee, S.-K.; Chung, M.-H.; Park, Y.-I.; Sung, C.-K.; Choi, J.-S.; Kim, K.-W. A ncovel angiogenic factor derived from *Aloe vera* gel: β-sitosterol, a plant sterol. *Angiogenesis* **1999**, *3*, 117–123. [CrossRef]
176. Choi, S.; Kim, K.-W.; Choi, J.-S.; Han, S.-T.; Park, Y.-I.; Lee, S.-K.; Kim, J.-S.; Chung, M.-H. Angiogenic activity of β-sitosterol in the ischaemia/reperfusion-damaged brain of mongolian gerbil. *Planta Med.* **2002**, *68*, 330–335. [CrossRef]
177. Liu, F.-W.; Liu, F.C.; Wang, Y.-R.; Tsai, H.-I.; Yu, H.-P. Aloin protects skin fibroblasts from heat stress-induced oxidative stress damage by regulating the oxidative defense system. *PLoS ONE* **2015**, *10*, e0143528. [CrossRef] [PubMed]
178. Shanmugam, D.K.; Madhavan, Y.; Manimaran, A.; Kaliaraj, G.S.; Mohanraj, K.G.; Kandhasamy, N.; Mosas, K.K.A. Efficacy of Graphene-Based Nanocomposite Gels as a Promising Wound Healing Biomaterial. *Gels* **2023**, *9*, 22. [CrossRef] [PubMed]
179. Teplicki, E.; Ma, Q.; Castillo, D.E.; Zarei, M.; Hustad, A.P.; Chen, J.; Li, J. The Effects of *Aloe vera* on Wound Healing in Cell Proliferation, Migration, and Viability. *Wounds* **2018**, *30*, 263–268. [PubMed]
180. Negahdari, S.; Galehdari, H.; Kesmati, M.; Rezaie, A.; Shariati, G. Wound healing activity of extracts and formulations of *Aloe vera*, henna, adiantum capillus-veneris, and myrrh on mouse dermal fibroblast cells. *Int. J. Prev. Med.* **2017**, *8*, 18. [CrossRef]
181. Wahedi, H.M.; Jeong, M.; Chae, J.K.; Gil Do, S.; Yoon, H.; Kim, S.Y. Aloesin from *Aloe vera* accelerates skin wound healing by modulating MAPK/Rho and Smad signaling pathways in vitro and in vivo. *Phytomedicine* **2017**, *28*, 19–26. [CrossRef]
182. Müller-Heupt, L.K.; Wiesmann, N.; Schröder, S.; Korkmaz, Y.; Vierengel, N.; Groß, J.; Dahm, R.; Deschner, J.; Opatz, T.; Brieger, J.; et al. Extracts of *Rheum palmatum* and *Aloe vera* Show Beneficial Properties for the Synergistic Improvement of Oral Wound Healing. *Pharmaceutics* **2022**, *14*, 2060. [CrossRef]
183. Hamman, J.H. Composition and Applications of *Aloe vera* leaf gel. *Molecules* **2008**, *13*, 1599–1616. [CrossRef] [PubMed]
184. Anilakumar, K.R.; Sudarshanakrishnam, K.R.; Chandramohan, G.; Ilaiyaraja, N.; Khanum, F.; Bawa, A.S. Effect of *Aloe vera* gel extract on antioxidant enzymes and azoxymethane-induced oxidative stress in rats. *Indian J. Exp. Biol.* **2010**, *48*, 837–842.
185. Hassanpour, H. Effect of *Aloe vera* gel coating on antioxidant capacity, antioxidant enzyme activities and decay in raspberry fruit. *LWT Food Sci. Technol.* **2015**, *60*, 495–501. [CrossRef]
186. Hormozi, M.; Assaei, R.; Boroujeni, M.B. The effect of *Aloe vera* on the expression of wound healing factors (TGFβ1 and bFGF) in mouse embryonic fibroblast cell: In vitro study. *Biomed. Pharmacother.* **2017**, *88*, 610–616. [CrossRef]
187. Ali, J.; Khan, A.W.; Kotta, S.; Ansari, S.H.; Sharma, R.K.; Kumar, A. Formulation development, optimization and evaluation of *Aloe vera* gel for wound healing. *Pharmacogn. Mag.* **2013**, *9*, 6–10. [CrossRef]
188. Curto, E.M.; Labelle, A.; Chandler, H.L. *Aloe vera*: An in vitro study of effects on corneal wound closure and collagenase activity. *Vet. Ophthamol.* **2014**, *17*, 403–410. [CrossRef]
189. De Oliveira, A.C.L.; Tabrez, S.; Shakil, S.; Khan, M.I.; Asghar, M.N.; Matias, B.D.; de Carvalho, R.M. Mutagenic, antioxidant and wound healing properties of *Aloe vera*. *J. Ethnopharmacol.* **2018**, *227*, 191–197. [CrossRef]
190. Zeng, W.M.; Barnes, C.W.; Hiro, M.E.; Parus, A.; Robson, M.C.; Payne, W.G. *Aloe vera*—Mechanisms of Action, Uses, and Potential Uses in Plastic Surgery and Wound Healing. *Surg. Sci.* **2020**, *11*, 312–328. [CrossRef]

191. Na, H.S.; Song, Y.R.; Kim, S.; Heo, J.Y.; Chung, H.Y.; Chung, J. Aloin Inhibits Interleukin (IL)-1β– Stimulated IL-8 Production in KB Cells. *J. Periodontol.* **2016**, *87*, 108–115. [CrossRef] [PubMed]
192. Reuter, J.; Jocher, A.; Stump, J.; Grossjohann, B.; Franke, G.; Schempp, C. Investigation of the anti-inflammatory potential of *Aloe vera* gel (97.5%) in the ultraviolet erythema test. *Ski. Pharmacol. Physiol.* **2008**, *21*, 106–110. [CrossRef] [PubMed]
193. Li, C.-Y.; Suzuki, K.; Hung, Y.-L.; Yang, M.-S.; Yu, C.-P.; Lin, S.-P.; Hou, Y.-C.; Fang, S.-H. Aloe metabolites prevent lps-induced sepsis and inflammatory response by inhibiting mitogen-activated protein kinase activation. *Am. J. Chin. Med.* **2017**, *45*, 847–861. [CrossRef] [PubMed]
194. Pulliero, A.; Profumo, A.; Izzotti, A.; Saccà, S.C. Release of *Aloe vera* Extracts from Therapeutic Lenses. *Appl. Sci.* **2020**, *10*, 9055. [CrossRef]
195. Chelu, M.; Moreno, J.C.; Atkinson, I.; Cusu, J.P.; Rusu, A.; Bratan, V.; Aricov, L.; Anastasescu, M.; Seciu-Grama, A.-M.; Musuc, A.M. Green synthesis of bioinspired chitosan-ZnO-based polysaccharide gums hydrogels with propolis extract as novel functional natural biomaterials. *Int. J. Biol. Macromol.* **2022**, *211*, 410–424. [CrossRef]
196. Chelu, M.; Musuc, A.M. Polymer Gels: Classification and Recent Developments in Biomedical Applications. *Gels* **2023**, *9*, 161. [CrossRef]
197. Miguel, S.P.; Ribeiro, M.P.; Coutinho, P.; Correia, I.J. Electrospun Polycaprolactone/*Aloe vera*_Chitosan Nanofibrous Asymmetric Membranes Aimed for Wound Healing Applications. *Polymers* **2017**, *9*, 183. [CrossRef]
198. International Agency for Research on Cancer. *Aloe vera*. In *Some Drugs and Herbal Products (Volume 108)*; International Agency for research on Cancer: Lyon, France, 2015. Available online: http://monographs.iarc.fr/ENG/Monographs/vol108/mono108.pdf (accessed on 20 May 2023).
199. Cosmetic Ingredient Review Expert Panel (CIR). Final report on the safety assessment of aloe andongensis extract, aloe andongensis leaf juice, aloe arborescens leaf extract, aloe arborescens leaf juice, aloe arborescens leaf protoplasts, aloe barbadensis flower extract, aloe barbadensis leaf, aloe barbadensis leaf extract, aloe barbadensis leaf juice, aloe barbadensis leaf polysaccharides, aloe barbadensis leaf water, aloe ferox leaf extract, aloe ferox leaf juice, and aloe ferox leaf juice extract. *Int. J. Toxicol.* **2007**, *26* (Suppl. S2), 1–50.
200. Boudreau, M.D.; Beland, F.A. An evaluation of the biological and toxicological properties of *Aloe barbadensis (miller)*, *Aloe vera*. *J. Environ. Sci. Health C Environ. Carcinog. Ecotoxicol. Rev.* **2006**, *24*, 103–154. [CrossRef]
201. Boudreau, M.D.; Beland, F.A.; Nichols, J.A.; Pogribna, M. Toxicology and carcinogenesis studies of a nondecolorized whole leaf extract of Aloe barbadensis Miller (*Aloe vera*) in F344/N rats and B6C3F1 mice (drinking water study). *Natl. Toxicol. Program Tech. Rep. Ser.* **2013**, *577*, 1–266.
202. Guo, X.; Mei, N. *Aloe vera*: A review of toxicity and adverse clinical effects. *J. Environ. Sci. Health Part C* **2016**, *34*, 77–96. [CrossRef] [PubMed]
203. Ryall, C.; Duarah, S.; Chen, S.; Yu, H.; Wen, J. Advancements in Skin Delivery of Natural Bioactive Products for Wound Management: A Brief Review of Two Decades. *Pharmaceutics* **2022**, *14*, 1072. [CrossRef] [PubMed]
204. Ferreira, A.S.; Macedo, C.; Silva, A.M.; Delerue-Matos, C.; Costa, P.; Rodrigues, F. Natural Products for the Prevention and Treatment of Oral Mucositis—A Review. *Int. J. Mol. Sci.* **2022**, *23*, 4385. [CrossRef] [PubMed]

gels

Review

Progress in Surface Modification of Titanium Implants by Hydrogel Coatings

Huangqin Chen [1,†], Rui Feng [1,†], Tian Xia [1], Zhehan Wen [1], Qing Li [1], Xin Qiu [1], Bin Huang [1,*] and Yuesheng Li [2,*]

1 Department of Stomatology, School of Stomatology and Ophthalmology, Xianning Medical College, Hubei University of Science and Technology, Xianning 437100, China; chenhuangqin79@163.com (H.C.)
2 Hubei Key Laboratory of Radiation Chemistry and Functional Materials, Non-Power Nuclear Technology Collaborative Innovation Center, Hubei University of Science and Technology, Xianning 437100, China
* Correspondence: huangbin914@163.com (B.H.); frank78929@163.com (Y.L.)
† These authors contributed equally to this work.

Abstract: Although titanium and titanium alloys have become the preferred materials for various medical implants, surface modification technology still needs to be strengthened in order to adapt to the complex physiological environment of the human body. Compared with physical or chemical modification methods, biochemical modification, such as the introduction of functional hydrogel coating on implants, can fix biomolecules such as proteins, peptides, growth factors, polysaccharides, or nucleotides on the surface of the implants, so that they can directly participate in biological processes; regulate cell adhesion, proliferation, migration, and differentiation; and improve the biological activity on the surface of the implants. This review begins with a look at common substrate materials for hydrogel coatings on implant surfaces, including natural polymers such as collagen, gelatin, chitosan, and alginate, and synthetic materials such as polyvinyl alcohol, polyacrylamide, polyethylene glycol, and polyacrylic acid. Then, the common construction methods of hydrogel coating (electrochemical method, sol–gel method and layer-by-layer self-assembly method) are introduced. Finally, five aspects of the enhancement effect of hydrogel coating on the surface bioactivity of titanium and titanium alloy implants are described: osseointegration, angiogenesis, macrophage polarization, antibacterial effects, and drug delivery. In this paper, we also summarize the latest research progress and point out the future research direction. After searching, no previous relevant literature reporting this information was found.

Keywords: hydrogel coating; titanium alloy; biochemical modification; application

Citation: Chen, H.; Feng, R.; Xia, T.; Wen, Z.; Li, Q.; Qiu, X.; Huang, B.; Li, Y. Progress in Surface Modification of Titanium Implants by Hydrogel Coatings. *Gels* **2023**, *9*, 423. https://doi.org/10.3390/gels9050423

Academic Editors: Adina Magdalena Musuc, Magdalena Mititelu and Mariana Chelu

Received: 5 May 2023
Revised: 16 May 2023
Accepted: 17 May 2023
Published: 18 May 2023

1. Introduction

In the 1940s, some scholars implanted pure titanium (Ti) into the femur of mice without causing adverse reactions, which proved that Ti had good biocompatibility [1]. Later, more and more scholars began to apply pure Ti in dental implants, joint prostheses, and other clinical fields. However, in the process of application, it was found that the low hardness and poor wear resistance of Ti did not meet the requirements of the force parts of knee joint and hip joint, promoting research into and development of titanium alloys. Ti6Al4V is an $\alpha + \beta$ alloy with higher hardness, better wear resistance, and better workability compared with Ti (Table 1). However, aluminum (Al) and vanadium (V) in Ti6Al4V alloy are harmful metal elements, which have the risk of releasing after implantation into the human body. There is an urgent need to develop new medical titanium alloys with better biocompatibility. Representative materials are Ti–Nb–Zr and Ti–Nb–Zr–ME (Me (metal)) systems, especially Ti–Nb–Zr–Si (TNZS) alloy, which not only has better biocompatibility, but also has improved corrosion resistance, and can better match with human bone tissue [2].

Table 1. Development of medical titanium and titanium alloys.

Classification	Time	Representative Material	Advantage	Disadvantage
α	1960s	Ti	Good biocompatibility	Low strength, poor wear resistance
α + β	1970s	Ti6Al4V	Higher hardness, better wear resistance, lower elastic modulus, better mechanical compatibility	Biological toxicity of metal ions Al and V
	1980s	Ti6Al7Nb Ti5Al2.5Fe	Better biocompatibility	Easy corrosion, biological toxicity of Al metal ions
β	1990s	Ti13Nb13Zr Ti12Mo6Zr2Fe Ti15Mo	The low modulus of elasticity is close to that of human bones, non-biological toxicity of metal ions	Biological activity, abrasion resistance, and corrosion resistance still need to be improved

At present, titanium and titanium alloys have become the preferred materials for medical metal products because of their lower density, higher specific strength, and better biocompatibility.

Although medical titanium and titanium alloys have outstanding properties, it is still necessary to strengthen the research on surface modification technology to match the complex physiological environment in human body (Figure 1) [3]. Now, construction of functional coatings on titanium and titanium alloys has attracted more and more attention. For example, titanium nitride [4], titanium aluminum nitride [5], and titanium dioxide [6] coatings significantly improve wear resistance and corrosion resistance of titanium and titanium alloys. The modification of Si–TiO_2 nanotubes on the Ti substrate generates a nanostructured and hydrophilic surface, which can promote cell growth. Moreover, the existence of the TiO_2 nanotubes and Si element improves the in vitro osteogenic differentiation of MC3T3-E1 cells and early bone-formation around the implanted screws [7]. Coating of silver (Ag), copper (Cu), zinc (Zn), and other antibacterial metal elements show excellent antibacterial properties [8,9]. Adding Cu with Ti-15Mo reduces the possibility of bacterial infection during biomedical implant surgeries [10].

Differently from physical or chemical modification methods such as micro-arc oxidation and sandblasting to prepare oxide film or rough surface on the surface of implants, biochemical modification of fixing specific proteins, peptides, growth factors, polysaccharides, nucleotides, and other biomolecules on the surface of implants can directly participate in biological processes and regulate cell adhesion, proliferation, migration, and differentiation [11,12]. Hydrogel is a type of hydrophilic three-dimensional network structure formed by natural or artificial synthetic polymer materials through the gelation process of sol, which is widely used in many fields such as tissue engineering, drug delivery, and biosensors. The functional hydrogel coatings on titanium implants can effectively coordinate the advantages of hydrogel (lubricity, biocompatibility, and controlled release) with those of implants (stiffness, strength, and toughness) [13], and change the electrochemical behavior of titanium implants and enhance corrosion resistance [14].

In this review, we will first introduce the classification of hydrogel coatings on the surface of titanium implants, including natural hydrogels and synthetic hydrogels, according to the composition of the hydrogel matrix. We will then introduce the common binding methods of hydrogel coatings and titanium implants, such as the electrochemical precipitation method, the sol–gel method, and the layer-by-layer self-assembly method. Subsequently, the improvement of titanium implants by hydrogel coating on osseointegration, angiogenesis, macrophage polarization, antibacterial, and drug delivery are summarized in detail. Finally, the possible problems and future development direction of hydrogel coatings are presented in order to provide reference for scientific research workers in related fields. After searching, no previous research reporting of this information was found.

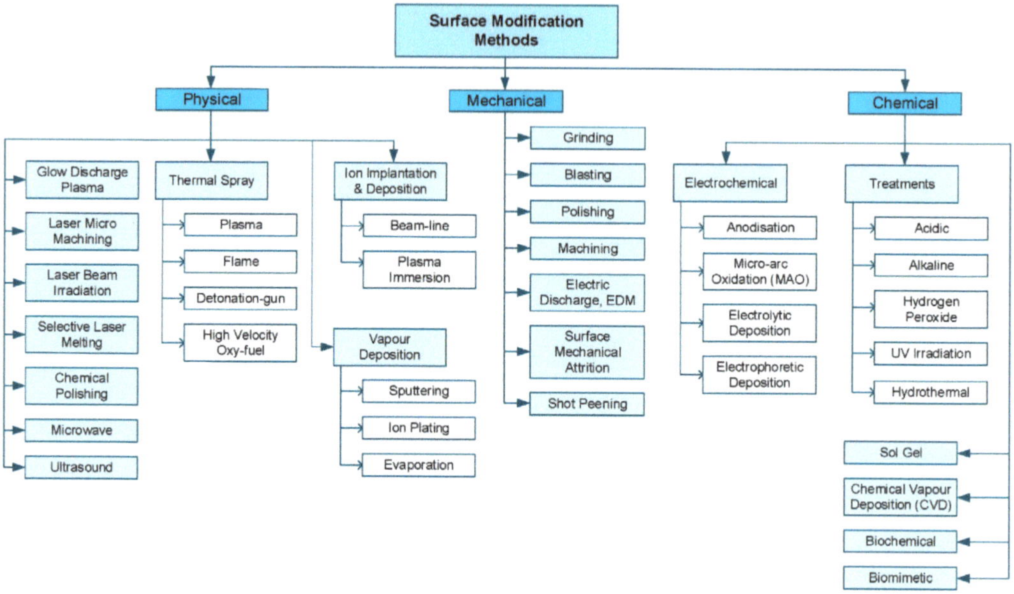

Figure 1. Surface modification methods of titanium and its alloys [3].

2. Classification of Hydrogel Coatings

According to the main components of the hydrogel matrix, hydrogel coatings can be divided into natural hydrogel coatings and synthetic hydrogel coatings (Table 2).

Table 2. Classification of hydrogel coatings.

Classification	Representative Material	Advantage	Reference
Natural hydrogel coatings	Collagen-based	Improve the attachment of the peri-implant soft tissue to titanium at early stages	[15]
		Enhance tissue vascularization and reduce inflammatory response	[16]
		Improve gingival connective tissue response to titanium implants	[17]
	Gelatin-based	Improve surface bio-activity	[18]
		Load with antibacterial agent curcumin	[19]
	Chitosan-based	Enhance the antibacterial activity and osteoinductive properties	[20]
		Develop a close bony apposition or the osseointegration of dental/craniofacial and orthopedic implants	[21]
		Provide a self-protective surface that prevents bacterial colonisation and implant-associated infections	[22]
		Great potential in implant anticorrosion	[23]
	Alginate-based	Improve the antibacterial effect and induce mineralization of dental implants	[24]
		Successively functionalize titanium surface	[25]

Table 2. *Cont.*

Classification	Representative Material	Advantage	Reference
Synthetic hydrogels coatings	Polyvinyl alcohol	Improve the calcium silicate coating-to-substrate adhesion.	[26]
	Polyacrylamide	Antimicrobial-loaded hydrogel coatings	[27]
	Polyethylene glycol	Lower albumin adsorption and presented a decreased fibroblast, *Streptococcus sanguinis* and *Lactobacillus salivarius* adhesion.	[28]
	Poly (lacto-glycolic acid)	Drug release	[29]
	Polyacrylic acid	Acts as both an effective bioactive surface and an effective anti-corrosion barrier	[14]

2.1. Natural Hydrogel Coating

Natural hydrogel is composed of natural biological materials which are highly similar to the extracellular matrix. It is considered as good biomimetic material in tissue engineering because of its complete bioactivity in promoting cell adhesion, proliferation, differentiation, and biodegradation. Common natural biomaterials are collagen and gelatin from animal protein, hyaluronic acid from animal epithelium and connective tissue, chitosan from shells of crustaceans, and alginate from the cytoplasm and cell wall.

2.1.1. Collagen-Based Hydrogel Coating

Collagen is the main component of the extracellular matrix in mammals, and is mainly distributed in the cornea, cartilage, bone, blood vessels, viscera, intervertebral discs, and dentin, and plays an important role in supporting and protecting the body and organs. Collagen has the advantages of non-cytotoxicity, good biocompatibility, easy absorption, small immune response, low antigenicity, etc. Coating titanium alloys with collagen promotes adhesion, proliferation, and differentiation of born-forming cells [30–33] as well as fibroblasts [34,35]. In comparison with uncoated commercially pure titanium, collagen coating significantly improves bone mineralization and maturation [36]. More rapid osteointegration will be achieved if the coating is combined with vitamins [37], phospholipid [38], or hydroxyapatite (HA) [39–41]. Besides osteogenesis, collagen coating can support the timely conversion of macrophages from the pro-inflammatory M1 to the pro-healing M2 phenotype, inhibiting inflammatory reaction and generating a beneficial osteoimmune microenvironment [42,43] (Figure 2). Simultaneously, collagen coating prominently facilitates angiogenesis of endothelial cells [42] and strengthens local blood supply restoration through sustained release of vascular endothelial growth factor (VEGF) [44].

2.1.2. Gelatin-Based Hydrogel Coating

Gelatin is a product of collagen hydrolysis but retains arginine–glycine–aspartic (RGD) cell adhesion peptide and protease degradation sites of collagen with lower immunogenicity. Gelatin-based hydrogel coatings enhance the integration between implant and tissue [45], and promote biological activity by loading various growth factors. For example, gelatin coating loaded with VEGF/bone morphogenetic protein 2 (BMP-2) shell-core microspheres promoted osteogenic differentiation and osseointegration effectively in 3D-printed porous titanium alloy [46]. In 2000, Van den Bulcke et al. introduced the methacryl group into modified gelatin for the first time to prepare methacryl amide gelatin, which gave the gelatin the property of photo-cross-linking under the photoinitiator and light [47]. When loaded with a short cationic antimicrobial peptide and synthetic silicate nanoparticles, the photo-cross-linked gelatin-based hydrogel coating demonstrated excellent antimicrobial activity and enhanced osteogenesis [48]. The addition of ginger inhibited the growth of *S. mutans* and *P. gingivalis* [49]. Methylacrylamide gelatin combined with photosensitizer and photocatalyst offers direct fibroblast activation [50] (Figure 3) and multi-mode photothermal and photodynamic antibacterial effects [51]. Furthermore, the allylated gelatin co-encapsulated human umbilical vein endothelial cells (HUVECs) and human mesenchymal stromal cells (hMSCs) support and achieve concurrent vasculogenic and osteogenic performance [52].

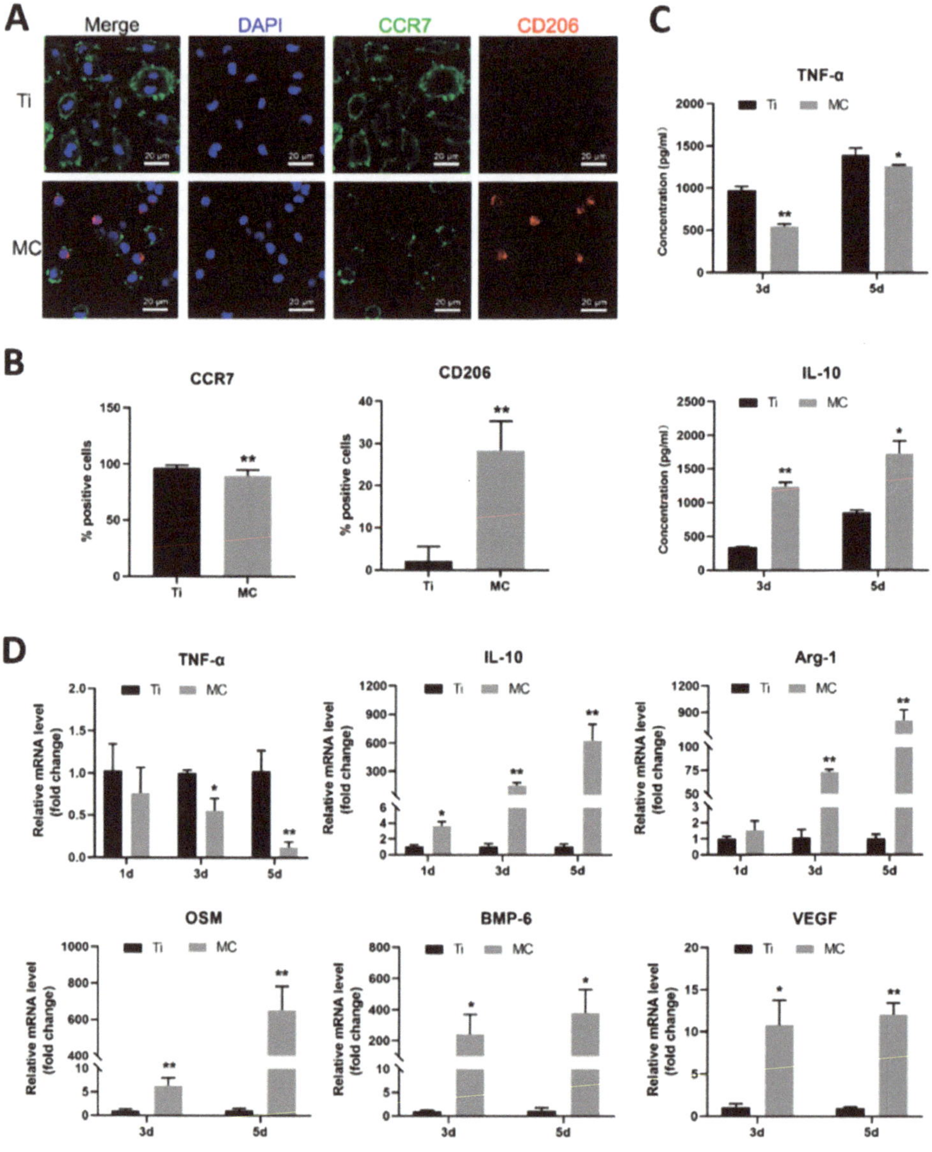

Figure 2. Functional evaluation for BMDM polarization through surface biomarkers (**A**,**B**), inflammatory factor secreting pattern (**C**), and related gene expression levels (**D**). CCR7 and TNF-α served as the M1-polarized markers; CD206, IL-10, and Arg-1 served as the M2-polarized markers; OSM, BMP-6, and VEGF served as the pro-regeneration biomarker. * $p < 0.05$; ** $p < 0.01$ [43].

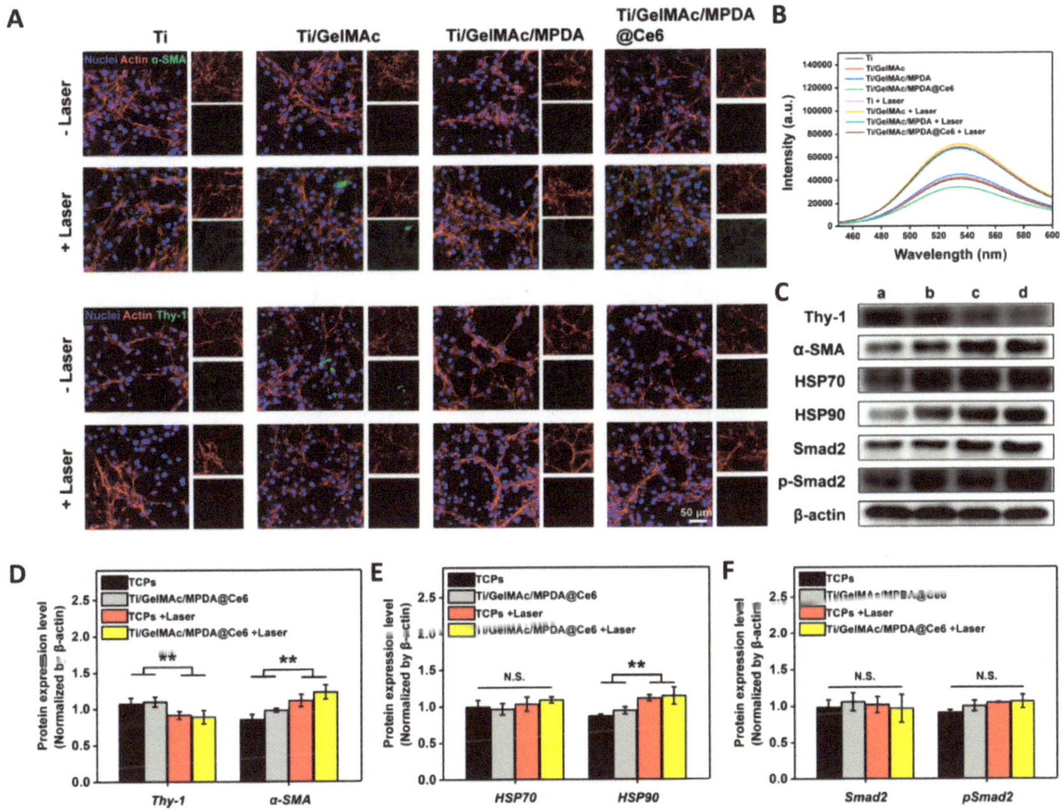

Figure 3. Fibroblast activation, or fibroblast-myofibroblast transition, of NIH/3T3 cells after irradiation [50]. (**A**) Representative fluorescence images of NIH/3T3 cells cultured with different samples for 2 days. The expression of α-SMA and Thy-1 was stained as green. (**B**) Cellular ATP level reflected by luminescence intensity after different treatments. (**C**) Western blotting detecting the expression of Thy-1, α-SMA, HSP70, HSP90, Smad-2, and p-Smad-2 after different treatments. a: TCPs; b: Ti/GelMAc/MPDA@Ce6; c: TCPs + Laser; d: Ti/GelMAc/MPDA@Ce6 + Laser. (**D–F**) Quantitative analysis according to Western blotting results. (n = 6, ** $p < 0.01$, N.S.: no significance). (For interpretation of the references to color in this figure legend, the reader is referred to the Web version of this article.)

2.1.3. Chitosan-Based Hydrogel Coating

Chitosan, the deacetylated chitin, is the only natural alkaline polysaccharide with charge. Due to their pH, ionic strength, and temperature sensitivity, chitosan-based hydrogels have good application prospects in the fields of targeting, sustained drug release, tissue engineering, and medical dressings [53]. Chitosan-based hydrogel coatings increase the antibacterial ability of the implant by loading antibacterial agents [54–56], or metal ions (Ag, Cu) [57–60]. Coatings give the implant the photocatalytic antibacterial effect by modifying or loading novel semiconductor materials, such as graphene [61], molybdenum disulfide [62], black phosphorus [63], and molybdenum diselenide [64]. They also promote osteogenesis through loading drugs, (for example, pitavastatin [65] and quercetin [66]), active substances (insulin growth factor binding protein-3 [67] and BMP-2 [68]) and inorganic matter (HA [69,70] and bioactive glass [71]). In addition, chitosan combines with polyanions such as gelatin [72,73], hyaluronic acid [74], and sodium alginate [75,76] to form a polyelectrolyte complex, promoting the surface functionalization of titanium. Modified carboxymethyl chitosan nanofibers, as a novel implant coating on titania nanotube arrays, inhibit bacterial colony formation and increase osteoblast cell survival [77] (Figure 4). Similarly, carboxymethyl chitosan loaded with silver nanoparticles enhances the antibacterial properties of the titanium alloy [78].

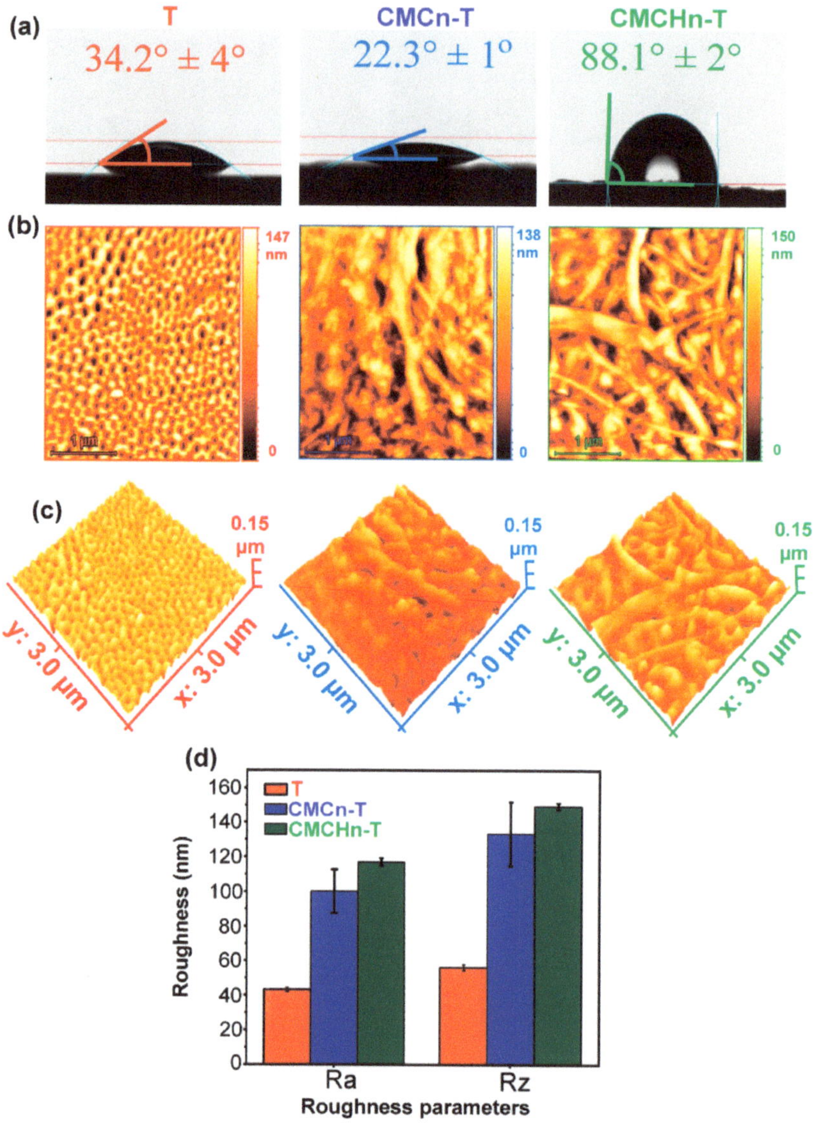

Figure 4. Wettability and topography of the implant's surface [77]. (**a**) Contact angle values and water droplet images, (**b**,**c**) 2D and 3D AFM images, and (**d**) average maximum height of the profile (Rz) and the arithmetic mean roughness (Ra) of the synthesized samples ($p < 0.05$).

2.1.4. Alginate-Based Hydrogel Coating

Alginate is a type of linear hydrophilic polysaccharide existing in brown algae. It forms hydrogels by ionic cross-linking with Ca^{2+} and other polyvalent inorganic cations. There are a large number of –OH and –COOH groups on the alginate skeleton, which can be modified by chemical or physical methods to achieve controlled release of cells or bioactive molecules in response to temperature, pH, and light [79]. Composite coating formed by alginate crosslinking with collagen enhances the cell adhesion of titanium implants [80,81]. Alginate and chitosan coating improves the biomineralization, the antibacterial activity, and corrosion resistance [75,82]. The addition of Ag further promotes the antibacterial ability and reduces the bacterial adhesion [83,84]. Alginate-based hydrogel coatings also provide sustained antibacterial activity by loading various antibacterial agents such as gentamicin [68], vancomycin [85], and chlorchloridine [86], and improving in vitro osteogenic differentiation as well as bone integration by loading BMP2 [87] and RGD [88].

2.2. *Synthetic Hydrogel Coatings*

The synthetic hydrogels have great application potential due to the wide source of raw materials, simple synthesis method, and controllable composition and structure. Synthetic hydrogels can be divided into the following functional groups: non-ionic hydrogels including polyvinyl alcohol (PVA), polyacrylamide (PAM), poly N-isopropylacrylamide (PNIPAm), polyethylene glycol (PEG), poly (lacto-glycolic acid) (PLGA), etc., and ionic hydrogels such as polyacrylic acid (PAA).

PVA hydrogels with porous titanium bases are being developed to repair or replace articular cartilage due to their high mechanical strength [89]. The Ti–hydrogel artificial cartilage material constructed with polydopamine (PDA), PVA, HA, or PAA as raw materials is an ideal high-strength and low-friction biomimetic cartilage material [90]. The novel "soft (PVA hydrogel layer)–hard (porous Ti6Al4V alloy substrate)" structure improves the surface wettability and tribological properties of Ti6Al4V alloy [91].

PAM hydrogel is an injectable soft-tissue-filling material. PAM-based hydrogels in combination with titanium-oxide nanotubes are also widely used as potential candidates for cartilage replacement [92]. PAM/PVA hydrogel on Ti6Al4V alloy configuration combines the good load-bearing capacity of the rigid substrate and the excellent lubrication of the hydrogel layer [93]. The cross-linked network porous structure of hydrogel is the main factor accounting for the low dynamic friction [94].

PNIPAm is a thermo-responsive polymer with lower critical solution temperature (LCST) around 32 °C. When the temperature is above the LCST, the polymer chains become hydrophobic and collapse, resulting in dense crosslinking networks in which the loaded molecules are more likely to be trapped, thus leading to slow release [95].

PEG is obtained by glycol dehydration polycondensation. The functional-group-hydroxyl at the end of the molecular chain is prone to chemical reactions and chemical modifications [96]. PLGA–PEG–PLGA hydrogels, polymerization of PLGA with PEG, are suitable for drug loading in vitro and sustained drug release in vivo, owing to the thermo-sensitive properties [97].

PAA hydrogels are three-dimensional macromolecules containing a large number of carboxyl groups that cannot move freely. When the pH value of the solution is different, it presents different degrees of shrinkage or swelling state. It can be used to prepare a simple and low-cost hydrogel-based bone adhesive to improve the osseointegration and anti-infection ability of the bone-implant interface [98].

3. Binding Method of Hydrogel Coating and Titanium Implant (Preparation Method of Hydrogel Coating)

Titanium implants have some disadvantages and one of the effective strategies is to prepare multifunctional hydrogel coatings on the surface. The most commonly used preparation method is sol–gel method.

3.1. *Electrochemical Methods*

The electrochemical methods for preparing hydrogel coatings mainly include electrochemical deposition and electrophoretic deposition. Electrochemical deposition refers to the process of forming coatings on the surface of metals or alloys in aqueous or non-aqueous solutions of inorganic salts and bio-active factors, which is a promising technique for surface modification of implants with various shapes, especially deformed structures [99,100]. The chitosan hydrogel coating, which is a versatile platform for Cu immobilization and precisely controlled synthesis via electrochemical deposition, has in vitro cell biocompatibility and catalyzed nitric-oxide-generation activity [101] (Figure 5). Electrophoretic deposition refers to the phenomenon of powder particles deposited from the suspension on the electrodes with opposite charges and certain shapes, relying on the action of the direct current [102]. The lanthanum- and silicate-substituted composite coating on a titanium implant achieved by the electrophoretic deposition method exhibited strong osteogenic ability [103]. Moreover, UV irradiation can be used as a crosslinking activator to deposit gentamicin-loaded agarose hydrogels, controlling the release of the loaded antibacterial agents while improving cell integration [104].

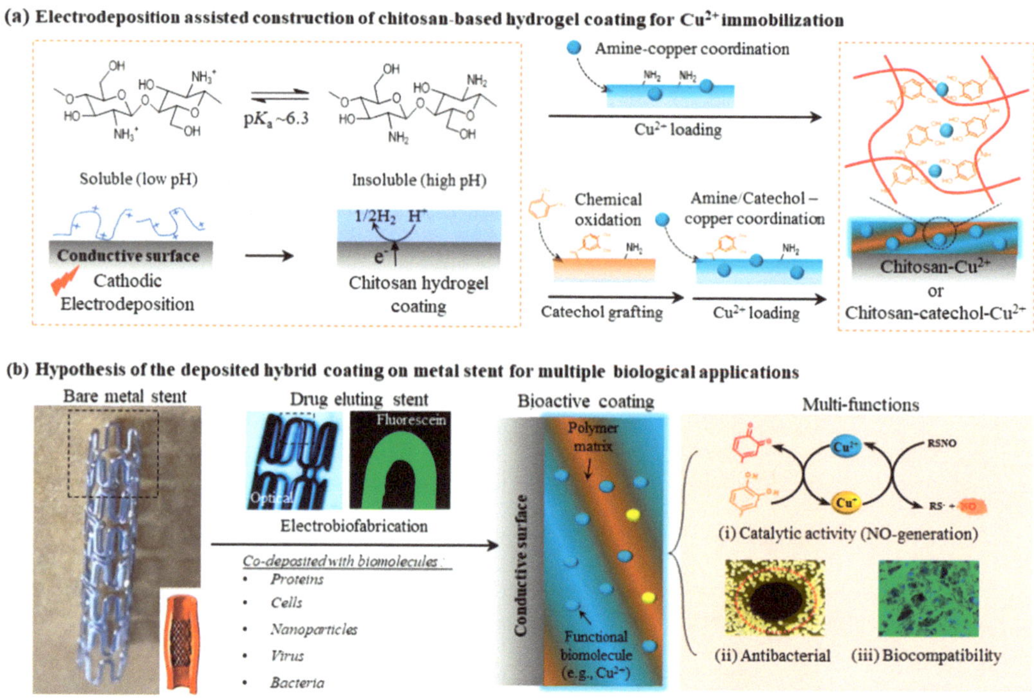

Figure 5. Electrochemical construction of a catechol-grafted chitosan film for Cu^{2+} incorporation [101].

3.2. Sol–Gel Method

Hydrogel coatings are mostly prepared by the sol–gel method. That is, monomers and coupling agents, as well as substances with different functions (initiators, loaded drugs, etc.), are dissolved in water to generate a free-radical polymerization reaction to form uncross-linked polymer chains. Then, the formulated aqueous solution is coated on the prepared matrix, and the polymer chain is cross-linked into a polymer network by the coupling agent. Finally, the polymer network is connected to the matrix by reacting with complementary functional groups on the matrix surface (Figure 6) [105].

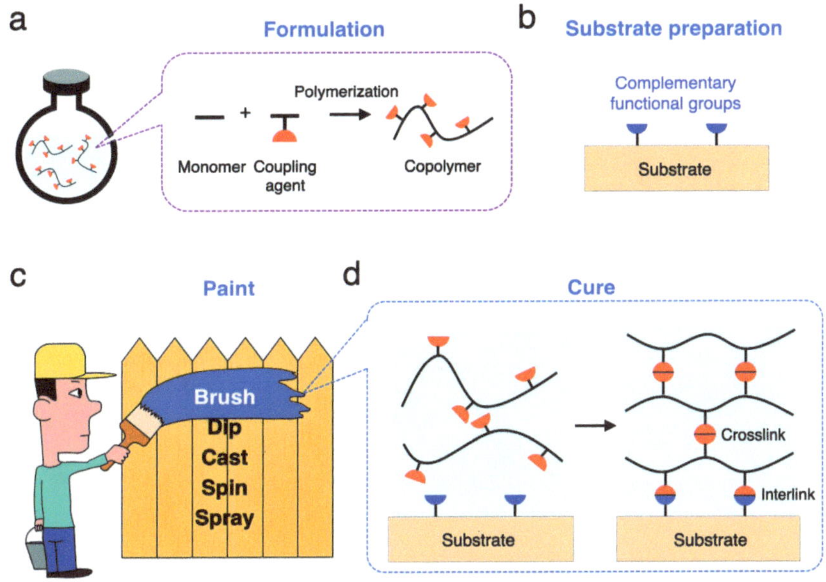

Figure 6. Principle of hydrogel coating [105]. (**a**) Formulation: monomer units and coupling agents copolymerize into polymer chains, but do not crosslink into a network, resulting in an aqueous solution. The solution may also contain other compounds for various functions but are not drawn here. (**b**) Substrate preparation: functional groups complementary to the coupling agents are imparted onto the surface of a substrate. (**c**) Paint: The aqueous solution is painted on the substrate by various operations. (**d**) Cure: The coupling agents react with each other to crosslink the polymer chains into network and react with the complementary functional groups to interlink the polymer network to the substrate.

Common methods of applying aqueous solution to the substrate are spraying [106], spin [26,107], and impregnated lift [52,108]. In order to enhance integration between implant material and hydrogel, a PDA layer was introduced onto the surface of the titanium alloy. Through chemical crosslinking between PDA and gelatin [45,51] or HRP/H_2O_2 catalysis [78], the hydrogel precursor could simply form a firm gel layer on the titanium alloy plate. Methylacryylated gelatin (GelMA) is a photo-cross-linked gelatin derivative. The photoinitiator [109] or catechol motifs [50,110] stabilize the GelMA hydrogel system and make the coating tightly adhere to titanium substrates after 365 nm UV exposure. The sol–gel method caused by ionizing radiation is a safe, simple operation with no polluting effects. Unfortunately, this method is rarely used in the preparation of hydrogel coatings on titanium and titanium alloys.

3.3. Layer-by-Layer Self-Assembly

Layer-by-layer self-assembly (LBL) is a popular surface modification method that uses electrostatic adsorption to self-assemble layers of materials with opposite charges into multilayer structures [111] (Figure 7). The assembly process is simple and gentle, and can maintain the biological activity of cytokines and achieve sustained release drug delivery [112]. However, it is necessary to pretreat the Ti surface with microarc oxidation, electrochemical deposition technology [113], etc., to firmly immobilize the multilayers. As one of the silyl reagents, 3-aminopropyl triethoxysilane is often used to aminofunctionalize titanium substrates, promoting covalent coupling to form precursor layers and facilitating the construction of future multilayer coatings [114]. The titanium alloy surfaces can also conjugate with dopamine as the base layer, which enables the deposition of gelatin molecules of hydrogel precursor [115]. PDA is a common mussel-inspired anchoring polymer and exhibits powerful reactivity to various bioactive molecules containing carboxyl groups, amino groups, and thiol groups. A multilayer type-I-collagen decorated nanoporous network was successfully developed on alkali-treated titanium surfaces via PDA coating and LBL [42]. The phase-transited lysozyme provides a new approach to achieving a high binding force that is superior to dopamine, and which forms an amyloid-like microfiber net that tightly adheres to Ti surfaces according to the transition process of lysozyme based on the β-sheet of lysozyme microfibers [116,117]. In addition, tannic acid is a low-cost plant polyphenol, which can bind materials tightly via hydrogen bonds, Michael addition reactions, Schiff base reactions, etc., due to the composition of a glucose core and a hydroxyl-rich phenolic shell, show great potential in LBL [118].

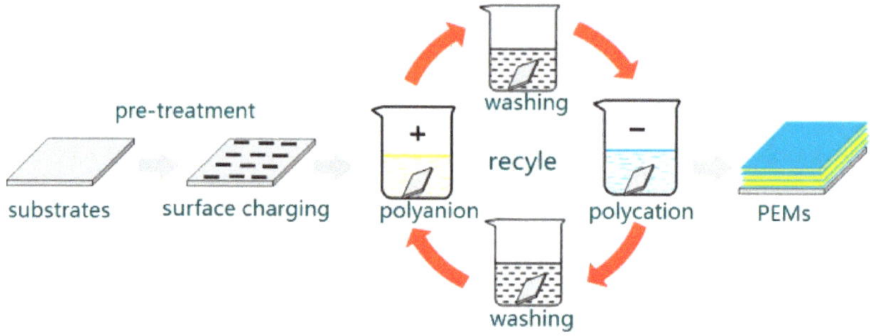

Figure 7. Layer-by-layer electrostatic self-assembly [112].

4. Characterization Methods of Surface Modification

Characterization methods for hydrogel coatings include nuclear magnetic resonance (NMR), Fourier transform infrared spectroscopy (FTIR), X-ray diffraction (XRD), and scanning electron microscopy (SEM). The position of the resonance signal on the NMR spectrum reflects the local structure of the sample molecules, such as functional groups. FTIR is the absorption spectrum generated by the absorption of specific wavelengths of infrared light during the vibrational energy level transition of bond-forming atoms in compound molecules, and is mainly used for structural analysis, qualitative identification, and quantitative analysis. XRD is the crystal structure analysis of hydrogel precursor polymer, such as fibroin protein, collagen, and other natural macromolecules containing crystal structure, or loaded nanoparticles such as biological glass and HA. Moreover, SEM and 3D optical profilometer are used to detect the morphological changes and thickness of hydrogel coating, respectively.

5. Application of the Hydrogel Coating

5.1. Osseointegration

Osseointegration is the direct contact between the implant and the bone tissue under the optical microscope, without fibrous connective tissue. Good osseointegration is a key factor in the long-term success of implants. Physical and chemical modification methods, such as changing surface properties [119] and loading inorganic substances, mainly indirectly affect cell behavior, with limitations in improving osteogenic activity [120,121]. Biochemical modifications caused by biomolecules [122–127] immobilized on the surface of titanium implants directly participate in biological processes and are more effective in inducing bone-formation, especially in poor bone conditions [128].

Hydrogel is a three-dimensional network cross-linked structure, which can not only simulate the extracellular matrix environment and develop bio-mimetic implants to design and repair bone defects [129], but also serve as drug carrier to carry and slowly release various active substances to promote bone-formation. Poloxamer-407 hydrogel loaded with simvastatin induces endogenous osteogenic growth factors and promotes bone ingrowth [130]. Pluronic F-127 hydrogel controls the release of 1α,25-Dihydroxyvitamin D3 as a bio-cap [131]. A non-toxic click hydrogel that rapidly polymerizes in situ provides localized controlled delivery of osteoprotective factor Semaphorin 3A [132]. Hydrogel containing BMP-2 facilitates dimensionally stable bone regeneration [133]. Dopamine-loaded RGD coatings on a vaterite-modified titanium surface successfully provided a solution to bone remodeling imbalance in osteoporosis by promoting osteoblasts and inhibiting osteoclasts at different concentrations [88] (Figure 8). Titanium implants loaded with human bone marrow mesenchymal stem cells (hBMSCs) show superior tissue ingrowth, and the synergic action of the bioactive hydrogel and hBMSCs increases both the bone deposition and integration [134]. In addition to these active ingredients, some inorganic substances such as a tri-calcium phosphate- [135], HA- [136], and silica-nanoparticle-loaded [137] hybrid hydrogels also improve the osteogenic ability of titanium implants, especially in combination with BMP-2 [138] or osteoblasts [139].

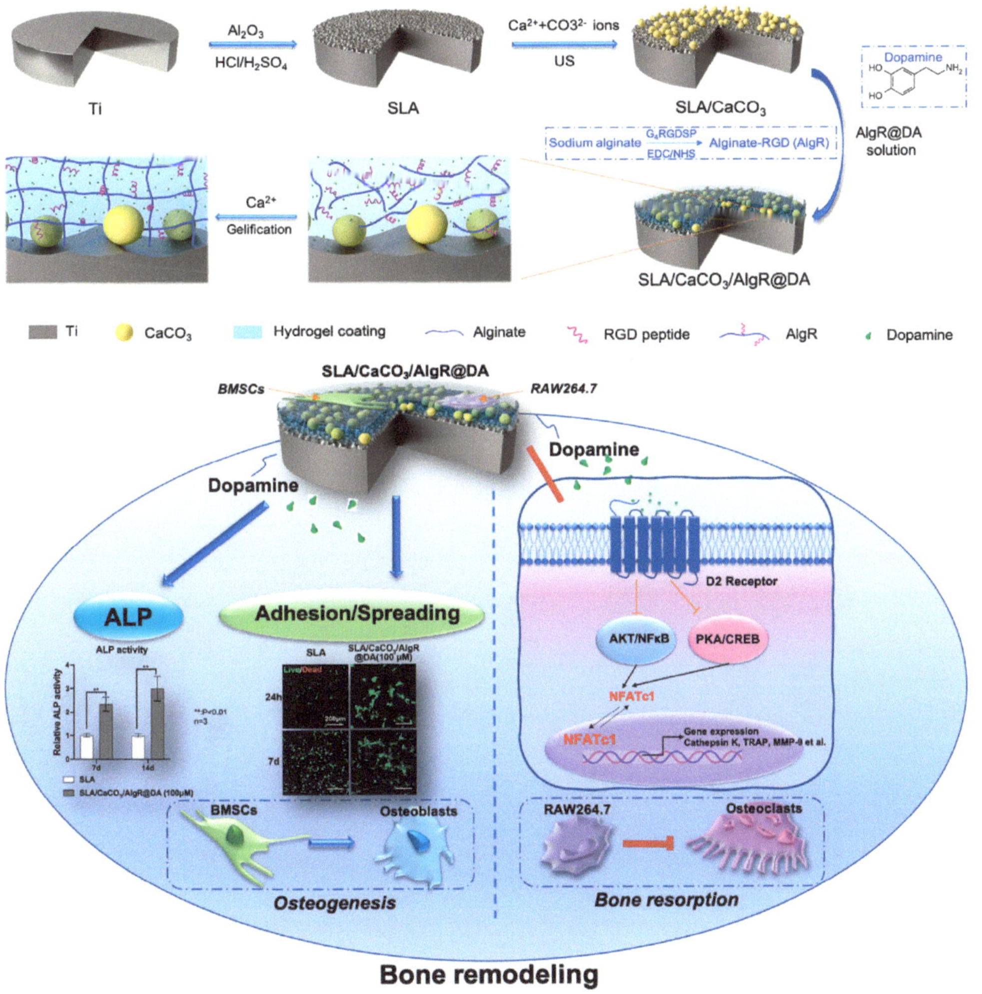

Figure 8. Hydrogel coatings on titanium bidirectionally regulate osteoclastic and osteogenic response behaviors [88].

5.2. Angiogenesis

Adequate blood supply plays an indispensable role in promoting bone regeneration, and angiogenesis promotion has become one of the key factors for the success of titanium implants. Hydrogels can act as carriers for drugs, growth factors, and cells, to promote angiogenesis around titanium implants. The combination of simvastatin-loaded hydrogel coating with porous titanium alloy significantly improved the formation of new blood vessels around rabbit tibial implants, providing an effective strategy for bone integration and bone growth [140]. The heat-sensitive collagen hydrogel/porous titanium alloy scaffold system equipped with VEGF, increased vascular permeability, promoted proliferation and induction of HUVECs, and aided in angiogenic-mediated bone regeneration [44] (Figure 9). The composite scaffold loaded with VEGF and BMP continuously provided angiogenic and osteogenic growth factors at the site of osseous defect, thus exhibiting higher bone integration capacity and new bone amount [46,141]. Combining cell-laden hydrogels with porous titanium alloys develops a vascularized bone implant. Co-encapsulating hMSCs with HUVECs [52] or endothelial progenitor cells (EPCs) [142], support HUVEC- and EPC-spreading and vascular-like network formation, along with osteogenesis of hMSCs.

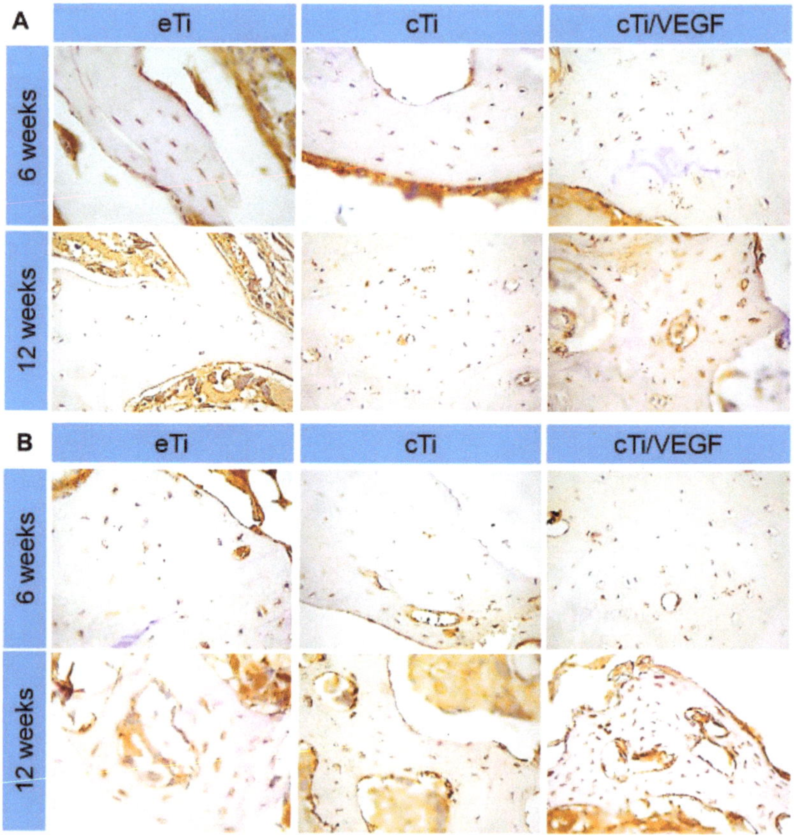

Figure 9. Sustained release of VEGF promotes COL I (**A**) and CD 31 (**B**) expression in bone-surrounding scaffolds 6 and 12 weeks after implantation (n = 3) [44].

5.3. Macrophage Polarization

Macrophage polarization is a reversible and modified dynamic process involving in the occurrence, development and outcome of many immune inflammatory diseases, including peri-implantitis. The introduction of hydrogel for "reprogramming" of the macrophage state is a novel strategy to induce resolution of inflammation [143]. Interleukin-4 (IL-4) is a common inflammatory factor, which can regulate the antigen-presenting ability of macrophages, inhibit the secretion of inflammatory factors such as IL-1 β and TNF- α, and promote the differentiation of macrophages into profibrotic macrophages to secrete TGF-β. IL-4-loading of a hydrogel system on titanium modulated pro-inflammatory reactions [110]. Hydrogels containing interferon-γ and IL-4 were able to modulate the transformation with a stronger effect than those containing only IL-4 [144] (Figure 10). Combination of IL-4 and cell adhesive motif (RGD) onto the Ti substrate synergistically generated a more favorable early-stage osteo-immune environment with superior osteogenic properties [145]. Dexamethasone, as a glucocorticoid, can also regulate macrophage polarization and plays an important role in the regression of inflammation. The novel DNA hydrogel on the titanium surface, as the platform for dexamethasone delivery, extends the half- life of the release profile [146]. Reactive oxygen species (ROS) produced by macrophages regulate a variety of physiological functions including endothelial cells growth, migration, and mesenchymal stem cells activation. Removing excessive ROS by a two-component hydrogel coating containing borate ester bond and thymosin β4 favors M1 to M2 phenotype switch of macrophages and inflammatory response regulation [147].

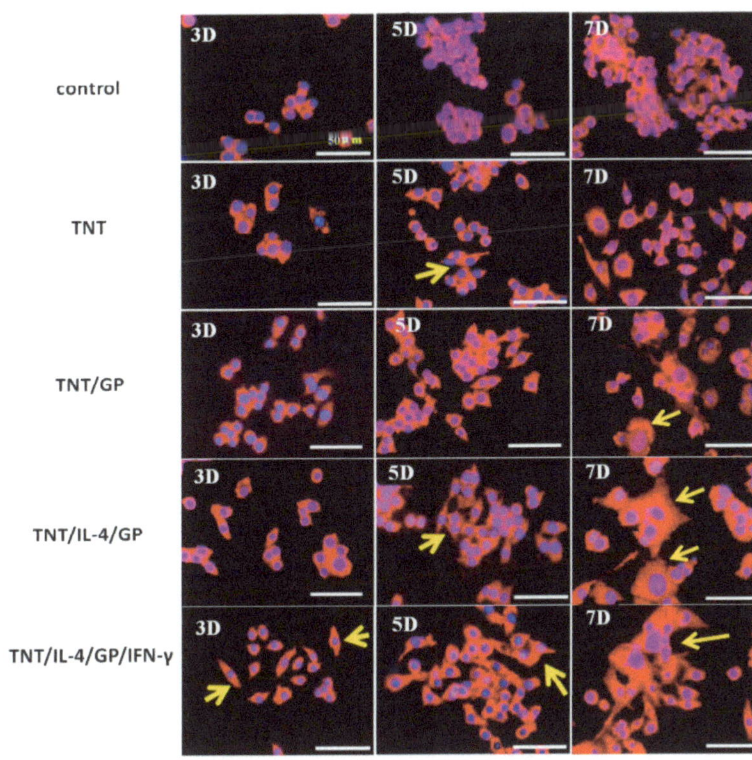

Figure 10. Fluorescence images of macrophage morphologies on dual-inflammatory cytokine (interferon-γ and IL-4)-coated TiO_2 nanotube surfaces [144]. Activated macrophages are indicated by yellow arrows.

5.4. Antibacterial

Bacterial biofilm formation can cause implant infection and osseointegration loss, resulting in loosening and dropping. Hydrogels with good biocompatibility and drug loading capability can slowly release various antibacterial components to prevent initial bacterial adhesion [13]. Designing and constructing a hydrogel drug-controlled release system by loading with antibacterial drugs such as gentamicin [104,108,148] or vancomycin [98,149,150] on a titanium surface is a frequently used strategy. Antibacterial peptides have garnered more attention as alternative antibacterial agents of implant coating due to their unique antibacterial mechanism [151,152]. Bacteriophage-loaded hydrogels also showed excellent antimicrobial activity in inhibiting attachment and colonization of multidrug-resistant *E. faecalis* surrounding and within femoral tissues [153]. Metal antibacterial agents are introduced into the implant hydrogel coating because of their broad-spectrum antibacterial properties and no drug resistance. Among them, silver ion is most commonly used [26,78,154,155]. Metal oxide antimicrobial agents such as zinc oxide [109] and calcium oxide [110] also show significant antibacterial ability in the coating, although Zn ion has renal absorption toxicity. Photodynamic therapy is a promising modality in antibacterial material design. The introduction of photosensitizer Chlorin e6 with laser-triggered ROS generation property exhibited a remarkable and rapid antibacterial activity when the laser power was 1 W cm^{-2} [50]. Coatings with semiconductor photocatalytic materials, such as bismuth [51] and red phosphorus [156], can produce ROS, kill bacteria and eradicate biofilm under light, which might provide a novel multimodal antibacterial and anti-biofilm treatment for infection.

5.5. Drug Delivery

Hydrogels have been widely used in various fields of medicine as vehicles to control the continuous release of drugs [157]. Loading cefuroxime, tetracycline, amoxicillin, or acetylsalicylic acid through hydrogel coating can improve the anti-infection effect of the implant [158]. Loading bone-metabolism-related drugs, proteins, peptides, and growth factors has demonstrated better osseointegration, especially in challenged degenerative conditions, such as osteoporosis, osteoarthritis, and osteogenesis imperfecta [159]. Similarly, hydrogels can load with cytokines to promote macrophage polarization [160] and angiogenesis [141], which have great potential for application in bone-tissue regeneration and repair.

6. Conclusions and Future Protects

In recent decades, metallic materials have been widely used in the field of biomaterials for their good mechanical properties and biocompatibility. Among them, the application prospects of biomedical titanium alloy are particularly remarkable. However, some disadvantages of titanium alloys limit their further application. Therefore, scientists have been working to explore improvements in the properties of titanium alloys. Hydrogel coatings can serve as ideal carriers to introduce drugs, peptides, metal ions, growth factors, and cells to effectively bio-modify titanium alloys. In this study, we have reviewed the popular matrix of hydrogel coatings, especially the natural materials such as collagen, gelatin, chitosan, and alginate. The usual modification methods are the electrochemical method, the sol–gel method, and layer-by-layer self-assembly. Hydrogel coatings significantly improve the properties of the titanium implant in osseointegration, angiogenesis, macrophage polarization, antibacterial effects, and drug delivery.

Although the improvement of titanium alloy caused by hydrogel coating is obvious, there are still some problems worth noting: (1) The mechanical properties of hydrogel coating are poor, and whether some inorganic fillers can be added to promote bone integration and mechanical properties needs further research. (2) The strong bond between the hydrogel coating and titanium alloy needs to be further strengthened. (3) Some semiconductor materials have excellent photocatalytic properties, and the introduction of semiconductor materials in hydrogel coatings is promising for photodynamic therapy. Therefore, future research should focus on these aspects to further improve the properties of hydrogel coating on titanium alloy.

Author Contributions: Conceptualization, B.H., Y.L., R.F. and H.C.; formal analysis, X.Q.; data curation, T.X.; writing—original draft preparation, R.F., H.C., Q.L. and Z.W.; writing—review and editing, B.H.; visualization. All authors have read and agreed to the published version of the manuscript.

Funding: This work was supported by the Key R&D Plan of Hubei Provincial Department of Science and Technology (Major Health Project, No. 2022BCE026), the Hubei Provincial Colleges and Universities Outstanding Young and Middle-aged Technological Innovation Team Project (No. T2020022), the Xianning City Key Program of Science & Technology (No. 2021GXYF021), the Special Fund for medical research of Hubei University of Science and Technology (2022YKY14), the Special fund of the School of Stomatology and Ophthalmology of Hubei University of Science and Technology (2020XZ32, 2020WG01), the Science Development Foundation of Hubei University of Science & Technology (No. 2020TD01, 2021ZX01, 2022FH09), and the Provincial College Students Innovation Training Program (S202210927030, S202210927031).

Institutional Review Board Statement: Not applicable.

Informed Consent Statement: Not applicable.

Data Availability Statement: Not applicable.

Conflicts of Interest: The authors declare no conflict of interest.

References

1. Bothe, R.T.; Beaton, L.E.; Davenport, H.A. Reaction of bone to multiple metallic implants. *Surg. Gynecol. Obstet.* **1940**, *71*, 598–602.
2. Bordbar-Khiabani, A.; Gasik, M. Electrochemical and biological characterization of Ti-Nb-Zr-Si alloy for orthopedic applications. *Sci. Rep.* **2023**, *13*, 2312. [CrossRef] [PubMed]
3. Alipal, J.; Mohd Pu'ad, N.A.S.; Nayan, N.H.M. An updated review on surface functionalisation of titanium and its alloys for implants applications. *Mater. Today Proc.* **2021**, *42*, 270–282. [CrossRef]
4. Del Castillo, R.; Chochlidakis, K.; Galindo-Moreno, P.; Ercoli, C. Titanium Nitride Coated Implant Abutments: From Technical Aspects and Soft tissue Biocompatibility to Clinical Applications. A Literature Review. *J. Prosthodont.* **2022**, *31*, 571–578. [CrossRef]
5. Prabhakar, V.; Chidambaranathan, A.S.; Balasubramanium, M. Effect of Cathodic Arc Plasma Deposition on Shear Bond Strength between Palladium Cobalt Chromium Coated with Titanium Nitride and Titanium Aluminium Nitride with Ceramic. *Contemp. Clin. Dent.* **2021**, *12*, 49–54. [PubMed]
6. Zhang, R.; Wan, Y.; Ai, X.; Zhang, D. Corrosion resistance and biological activity of TiO_2 implant coatings produced in oxygen-rich environments. *Proc. Inst. Mech. Eng. Part H J. Eng. Med.* **2017**, *231*, 20–27. [CrossRef] [PubMed]
7. Zhao, X.; You, L.; Wang, T.; Li, B. Enhanced Osseointegration of Titanium Implants by Surface Modification with Silicon-doped Titania Nanotubes. *Int. J. Nanomed.* **2020**, *15*, 8583–8594. [CrossRef]
8. Shimabukuro, M. Antibacterial Property and Biocompatibility of Silver, Copper, and Zinc in Titanium Dioxide Layers Incorporated by One-Step Micro-Arc Oxidation: A Review. *Antibiotics* **2020**, *9*, 716. [CrossRef] [PubMed]
9. Alshimaysawee, S.; Fadhel Obaid, R.; Al-Gazally, M.E.; Bathaei, M.S. Recent Advancements in Metallic Drug-Eluting Implants. *Pharmaceutics* **2023**, *15*, 223. [CrossRef]
10. Yuan, Y.X.; Luo, R.D.; Ren, J.K.; He, Z.Y. Design of a new Ti-Mo-Cu alloy with excellent mechanical and antibacterial properties as implant materials. *Mater. Lett.* **2022**, *306*, 130875. [CrossRef]
11. Lupi, S.M.; Torchia, M.; Rizzo, S. Biochemical Modification of Titanium Oral Implants: Evidence from in Vivo Studies. *Materials* **2021**, *14*, 2798. [CrossRef] [PubMed]
12. Zheng, Q.C.; Mao, L.L.; Shi, Y.T.; Hu, Y.H. Biocompatibility of Ti-6Al-4V titanium alloy implants with laser microgrooved surfaces. *Mater. Technol.* **2022**, *37*, 2039–2048. [CrossRef]
13. Bohara, S.; Suthakorn, J. Surface coating of orthopedic implant to enhance the osseointegration and reduction of bacterial colonization: A review. *Biomater. Res.* **2022**, *26*, 26. [CrossRef] [PubMed]
14. De Giglio, E.; Cometa, S.; Cioffi, N.; Sabbatini, L. Analytical investigations of poly (acrylic acid) coatings electrodeposited on titanium-based implants: A versatile approach to biocompatibility enhancement. *Anal. Bioanal. Chem.* **2007**, *389*, 2055–2063. [CrossRef]
15. Nagai, M.; Hayakawa, T.; Fukatsu, A.; Kato, T. In vitro study of collagen coating of titanium implants for initial cell attachment. *Dent. Mater. J.* **2002**, *21*, 250–260. [CrossRef]
16. Hauser, J.; Ring, A.; Schaffran, A.; Langer, S. In vivo analysis of tissue response to plasma-treated collagen-I-coated titanium alloys. *Eur. Surg. Res.* **2009**, *43*, 262–268. [CrossRef]
17. Raita, Y.; Komatsu, K.; Hayakawa, T. Pilot study of gingival connective tissue responses to 3-dimensional collagen nanofiber-coated dental implants. *Dent. Mater. J.* **2015**, *34*, 847–854. [CrossRef] [PubMed]
18. Vanderleyden, E.; Van Bael, S.; Chai, Y.C.; Dubruel, P. Gelatin functionalised porous titanium alloy implants for orthopaedic applications. *Mat. Sci. Eng. C Mater.* **2014**, *42*, 396–404. [CrossRef]

19. Mahin, T.; Torab, A.; Negahdari, R.; Sharifi, S. The Antibacterial Effects of Healing Abutments Coated with Gelatin-curcumin Nanocomposite. *Pharm. Nanotechnol.* **2023**. [CrossRef]
20. López-Valverde, N.; Aragoneses, J.; López-Valverde, A.; Aragoneses, J.M. Role of chitosan in titanium coatings. trends and new generations of coatings. *Front. Bioeng. Biotechnol.* **2022**, *10*, 907589. [CrossRef]
21. Bumgardner, J.D.; Chesnutt, B.M.; Yuan, Y.; Ong, J.L. The integration of chitosan-coated titanium in bone: An in vivo study in rabbits. *Implant* **2007**, *16*, 66–79. [CrossRef] [PubMed]
22. Peng, Z.X.; Ao, H.Y.; Wang, L.; Tang, T.T. Quaternised chitosan coating on titanium provides a self-protective surface that prevents bacterial colonisation and implant-associated infections. *RSC Adv.* **2015**, *5*, 54304–54311. [CrossRef]
23. Jian, Y.H.; Yang, C.; Zhang, J.X.; Du, Y.M. One-step electrodeposition of Janus chitosan coating for metallic implants with anti-corrosion properties. *Colloid Surf.* **2022**, *641*, 128498. [CrossRef]
24. Guo, C.C.; Cui, W.D.; Wang, X.W.; Chen, J.L. Poly-L-lysine/Sodium Alginate Coating Loading Nanosilver for Improving the Antibacterial Effect and Inducing Mineralization of Dental Implants. *ACS Omega* **2020**, *5*, 10562–10571. [CrossRef] [PubMed]
25. Yuan, N.; Jia, L.; Geng, Z.; Liu, Y. The Incorporation of Strontium in a Sodium Alginate Coating on Titanium Surfaces for Improved Biological Properties. *Biomed. Res. Int.* **2017**, *2017*, 9867819. [CrossRef] [PubMed]
26. Abdul Azam, F.A.; Ismail, H.; Abdul Hamid, M.A. Characterizations on Morphology and Adhesion of Calcium Silicate Coating on Ti6Al4V Substrate. *Key Eng. Mater.* **2016**, *694*, 83–87. [CrossRef]
27. Sille, I.E.; Pissinis, D.E.; Fagali, N.S.; Schilardi, P.L. Antimicrobial-Loaded Polyacrylamide Hydrogels Supported on Titanium as Reservoir for Local Drug Delivery. *Pathogens* **2023**, *12*, 202. [CrossRef]
28. Buxadera-Palomero, J.; Calvo, C.; Torrent-Camarero, S.; Rodríguez, D. Biofunctional polyethylene glycol coatings on titanium: An in vitro-based comparison of functionalization methods. *Colloid Surf.* **2017**, *152*, 367–375. [CrossRef]
29. Xiao, D.Q.; Liu, Q.; Weng, J. Room-temperature attachment of PLGA microspheres to titanium surfaces for implant-based drug release. *Appl. Surf. Sci.* **2014**, *30*, 112–118. [CrossRef]
30. Iafiscol, M.; Quirici, N.; Foltran, I.; Rimondini, L. Electrospun collagen mimicking the reconstituted extracellular matrix improves osteoblastic differentiation onto titanium surfaces. *J. Nanosci. Nanotechnol.* **2013**, *13*, 4720–4726. [CrossRef]
31. Becker, D.; Geissler, U.; Hempel, U.; Bierbaum, S.; Scharnweber, D.; Worch, H.; Wenzel, K.W. Proliferation and differentiation of rat calvarial osteoblasts on type I collagen-coated titanium alloy. *J. Biomed. Mater. Res.* **2002**, *59*, 516–527. [CrossRef] [PubMed]
32. Bierbaum, S.; Hempel, U.; Geissler UHanke, T.; Scharnweber, D.; Wenzel, K.W.; Worch, H. Modification of Ti6Al4V surfaces using collagen i, iii, and fibronectin. Ii. Influence on osteoblast responses. *J. Biomed. Mater. Res. A* **2003**, *67*, 431–438. [CrossRef] [PubMed]
33. Geissler, U.; Hempel, U.; Wolf, C.; Scharnweber, D.; Worch, H.; Wenzel, K. Collagen type I-coating of Ti6Al4V promotes adhesion of osteoblasts. *J. Biomed. Mater. Res.* **2000**, *51*, 752–760. [CrossRef]
34. Ritz, U.; Nusselt, T.; Sewing, A.; Hofmann, A. The effect of different collagen modifications for titanium and titanium nitrite surfaces on functions of gingival fibroblasts. *Clin. Oral. Investig.* **2017**, *21*, 255–265. [CrossRef] [PubMed]
35. Sharan, J.; Koul, V.; Dinda, A.K.; Singh, M.P. Bio-functionalization of grade V titanium alloy with type I human collagen for enhancing and promoting human periodontal fibroblast cell adhesion—An in-vitro study. *Colloid Surf.* **2018**, *161*, 1–9. [CrossRef]
36. Morra, M.; Cassinelli, C.; Meda, L.; Giardino, R. Surface analysis and effects on interfacial bone microhardness of collagen-coated titanium implants: A rabbit model. *Int. J. Oral Maxillofac. Implant.* **2005**, *20*, 23–30.
37. Ciobanu, G.; Ciobanu, O. Investigation on the effect of collagen and vitamins on biomimetic hydroxyapatite coating formation on titanium surfaces. *Mat. Sci. Eng. C Mater.* **2013**, *33*, 1683–1688. [CrossRef]
38. Ruiz, G.C.M.; Cruz, M.A.E.; Faria, A.N.; Ramos, A.P. Biomimetic collagen/phospholipid coatings improve formation of hydroxyapatite nanoparticles on titanium. *Mat. Sci. Eng. C Mater.* **2017**, *77*, 102–110. [CrossRef]
39. Patty, D.J.; Nugraheni, A.D.; Dewi Ana, I.; Yusuf, Y. Mechanical Characteristics and Bioactivity of Nanocomposite Hydroxyapatite/Collagen Coated Titanium for Bone Tissue Engineering. *Bioengineering* **2022**, *9*, 784. [CrossRef]
40. Pokorný, M.; Suchý, T.; Kotzianová, A.; Čejka, Z. Surface Treatment of Acetabular Cups with a Direct Deposition of a Composite Nanostructured Layer Using a High Electrostatic Field. *Molecules* **2020**, *25*, 1173. [CrossRef]
41. Iwanami-Kadowaki, K.; Uchikoshi, T.; Uezono, M.; Moriyama, K. Development of novel bone-like nanocomposite coating of hydroxyapatite/collagen on titanium by modified electrophoretic deposition. *J. Biomed. Mater. Res. A* **2021**, *109*, 1905–1911. [CrossRef] [PubMed]
42. Zhao, Y.; Bai, L.; Zhang, Y.; Hang, R. Type I collagen decorated nanoporous network on titanium implant surface promotes osseointegration through mediating immunomodulation, angiogenesis, and osteogenesis. *Biomaterials* **2022**, *288*, 121684. [CrossRef] [PubMed]
43. Shao, J.; Weng, L.; Li, J.; Lin, J. Regulation of Macrophage Polarization by Mineralized Collagen Coating to Accelerate the Osteogenic Differentiation of Mesenchymal Stem Cells. *ACS Biomater. Sci. Eng.* **2022**, *8*, 610–619. [CrossRef] [PubMed]
44. Li, Y.; Liu, Y.; Bai, H.; Huang, L. Sustained Release of VEGF to Promote Angiogenesis and Osteointegration of Three-Dimensional Printed Biomimetic Titanium Alloy Implants. *Front. Bioeng. Biotechnol.* **2021**, *9*, 757767. [CrossRef] [PubMed]
45. Dinh, T.N.; Hou, S.; Park, S.; Jeong, K.J. Gelatin Hydrogel Combined with Polydopamine Coating to Enhance Tissue Integration of Medical Implants. *ACS Biomater. Sci. Eng.* **2018**, *4*, 3471–3477. [CrossRef] [PubMed]
46. Liu, Z.; Xu, Z.; Wang, X.; Jia, R. Construction and osteogenic effects of 3D-printed porous titanium alloy loaded with VEGF/BMP-2 shell-core microspheres in a sustained-release system. *Front. Bioeng. Biotechnol.* **2022**, *10*, 1028278. [CrossRef] [PubMed]

47. Van den Bulcke, A.I.; Bogdanov, B.; de Rooze, N.; Schacht, E.H.; Cornelissen, M.; Berghmans, H. Structural and rheological properties of methaerylamide modified gelatin hydrogels. *Biomacromolecules* **2000**, *1*, 31–38. [CrossRef]
48. Cheng, H.; Yue, K.; Kazemzadeh-Narbat, M.; Khademhosseini, A. Mussel-Inspired Multifunctional Hydrogel Coating for Prevention of Infections and Enhanced Osteogenesis. *ACS Appl. Mater. Interfaces* **2017**, *9*, 11428–11439. [CrossRef] [PubMed]
49. Kim, S.Y.; Choi, A.J.; Park, J.E.; Lee, M.H. Antibacterial Activity and Biocompatibility with the Concentration of Ginger Fraction in Biodegradable Gelatin Methacryloyl (GelMA) Hydrogel Coating for Medical Implants. *Polymers* **2022**, *14*, 5317. [CrossRef] [PubMed]
50. He, Y.; Leng, J.; Li, K.; Cai, K. A multifunctional hydrogel coating to direct fibroblast activation and infected wound healing via simultaneously controllable photobiomodulation and photodynamic therapies. *Biomaterials* **2021**, *278*, 121164. [CrossRef]
51. Ding, Y.; Ma, R.; Liu, G.; Cai, K. Fabrication of a New Hyaluronic Acid/Gelatin Nanocomposite Hydrogel Coating on Titanium-Based Implants for Treating Biofilm Infection and Excessive Inflammatory Response. *ACS Appl. Mater. Interfaces* **2023**, *15*, 13783–13801. [CrossRef] [PubMed]
52. Li, J.; Cui, X.; Lindberg, G.C.J.; Woodfield, T.B.F. Hybrid fabrication of photo-clickable vascular hydrogels with additive manufactured titanium implants for enhanced osseointegration and vascularized bone formation. *Biofabrication* **2022**, *14*, 034103. [CrossRef]
53. Taokaew, S.; Kaewkong, W.; Kriangkrai, W. Recent Development of Functional Chitosan-Based Hydrogels for Pharmaceutical and Biomedical Applications. *Gels* **2023**, *9*, 277. [CrossRef]
54. Stevanović, M.; Đošić, M.; Janković, A.; Mišković-Stanković, V. Gentamicin-Loaded Bioactive Hydroxyapatite/Chitosan Composite Coating Electrodeposited on Titanium. *ACS Biomater. Sci. Eng.* **2018**, *4*, 3994–4007. [CrossRef] [PubMed]
55. Zarghami, V.; Ghorbani, M.; Bagheri, K.P.; Shokrgozar, M.A. Improving bactericidal performance of implant composite coatings by synergism between Melittin and tetracycline. *J. Mater. Sci. Mater. Med.* **2022**, *33*, 46–58. [CrossRef] [PubMed]
56. Asadi, S.; Mortezagholi, B.; Hadizadeh, A.; Chaiyasut, C. Ciprofloxacin-Loaded Titanium Nanotubes Coated with Chitosan: A Promising Formulation with Sustained Release and Enhanced Antibacterial Properties. *Pharmaceutics* **2022**, *14*, 1359. [CrossRef] [PubMed]
57. Divakar, D.D.; Jastaniyah, N.T.; Altamimi, H.G.; Haleem, S. Enhanced antimicrobial activity of naturally derived bioactive molecule chitosan conjugated silver nanoparticle against dental implant pathogens. *Int. J. Biol. Macromol.* **2018**, *108*, 790–797. [CrossRef] [PubMed]
58. Pawłowski, Ł.; Wawrzyniak, J.; Banach-Kopeć, A.; Zieliński, A. Antibacterial properties of laser-encapsulated titanium oxide nanotubes decorated with nanosilver and covered with chitosan/Eudragit polymers. *Biomater. Adv.* **2022**, *138*, 212950. [CrossRef]
59. Zhang, T.; Qin, X.; Gao, Y.; Yin, P. Functional chitosan gel coating enhances antimicrobial properties and osteogenesis of titanium alloy under persistent chronic inflammation. *Front. Bioeng. Biotechnol.* **2023**, *11*, 1118487. [CrossRef] [PubMed]
60. Han, J.; Hassani Besheli, N.; Deng, D.; Yang, F. Tailoring Copper-Doped Bioactive Glass/Chitosan Coatings with Angiogenic and Antibacterial Properties. *Tissue Eng. Part C Methods* **2022**, *28*, 314–324. [CrossRef]
61. Stevanović, M.; Djošić, M.; Janković, A.; Mišković-Stanković, V. Antibacterial graphene-based hydroxyapatite/chitosan coating with gentamicin for potential applications in bone tissue engineering. *J. Biomed. Mater. Res. A* **2020**, *108*, 2175–2189. [CrossRef]
62. Zhu, M.; Liu, X.; Tan, L.; Wu, S. Photo-responsive chitosan/Ag/MoS_2 for rapid bacteria-killing. *J. Hazard. Mater.* **2020**, *383*, 121122. [CrossRef]
63. Bose, S.; Surendhiran, D.; Chun, B.S.; Kang, H.W. Facile synthesis of black phosphorus-zinc oxide nanohybrids for antibacterial coating of titanium surface. *Colloid Surf.* **2022**, *219*, 112807. [CrossRef]
64. Chai, M.; An, M.; Zhang, X. Construction of a TiO_2/$MoSe_2$/CHI coating on dental implants for combating Streptococcus mutans infection. *Mat. Sci. Eng. C Mater.* **2021**, *129*, 112416. [CrossRef]
65. Chen, W.; Xie, G.; Lu, Y.; Bao, J. An improved osseointegration of metal implants by pitavastatin loaded multilayer films with osteogenic and angiogenic properties. *Biomaterials* **2022**, *280*, 121260. [CrossRef]
66. Wang, B.; Chen, L.; Xie, J.; Yang, L. Coating Polyelectrolyte Multilayers Loaded with Quercetin on Titanium Surfaces by Layer-By-Layer Assembly Technique to Improve Surface Osteogenesis Under Osteoporotic Condition. *J. Biomed. Nanotechnol.* **2021**, *17*, 1392–1403. [CrossRef] [PubMed]
67. Takanche, J.S.; Kim, J.E.; Jang, S.; Yi, H.K. Insulin growth factor binding protein-3 enhances dental implant osseointegration against methylglyoxal-induced bone deterioration in a rat model. *J. Periodontal. Implant Sci.* **2022**, *52*, 155–169. [CrossRef]
68. Tao, B.; Deng, Y.; Song, L.; Cai, K. BMP2-loaded titania nanotubes coating with pH-responsive multilayers for bacterial infections inhibition and osteogenic activity improvement. *Colloid Surf.* **2019**, *177*, 242–252. [CrossRef] [PubMed]
69. Zhang, T.; Zhang, X.; Mao, M.; Sun, H. Chitosan/hydroxyapatite composite coatings on porous Ti_6Al_4V titanium implants: In vitro and in vivo studies. *J. Periodontal. Implant Sci.* **2020**, *50*, 392–405. [CrossRef] [PubMed]
70. Gaafar, M.S.; Yakout, S.M.; Barakat, Y.F.; Sharmoukh, W. Electrophoretic deposition of hydroxyapatite/chitosan nanocomposites: The effect of dispersing agents on the coating properties. *RSC Adv.* **2022**, *12*, 27564–27581. [CrossRef]
71. Rastegari, S.; Salahinejad, E. Surface modification of Ti-6Al-4V alloy for osseointegration by alkaline treatment and chitosan-matrix glass-reinforced nanocomposite coating. *Carbohyd. Polym.* **2019**, *205*, 302–311. [CrossRef] [PubMed]
72. Ma, K.; Cai, X.; Zhou, Y.; Jiang, T. In Vitro and In Vivo Evaluation of Tetracycline Loaded Chitosan-Gelatin Nanosphere Coatings for Titanium Surface Functionalization. *Macromol. Biosci.* **2017**, *17*, 201600130. [CrossRef] [PubMed]

73. Zhao, D.; Dong, H.; Niu, Y.; Zhang, Z. Electrophoretic deposition of novel semi-permeable coatings on 3D-printed Ti-Nb alloy meshes for guided alveolar bone regeneration. *Dent. Mater.* **2022**, *38*, 431–443. [CrossRef] [PubMed]
74. Tang, J.; Chen, L.; Yan, D.; Shen, L. Surface Functionalization with Proanthocyanidins Provides an Anti-Oxidant Defense Mechanism That Improves the Long-Term Stability and Osteogenesis of Titanium Implants. *Int. J. Nanomed.* **2020**, *15*, 1643–1659. [CrossRef]
75. Vakili, N.; Asefnejad, A. Titanium coating: Introducing an antibacterial and bioactive chitosan-alginate film on titanium by spin coating. *Biomed. Eng. Biomed. Tech.* **2020**, *65*, 621–630. [CrossRef] [PubMed]
76. Jabłoński, P.; Kyzioł, A.; Pawcenis, D.; Kyzioł, K. Electrostatic self-assembly approach in the deposition of bio-functional chitosan-based layers enriched with caffeic acid on Ti-6Al-7Nb alloys by alternate immersion. *Biomater. Adv.* **2022**, *136*, 212791. [CrossRef] [PubMed]
77. Rahnamaee, S.Y.; Dehnavi, S.M.; Bagheri, R.; Karimi, A. Boosting bone cell growth using nanofibrous carboxymethylated cellulose and chitosan on titanium dioxide nanotube array with dual surface charges as a novel multifunctional bioimplant surface. *Int. J. Biol. Macromol.* **2023**, *228*, 570–581. [CrossRef]
78. Ren, Y.; Qin, X.; Barbeck, M.; Liu, C. Mussel-Inspired Carboxymethyl Chitosan Hydrogel Coating of Titanium Alloy with Antibacterial and Bioactive Properties. *Materials* **2021**, *14*, 6901. [CrossRef]
79. Abasalizadeh, F.; Moghaddam, S.V.; Alizadeh, E.; Akbarzadeh, A. Alginate-based hydrogels as drug delivery vehicles in cancer treatment and their applications in wound dressing and 3D bioprinting. *J. Biol. Eng.* **2020**, *14*, 8. [CrossRef]
80. Vrana, N.E.; Dupret-Bories, A.; Bach, C.; Lavalle, P. Modification of macroporous titanium tracheal implants with biodegradable structures: Tracking in vivo integration for determination of optimal in situ epithelialization conditions. *Biotechnol. Bioeng.* **2012**, *109*, 2134–2146. [CrossRef]
81. Gregurec, D.; Wang, G.; Pires, R.H.; Moya, S.E. Bioinspired titanium coatings: Self-assembly of collagen-alginate films for enhanced osseointegration. *J. Mater. Chem.* **2016**, *4*, 1978–1986. [CrossRef] [PubMed]
82. Shaygani, H.; Seifi, S.; Shamloo, A.; Ebrahimi, S. Novel bilayer coating on gentamicin-loaded titanium nanotube for orthopedic implants applications. *Int. J. Pharm.* **2023**, *636*, 122764. [CrossRef] [PubMed]
83. Duan, Y.; Wu, Y.; Yan, R.; Ma, H. Chitosan-sodium alginate-based coatings for self-strengthening anticorrosion and antibacterial protection of titanium substrate in artificial saliva. *Int. J. Biol. Macromol.* **2021**, *184*, 109–117. [CrossRef] [PubMed]
84. Yuan, Z.; Liu, P.; Hao, Y.; Cai, K. Construction of Ag-incorporated coating on Ti substrates for inhibited bacterial growth and enhanced osteoblast response. *Colloid Surf.* **2018**, *171*, 597–605. [CrossRef] [PubMed]
85. Lian, Q.; Zheng, S.; Shi, Z.; Cheng, H. Using a degradable three-layer sandwich-type coating to prevent titanium implant infection with the combined efficient bactericidal ability and fast immune remodeling property. *Acta Biomater.* **2022**, *154*, 650–666. [CrossRef]
86. Perni, S.; Alotaibi, H.F.; Yergeshov, A.A.; Prokopovich, P. Long acting anti-infection constructs on titanium. *J. Control. Release* **2020**, *326*, 91–105. [CrossRef] [PubMed]
87. Chen, T.; Wang, S.; He, F.; Zou, S. Promotion of Osseointegration Using Protamine/Alginate/Bone Morphogenic Protein 2 Biofunctionalized Composite Coating on Nanopolymorphic Titanium Surfaces. *J. Biomed. Nanotechnol.* **2018**, *14*, 933–945. [CrossRef]
88. Wang, M.; Wang, C.; Zhang, Y.; Lin, Y. Controlled release of dopamine coatings on titanium bidirectionally regulate osteoclastic and osteogenic response behaviors. *Mat. Sci. Eng. C Mater.* **2021**, *129*, 112376. [CrossRef]
89. Chen, T.; Brial, C.; McCarthy, M.; Maher, S.A. Synthetic PVA Osteochondral Implants for the Knee Joint: Mechanical Characteristics during Simulated Gait. *Am. J. Sport. Med.* **2021**, *49*, 2933–2941. [CrossRef]
90. Chen, K.; Liu, S.Y.; Wu, X.F.; Zhang, D.K. Mussel-inspired construction of Ti_6Al_4V-hydrogel artificial cartilage material with high strength and low friction. *Mater. Lett.* **2020**, *265*, 127421. [CrossRef]
91. Cui, L.L.; Chen, J.Y.; Yan, C.Q.; Xiong, D.S. Articular Cartilage Inspired the Construction of LTi-DA-PVA Composite Structure with Excellent Surface Wettability and Low Friction Performance. *Tribol. Lett.* **2021**, *69*, 41. [CrossRef]
92. Awasthi, S.; Gaur, J.K.; Pandey, S.K.; Srivastava, C. High-Strength, Strongly Bonded Nanocomposite Hydrogels for Cartilage Repair. *ACS Appl. Mater. Interfaces* **2021**, *13*, 24505–24523. [CrossRef] [PubMed]
93. Shi, Y.; Liu, J.; Li, J.L.; Xiong, D.S.; Dini, D. Improved mechanical and tribological properties of PAAm/PVA hydrogel-Ti_6Al_4V alloy configuration for cartilage repair. *J. Polym. Res.* **2022**, *29*, 515. [CrossRef]
94. Deng, Y.L.; Sun, J.J.; Ni, X.Y.; Yu, B. Tribological properties of hierarchical structure artificial joints with poly acrylic acid (AA)-poly acrylamide (AAm) hydrogel and Ti_6Al_4V substrate. *J. Polym. Res.* **2020**, *27*, 157. [CrossRef]
95. Chen, Y.H.; Cheng, W.H.; Teng, L.J.; Wang, Y.J. Graphene Oxide Hybrid Supramolecular Hydrogels with Self-Healable, Bioadhesive and Stimuli-Responsive Properties and Drug Delivery Application. *Macromol. Mater. Eng.* **2018**, *303*, 1700660. [CrossRef]
96. Xupeng, G.; Fangyuan, Y.; Ming, C. Drug release kinetics of rifampicin from composite gel coating on surface of titanium alloy. *Orthop. J. China* **2018**, *26*, 649–654.
97. Jing, Z.; Ni, R.; Wang, J.; Liu, Z. Practical strategy to construct anti-osteosarcoma bone substitutes by loading cisplatin into 3D-printed titanium alloy implants using a thermosensitive hydrogel. *Bioact. Mater.* **2021**, *6*, 4542–4557. [CrossRef] [PubMed]
98. Wang, X.Y.; Fang, X.; Gao, X.; Qin, Y.G. Strong adhesive and drug-loaded hydrogels for enhancing bone-implant interface fixation and anti-infection properties. *Colloid Surf.* **2022**, *219*, 112817. [CrossRef] [PubMed]

99. Huang, T.; Yu, Z.; Yu, Q.; Yang, G. Electrochemical deposition of lithium coating on titanium implant with enhanced early stage osseointegration. *J. Biomed. Mater. Res.* **2022**, *110*, 2399–2410. [CrossRef] [PubMed]
100. Lu, M.; Chen, H.; Yuan, B.; Zhang, X. Electrochemical Deposition of Nanostructured Hydroxyapatite Coating on Titanium with Enhanced Early Stage Osteogenic Activity and Osseointegration. *Int. J. Nanomed.* **2020**, *15*, 6605–6618. [CrossRef] [PubMed]
101. Wang, B.; Hua, J.; You, R.; Ma, L. Electrochemically deposition of catechol-chitosan hydrogel coating on coronary stent with robust copper ions immobilization capability and improved interfacial biological activity. *Int. J. Biol. Macromol.* **2021**, *181*, 435–443. [CrossRef] [PubMed]
102. Croes, M.; Bakhshandeh, S.; van Hengel, I.A.J.; Amin Yavari, S. Antibacterial and immunogenic behavior of silver coatings on additively manufactured porous titanium. *Acta Biomater.* **2018**, *81*, 315–327. [CrossRef]
103. Yin, X.; Yan, L.; Jun Hao, D.; Liu, Z. Calcium alginate template-mineral substituted hydroxyapatite hydrogel coated titanium implant for tibia bone regeneration. *Int. J. Pharm.* **2020**, *582*, 119303. [CrossRef]
104. Soylu, H.M.; Chevallier, P.; Copes, F.; Mantovani, D. A Novel Strategy to Coat Dopamine-Functionalized Titanium Surfaces with Agarose-Based Hydrogels for the Controlled Release of Gentamicin. *Front. Cell. Infect. Microbiol.* **2021**, *11*, 678081. [CrossRef]
105. Yao, X.; Liu, J.; Yang, C.; Suo, Z. Hydrogel Paint. *Adv. Mater.* **2019**, *31*, e1903062. [CrossRef] [PubMed]
106. Tolle, C.; Riedel, J.; Mikolai, C.; Menzel, H. Biocompatible Coatings from Smart Biopolymer Nanoparticles for Enzymatically Induced Drug Release. *Biomolecules* **2018**, *8*, 103. [CrossRef] [PubMed]
107. Tsikopoulos, K.; Bidossi, A.; Drago, L.; Papaioannidou, P. Is Implant Coating with Tyrosol- and Antibiotic-loaded Hydrogel Effective in Reducing Cutibacterium (Propionibacterium) acnes Biofilm Formation? A Preliminary in Vitro Study. *Clin. Orthop. Relat. Res.* **2019**, *477*, 1736–1746. [CrossRef]
108. Wu, Y.; Hu, F.; Yang, X.; Zhang, X. Titanium surface polyethylene glycol hydrogel and gentamicin-loaded cross-linked starch microspheres release system for anti-infective drugs. *J. Drug Target.* **2023**, *31*, 217–224. [CrossRef] [PubMed]
109. Leng, J.; He, Y.; Yuan, Z.; Cai, K. Enzymatically-degradable hydrogel coatings on titanium for bacterial infection inhibition and enhanced soft tissue compatibility via a self-adaptive strategy. *Bioact. Mater.* **2021**, *6*, 4670–4685. [CrossRef]
110. He, Y.; Li, K.; Yang, X.; Cai, K. Calcium Peroxide Nanoparticles-Embedded Coatings on Anti-Inflammatory TiO_2 Nanotubes for Bacteria Elimination and Inflammatory Environment Amelioration. *Small* **2021**, *17*, e2102907. [CrossRef] [PubMed]
111. Stoetzel, S.; Malhan, D.; Wild, U.; El Khassawna, T. Osteocytes Influence on Bone Matrix Integrity Affects Biomechanical Competence at Bone-Implant Interface of Bioactive-Coated Titanium Implants in Rat Tibiae. *Int. J. Mol. Sci.* **2021**, *23*, 374. [CrossRef]
112. Shi, Q.; Qian, Z.; Liu, D.; Liu, H. Surface Modification of Dental Titanium Implant by Layer-by-Layer Electrostatic Self-Assembly. *Front. Physiol.* **2017**, *8*, 574. [CrossRef]
113. Wang, D.; Chen, M.W.; Wei, Y.J.; Cai, K.Y. Construction of Wogonin Nanoparticle-Containing Strontium-Doped Nanoporous Structure on Titanium Surface to Promote Osteoporosis Fracture Repair. *Adv. Healthc. Mater.* **2022**, *11*, e2201405. [CrossRef]
114. Lv, H.; Chen, Z.; Yang, X.; Gao, P. Layer-by-layer self-assembly of minocycline-loaded chitosan/alginate multilayer on titanium substrates to inhibit biofilm formation. *J. Dent.* **2014**, *42*, 1464–1472. [CrossRef]
115. Chen, M.; Huang, L.; Shen, X.; Hu, Y. Construction of multilayered molecular reservoirs on a titanium alloy implant for combinational drug delivery to promote osseointegration in osteoporotic conditions. *Acta Biomater.* **2020**, *105*, 304–318. [CrossRef]
116. Zhong, X.; Song, Y.; Yang, P.; Li, C. Titanium Surface Priming with Phase-Transited Lysozyme to Establish a Silver Nanoparticle-Loaded Chitosan/Hyaluronic Acid Antibacterial Multilayer via Layer-by-Layer Self-Assembly. *PLoS ONE* **2016**, *11*, e0146957. [CrossRef] [PubMed]
117. Song, Y.; Ma, A.; Ning, J.; Li, C. Loading icariin on titanium surfaces by phase-transited lysozyme priming and layer-by-layer self-assembly of hyaluronic acid/chitosan to improve surface osteogenesis ability. *Int. J. Nanomed.* **2018**, *13*, 6751–6767. [CrossRef]
118. Han, M.; Dong, Z.; Li, J.; Li, J. Mussel-inspired self-assembly engineered implant coatings for synergistic anti-infection and osteogenesis acceleration. *J. Mater. Chem.* **2021**, *9*, 8501–8511. [CrossRef] [PubMed]
119. Iwata, N.; Nozaki, K.; Horiuchi, N.; Nagai, A. Effects of controlled micro-/nanosurfaces on osteoblast proliferation. *J. Biomed. Mater. Res.* **2017**, *105*, 2589–2596. [CrossRef] [PubMed]
120. Yu, Y.; Jin, G.; Xue, Y.; Sun, J. Multifunctions of dual Zn/Mg ion co-implanted titanium on osteogenesis, angiogenesis and bacteria inhibition for dental implants. *Acta Biomater.* **2017**, *49*, 590–603. [CrossRef] [PubMed]
121. Lee, S.; Chang, Y.Y.; Lee, J.; Shin, H. Surface engineering of titanium alloy using metal-polyphenol network coating with magnesium ions for improved osseointegration. *Biomater. Sci.* **2020**, *8*, 3404–3417. [CrossRef] [PubMed]
122. Wu, Y.; Tang, H.; Liu, L.; Wang, A. Biomimetic titanium implant coated with extracellular matrix enhances and accelerates osteogenesis. *Nanomedicine* **2020**, *15*, 1779–1793. [CrossRef] [PubMed]
123. Lu, X.; Xiong, S.; Chen, Y.; Yang, B. Effects of statherin on the biological properties of titanium metals subjected to different surface modification. *Colloid Surf.* **2020**, *188*, 110783. [CrossRef] [PubMed]
124. Scarano, A.; Lorusso, F.; Orsini, T.; Valbonetti, L. Biomimetic Surfaces Coated with Covalently Immobilized Collagen Type I: An X-ray Photoelectron Spectroscopy, Atomic Force Microscopy, Micro-CT and Histomorphometrical Study in Rabbits. *Int. J. Mol. Sci.* **2019**, *20*, 724. [CrossRef] [PubMed]
125. Liu, J.; Tang, Y.; Yang, W.; Cai, K. Functionalization of titanium substrate with multifunctional peptide OGP-NAC for the regulation of osteoimmunology. *Biomater. Sci.* **2019**, *7*, 1463–1476. [CrossRef] [PubMed]

126. Zhang, G.; Zhang, X.; Yang, Y.; Zhang, X. Dual light-induced in situ antibacterial activities of biocompatibleTiO_2/MoS_2/PDA/RGD nanorod arrays on titanium. *Biomater. Sci.* **2020**, *8*, 391–404. [CrossRef] [PubMed]
127. Teng, F.Y.; Tai, I.C.; Ho, M.L.; Tseng, C.C. Controlled release of BMP-2 from titanium with electrodeposition modification enhancing critical size bone formation. *Mat. Sci. Eng. C Mater.* **2019**, *105*, 109879. [CrossRef] [PubMed]
128. Zhao, H.; Huang, Y.; Zhang, W.; Shi, Q. Mussel-Inspired Peptide Coatings on Titanium Implant to Improve Osseointegration in Osteoporotic Condition. *ACS Biomater. Sci. Eng.* **2018**, *4*, 2505–2515. [CrossRef]
129. Ma, L.; Wang, X.; Zhou, Y.; Zhang, Y. Biomimetic Ti-6Al-4V alloy/gelatin methacrylate hybrid scaffold with enhanced osteogenic and angiogenic capabilities for large bone defect restoration. *Bioact. Mater.* **2021**, *6*, 3437–3448. [CrossRef]
130. Zhang, W.; Sun, C.G.; Song, C.L. 3D printed porous titanium cages filled with simvastatin hydrogel promotes bone ingrowth and spinal fusion in rhesus macaques. *Biomater. Sci.* **2020**, *8*, 4147–4156. [CrossRef]
131. He, P.; Zhang, H.; Li, Y.; Yang, S. 1α,25-Dihydroxyvitamin D3-loaded hierarchical titanium scaffold enhanced early osseointegration. *Mat. Sci. Eng. C Mater.* **2020**, *109*, 110551. [CrossRef] [PubMed]
132. Deng, J.; Cohen, D.J.; Sabalewski, E.L.; Boyan, B.D. Semaphorin 3A delivered by a rapidly polymerizing click hydrogel overcomes impaired implant osseointegration in a rat type 2 diabetes model. *Acta Biomater.* **2023**, *157*, 236–251. [CrossRef] [PubMed]
133. Vaquette, C.; Mitchell, J.; Fernandez-Medina, T.; Ivanovski, S. Resorbable additively manufactured scaffold imparts dimensional stability to extraskeletally regenerated bone. *Biomaterials* **2021**, *269*, 120671. [CrossRef]
134. Lovati, A.B.; Lopa, S.; Talò, G.; Moretti, M. In vivo evaluation of bone deposition in macroporous titanium implants loaded with mesenchymal stem cells and strontium-enriched hydrogel. *J. Biomed. Mater. Res.* **2015**, *103*, 448–456. [CrossRef] [PubMed]
135. Kang, H.J.; Hossain, M.; Park, S.S.; Im, S.B.; Lee, B.T. Microstructures and biological properties of 3D-printed titanium intervertebral spacer with the tri-calcium phosphate loaded demineralized bone matrix hydrogel. *Mater. Lett.* **2021**, *303*, 130519. [CrossRef]
136. Kwon, K.A.; Juhasz, J.A.; Brooks, R.A.; Best, S.M. Bioactive conformable hydrogel-carbonated hydroxyapatite nanocomposite coatings on Ti-6Al-4V substrates. *Mater. Technol.* **2020**, *35*, 727–733. [CrossRef]
137. Alavi, S.E.; Panah, N.; Page, F.; Gholami, M. Hydrogel-based therapeutic coatings for dental implants. *Eur. Polym. J.* **2022**, *181*, 111652. [CrossRef]
138. Lee, J.H.; Jin, Y.Z. Recombinant human bone morphogenetic protein-2 loaded porous beta-tricalcium phosphate microsphere-hyaluronic acid composites promoted osseointegration around titanium implants. *Int. J. Polym. Mater. Polym. Biomater.* **2019**, *68*, 368–374. [CrossRef]
139. Kumar, A.; Nune, K.C.; Misra, R.D.K. Design and biological functionality of a novel hybrid Ti-6Al-4V/hydrogel system for reconstruction of bone defects. *J. Tissue Eng. Regen.* **2018**, *12*, 1133–1144. [CrossRef] [PubMed]
140. Liu, H.; Li, W.; Liu, C.; Song, C.L. Incorporating simvastatin/poloxamer 407 hydrogel into 3D-printed porous Ti6Al4V scaffolds for the promotion of angiogenesis, osseointegration and bone ingrowth. *Biofabrication* **2016**, *8*, 045012. [CrossRef]
141. Che, Z.J.; Sun, Y.F.; Luo, W.B.; Huang, L.F. Bifunctionalized hydrogels promote angiogenesis and osseointegration at the interface of three-dimensionally printed porous titanium scaffolds. *Mater. Des.* **2022**, *223*, 111118. [CrossRef]
142. Zhao, H.; Shen, S.; Zhao, L.; Zhuo, N. 3D printing of dual-cell delivery titanium alloy scaffolds for improving osseointegration through enhancing angiogenesis and osteogenesis. *BMC Musculoskelet. Disord.* **2021**, *22*, 734. [CrossRef]
143. Li, B.E.; Zhang, L.; Wang, D.H.; Zhao, X.Y. Thermo-sensitive hydrogel on anodized titanium surface to regulate immune response. *Surf. Coat. Technol.* **2021**, *405*, 126624. [CrossRef]
144. Gao, L.; Li, M.; Yin, L.; Feng, B. Dual-inflammatory cytokines on TiO_2 nanotube-coated surfaces used for regulating macrophage polarization in bone implants. *J. Biomed. Mater. Res.* **2018**, *106*, 1878–1886. [CrossRef] [PubMed]
145. Li, M.; Wei, F.; Yin, X.; Zhou, Y. Synergistic regulation of osteoimmune microenvironment by IL-4 and RGD to accelerate osteogenesis. *Mat. Sci. Eng. C Mater.* **2020**, *109*, 110508. [CrossRef]
146. Chen, F.; He, Y.; Li, Z.; Zhang, Y. A novel tunable, highly biocompatible and injectable DNA-chitosan hybrid hydrogel fabricated by electrostatic interaction between chitosan and DNA backbone. *Int. J. Pharm.* **2021**, *606*, 120938. [CrossRef] [PubMed]
147. Li, X.; Xu, K.; He, Y.; Cai, K. ROS-responsive hydrogel coating modified titanium promotes vascularization and osteointegration of bone defects by orchestrating immunomodulation. *Biomaterials* **2022**, *287*, 121683. [CrossRef]
148. Sun, C.K.; Ke, C.J.; Lin, Y.W.; Sun, J.S. Transglutaminase Cross-Linked Gelatin-Alginate-Antibacterial Hydrogel as the Drug Delivery-Coatings for Implant-Related Infections. *Polymers* **2021**, *13*, 414. [CrossRef]
149. Huang, H.; Wu, Z.; Yang, Z.; Xie, X. In vitro application of drug-loaded hydrogel combined with 3D-printed porous scaffolds. *Biomed. Mater.* **2022**, *17*, 065019. [CrossRef] [PubMed]
150. Boot, W.; Vogely, H.C.; Jiao, C.; Gawlitta, D. Prophylaxis of implant-related infections by local release of vancomycin from a hydrogel in rabbits. *Eur. Cells Mater.* **2020**, *39*, 108–120. [CrossRef]
151. Zhang, X.; Wang, W.; Chen, J.; Lai, M. Peptide GL13K releasing hydrogel functionalized micro/nanostructured titanium enhances its osteogenic and antibacterial activity. *J. Biomat. Sci. Polym.* **2022**, *18*, 1–17. [CrossRef]
152. Lin, H.Y.; Chiou, W.S.; Shiue, S.J.; Cheng, J.K. Non-RGD peptide H-ckrwwkwirw-NH2 grafting accentuates antibacterial and osteoinductive properties of biopolymer coating. *Soft Mater.* **2020**, *18*, 487–498. [CrossRef]
153. Barros, J.A.R.; Melo, L.D.R.; Silva, R.A.R.D.; Monteiro, F.J. Encapsulated bacteriophages in alginate-nanohydroxyapatite hydrogel as a novel delivery system to prevent orthopedic implant-associated infections. *Nanomed. Nanotechnol. Biol. Med.* **2020**, *24*, 102145. [CrossRef] [PubMed]

154. Qiao, S.; Wu, D.; Li, Z.; Gu, Y. The combination of multi-functional ingredients-loaded hydrogels and three-dimensional printed porous titanium alloys for infective bone defect treatment. *J. Tissue Eng.* **2020**, *11*, 2041731420965797. [CrossRef] [PubMed]
155. Li, Z.; Zhao, Y.; Wang, Z.; Wang, J. Engineering Multifunctional Hydrogel-Integrated 3D Printed Bioactive Prosthetic Interfaces for Osteoporotic Osseointegration. *Adv. Healthc. Mater.* **2022**, *11*, e2102535. [CrossRef] [PubMed]
156. Li, Y.; Liu, X.; Li, B.; Wu, S. Near-Infrared Light Triggered Phototherapy and Immunotherapy for Elimination of Methicillin-Resistant Staphylococcus aureus Biofilm Infection on Bone Implant. *ACS Nano* **2020**, *14*, 8157–8170. [CrossRef] [PubMed]
157. Chen, H.; Qiu, X.; Xia, T.; Li, Y. Mesoporous Materials Make Hydrogels More Powerful in Biomedicine. *Gels* **2023**, *9*, 207. [CrossRef] [PubMed]
158. Andrade Del Olmo, J.; Pérez-Álvarez, L.; Sáez Martínez, V.; Alonso, J.M. Multifunctional antibacterial chitosan-based hydrogel coatings on Ti6Al4V biomaterial for biomedical implant applications. *Int. J. Biol. Macromol.* **2023**, *231*, 123328. [CrossRef] [PubMed]
159. Barik, A.; Chakravorty, N. Targeted Drug Delivery from Titanium Implants: A Review of Challenges and Approaches. *Adv. Exp. Med. Biol.* **2020**, *1251*, 1–17.
160. Ding, Y.; Liu, G.; Liu, S.; Cai, K. A multi-function hydrogel-coating engineered implant for rescuing biofilm infection and boosting osseointegration by macrophage-related immunomodulation. *Adv. Healthc. Mater.* **2023**, *4*, e2300722. [CrossRef]

MDPI AG
Grosspeteranlage 5
4052 Basel
Switzerland
Tel.: +41 61 683 77 34

Gels Editorial Office
E-mail: gels@mdpi.com
www.mdpi.com/journal/gels

www.ingramcontent.com/pod-product-compliance
Lightning Source LLC
Chambersburg PA
CBHW041101270926
51850CB00025B/3177
* 9 7 8 3 7 2 5 8 2 7 3 8 1 *